Birmingham Archaeology Monograph Series 15

'The Homes of our Metal Manufactures. Messrs R.W. Winfield and Co's Cambridge Street Works & Rolling Mills, Birmingham'

Archaeological Excavations at the Library of Birmingham, Cambridge Street

Edited by

Chris Hewitson

Principal contributors

Chris Hewitson, Will Mitchell, Gerry McDonnell and Ray Shill

with

Emma Collins, David Dungworth, Sam Hepburn, Steve Litherland, Erica Macey-Bracken, Leonie Taibi and Christine Winter

Illustrations by

Nigel Dodds and Chris Hewitson

BAR British Series 579

2013

Published in 2016 by
BAR Publishing, Oxford

BAR British Series 579

Birmingham Archaeology Monograph Series 15
*'The Homes of our Metal Manufactures. Messrs R.W. Winfield and Co's Cambridge Street Works & Rolling Mills, Birmingham'**

* quoted in Martineau and Smith's Hardware Trade Journal (1887, 9) The works of Messrs R W Winfield and Co. in 1887 from 'The homes of our metal manufactures. Messrs R W Winfield and Co's Cambridge Street works and rolling mills, Birmingham', dated January 31st, 1887.

ISBN 9781407310992 paperback
ISBN 9781407322582 e-format
DOI https://doi.org/10.30861/9781407310992
A catalogue record for this book is available from the British Library

BAR Publishing is the trading name of British Archaeological Reports (Oxford) Ltd. British Archaeological Reports was first incorporated in 1974 to publish the BAR Series, International and British. In 1992 Hadrian Books Ltd became part of the BAR group. This volume was originally published by Archaeopress in conjunction with British Archaeological Reports (Oxford) Ltd / Hadrian Books Ltd, the Series principal publisher, in 2013. This present volume is published by BAR Publishing, 2016.

BAR

PUBLISHING

BAR titles are available from:

BAR Publishing
122 Banbury Rd, Oxford, OX2 7BP, UK
EMAIL info@barpublishing.com
PHONE +44 (0)1865 310431
FAX +44 (0)1865 316916
www.barpublishing.com

CONTENTS

LIST OF FIGURES

LIST OF PLATES

LIST OF TABLES

CHAPTER 1:
INTRODUCTION AND BACKGROUND TO THE WORK

Chris Hewitson

BACKGROUND TO THE WORK

With the redevelopment of the former car park adjacent to Baskerville House as part of the Library of Birmingham project, the opportunity arose to examine some of the most complete remains of the 19th-century industrialisation in Birmingham. Birmingham Archaeology of the University of Birmingham, in association with Carillion and the Birmingham City Council, undertook an archaeological excavation, before the construction of the new Library of Birmingham, in an area between Cambridge Street and Centenary Square, Broad Street in the city centre (hereinafter referred to as the site; Fig. 1.1, Plate 1.1).

LOCATION, TOPOGRAPHY AND GEOLOGY

The site lay within the city centre of Birmingham on the south side of Cambridge Street, Birmingham between the buildings of the Repertory Theatre to the east, Baskerville House to the west and Centenary Square to the south, centred on NGR SP 0631 8687.

The site lies in the middle of a ridge of higher ground running southwest – northeast. The Birmingham and Fazeley canal and redundant Newhall Arm are located to the north, and the Birmingham and Worcester canals located to the west. The early canals were contour canals, which skirt around the higher ground on which the site is located.

The underlying geology consists of a ridge of Bromsgrove Sandstone overlain by glacio-fluvial deposits. Immediately prior to excavation the site was tarmac hard standing (used as a public car park), landscaped raise garden beds, and an open public plaza.

ARCHAEOLOGICAL BACKGROUND

The project began in October 2006, when an archaeological desk-based assessment (Lobb 2006), consisting of analysis of the available documentary and cartographic evidence, concluded that there was the potential for finding well-preserved archaeological remains on the site. The buildings that once occupied the site included the house of John Baskerville, two former canal wharfs, and the Union Rolling Mill, which became part of Winfield's Cambridge Street Works. The Cambridge Street Works comprised metal rolling, gas fitting, brass founding, tube and wire manufacture. Subsequently an archaeological evaluation in 2008 of four trenches in the car park revealed well-preserved archaeological remains of the brass works and the canal basins (Ramsey 2008). Geotechnical test pits excavated in January and February 2009, in both the car park and the adjoining part of Centenary Square, confirmed these results (Krawiec 2009).

AIMS AND OBJECTIVES

The aims and objectives of the excavation were stated in a brief produced by Birmingham City Council (Appendix 6), and a Written Scheme of Investigation (Birmingham Archaeology 2009) in accordance with guidelines laid down in Planning Policy Guidance Note 16 (DoE 1990).

The objective of the 2010 excavation was to identify archaeological remains, and to preserve those remains by record. More specific research aims were:

- To recover and analyse industrial residues of all kinds.
- To identify and date phases of development of the site and the functions of different parts of it.
- To relate the excavated structures to the documentary record.
- To relate the historic development of the site to that found by excavation on nearby sites and to other sites of a similar type nearby.
- To interpret and discuss the remains in their regional, national and international context.

In the light of the above assessment of potential, the revised research aims of the project were to examine the excavated remains of the brass works and understand how they functioned in its many contexts. These wider research themes included:

- The site, both within and immediately outside its boundaries.
- Birmingham and the Black Country, its people, industry and economy.
- The realm of Britain, the Empire, Europe, America and the World.

Within this context, the research undertaken concentrated on three clear areas:

Industrial function: how the brass works functioned, technological development and innovation, and the nature of the product itself.

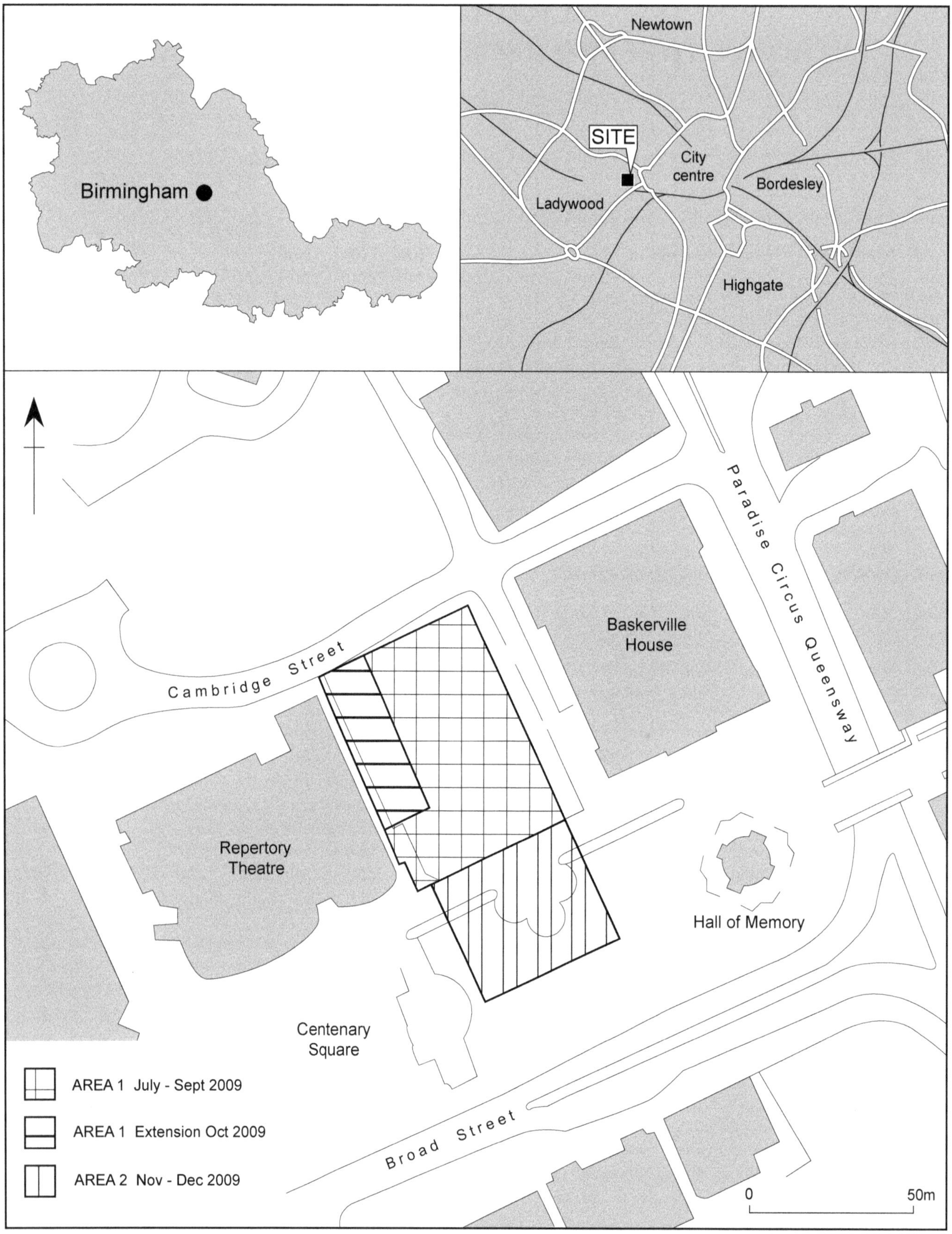

FIGURE 1.1 THE LOCATION OF THE EXCAVATIONS AT CAMBRIDGE STREET

Plate 1.1 Overall view of Area 1, facing southeast (Image by Aerial-Cam)

Economic function - how the brass works grew, developed and eventually declined. The economic locale of the brass industry in Birmingham: what the brass works produced and who consumed the goods.

Social function: how people in the brass works interacted and operated, the hierarchical structure.

Work throughout the excavation phase was supplemented by detailed historical research by local historian, Ray Shill, this was undertaken with the object of informing the process of the investigation.

ARRANGEMENT OF THE REPORT

The report outlines the final results of the excavation that was carried out between July and December 2009 and includes the archaeological, environmental and artefactual evidence recovered. The structure of the report is based on guidelines provided by English Heritage (1991; 2006).

The following chapters describe the results of the historic research and archaeological investigations. Chapter 2 describes the history and geography of the brass metal trade with particular reference to Birmingham. Chapter 3 details the history of the site before the foundation of the works and the initial laying out of the site. Chapters 4 and 5 – explore the history and archaeology of the Cambridge Street works, including a full description of the function of the works based on the archaeological evidence. Chapter 6 describes the artefactual evidence. Chapter 7 is a detailed discussion of the archaeo-metallurgical evidence conducted on the site. Chapter 8 aims to produce a synthesis of evidence for the development of the industrial features of the Birmingham brass industry. Chapter 9 examines the social history of the Birmingham brass trade and provides an overall discussion of the works.

THE PHASING

The phases and results are based upon the primary documentary evidence, stratigraphic relationships identified on site, pottery dating and the material evidence. Cartographic sources have been used extensively to interpret the archaeological evidence. Each chapter places the historical information alongside the archaeological results in an attempt to combine the available evidence into a coherent whole.

The excavation identified six phases of activity pre-dating, during and after the completion of the brass metal works (Fig. 1.2). Due to the long life and re-use of elements of the buildings, however, it was not always possible to confirm specific phases. However, these six phases broadly relate to the following sequence and correspond closely with the historical record.

Phase 1: **18th century** Occupation of the site associated with Easy Hill House
Phase 2: **1810–1820** Construction of canal basins

Figure 1.2 The phased development of the excavations

Phase 3: **1820–1830** Construction of the original Union Rolling Mill
Phase 4: **1830–1900** Major expansion of the Cambridge Street works
Phase 5: 1900–1920 Change in ownership of the works
Phase 6: **1920–1936** Demolition of the works

CHAPTER 2: HISTORICAL OVERVIEW

Chris Hewitson and Ray Shill

The non-ferrous processing industry has not received the same level of archaeological focus that the ferrous processing industries have received, and large-scale investigation and excavation are limited to two areas. These are the brass industry in Bristol, specifically the Warmley Brass Works (Day 1973; Day 1995; van Laun and Day 1995) and the copper and brass industry in Swansea, Wales known as Copperopolis (Hughes 2000; 2004). These remains relate to the primary smelting remains with some remains of secondary production of rolling and wire drawing. In contrast the Cambridge Street works relates to secondary and tertiary processing of wire and brass manufacture.

THE CAMBRIDGE STREET BRASS WORKS – AN OVERVIEW OF ITS HISTORY

This represents an overview of the history of the works. The individual phases of the works will be discussed in greater detail in the following chapters. The main archaeological evidence relates to Phases 3 and 4 between 1820 and 1900, with some alteration in Phase 5 in 1900–1920. This is a period the economic expansion of Britain and the Empire resulted in dramatic change in terms of the economy, industry and social aspects of Birmingham.

Phase 1

1745	In the early 18th century the Cambridge Street site lay north of the Harborne Road on the western fringes of Birmingham. From 1745 it formed part of the gardens of Easy Hill House within John Baskerville's estate.
1788–1791	From 1788 Easy Hill House was occupied by John Ryland but was destroyed in the Priestly riots of 1791. This left it as a ruin and a vacant plot of land.

Phase 2

1810–1811	Thomas Gibson and Shore acquired the Baskerville estate and the canal wharfs and Gibson's Arm and Baskerville Wharf were constructed.
1820	Industrial buildings began to occupy the former Baskerville Estate including those of RW Winfield, who established on the corner of Cambridge Street and Easy Row as a brass founder and brass-tube maker.

Phase 3

1824/5	Baskerville wharf was fully laid out with four canal arms. This can be seen on the Earl of Dartmouth's map of 1824–1825 (Fig. 2.1).
1824	Daniel and Joseph Ledsam, William Potts, Matthew Dixon and Robert W Winfield used the Cambridge Street site to set up the Union Rolling Mills.
1827–1829	Winfield took out patent for the production of brass bedsteads. By 1829 the Union Rolling Mills site was occupied solely by Robert Winfield and his company and was located at the Cambridge Street Works. This became known as RW Winfield and Co.

Phase 4

1840–1850s	From the 1840s Robert Winfield developed a bedstead works south of Gibson's Arm visible on Pigott-Smith's Map (1850–1855; Fig. 2.1).
1853	In 1853 Winfield acquired the lease of the Union Rolling Mills and combined the rolling, wire and tubedrawing plant into one operation. The Cambridge Street works was part of a larger complex of industrial buildings, trading as RW Winfield and Sons, that stretched further along Cambridge Street to the west, under the present day REP theatre, and east as far as Easy Hill Road (Pigott-Smith 1850-1855; Fig. 2.1).
1869	Robert Winfield died in 1869 and the company went through a variety of partnership changes.
1887	It enrolled as a limited liability company and traded as Winfields Limited.
1897–1898	Winfield's Limited went into receivership and was auctioned and parts were sold off separately. The sales plans reveal the extent of the works in the late- 19th century.

Phase 5

1900	The Rolling Mill taken over by ICI Metals Ltd but, the rest of the works were operated by a number of companies.
1903	Establishment of the Birmingham Aluminium Works in part of the former bedstead works.

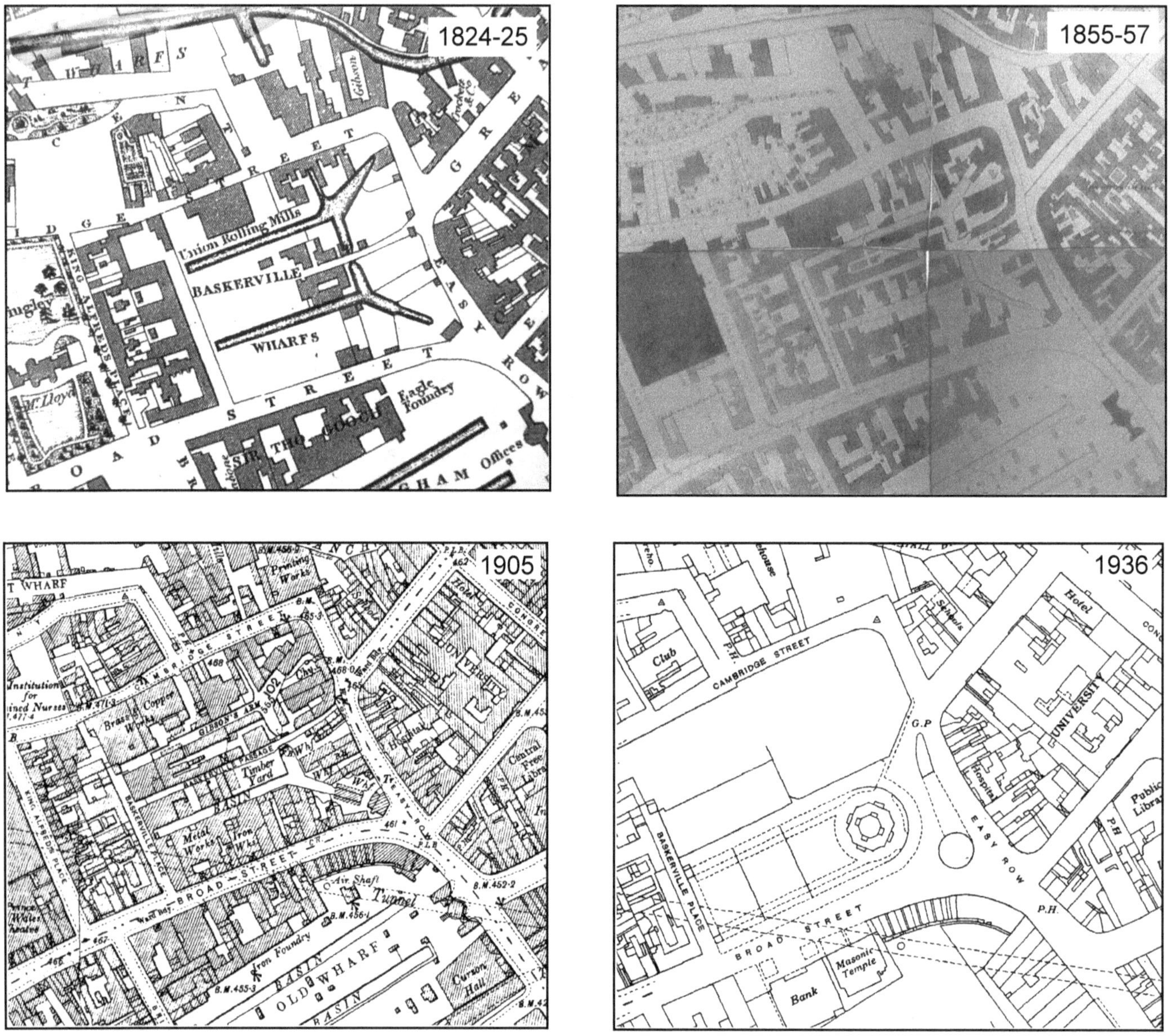

FIGURE 2.1 THE CARTOGRAPHIC DEVELOPMENT OF THE CAMBRIDGE STREET WORKS, 1824–1936 (MAPS ARE COURTESY OF BIRMINGHAM ARCHIVES AND HERITAGE)

1914	Birmingham Aluminium Works takes over the southern half of Gibson's Wharf.
1914-1918	The Cambridge Street works was taken over by the Ministry of Munitions during the Great War as part of the war effort (Fig. 2.1).
Phase 6	
1922	End of production on south of Gibson's Arm and beginning of site clearance (Fig. 2.1).
1922–1926	Centenary Square was developed.
1936	The Rolling Mill ceased production in 1936 when it transferred to a site near Icknield Port Loop, by the Edgbaston reservoir. Gibson's Arm Canal was filled in and the Beam engine scrapped.

THE NATIONAL DEVELOPMENT OF THE BRASS INDUSTRY

Chris Hewitson

The Cambridge Street works was part of a regional and national brass industry which underwent dramatic technological developments that transformed the industry in the 18th and 19th centuries. The British brass industry developed from the 17th century and was reliant on a number of areas of operation; mineral resources from Cornwall and Anglesey in the form of copper deposits, and zinc from the Mendip Hills, Derbyshire and Flintshire. Primary smelting of the ores developed in Swansea and Neath as well as in north Wales around Holywell, whilst secondary processing to make brass occurred in Bristol, Cheadle and London before transferring to Birmingham. Birmingham, London and Bristol were to maintain their

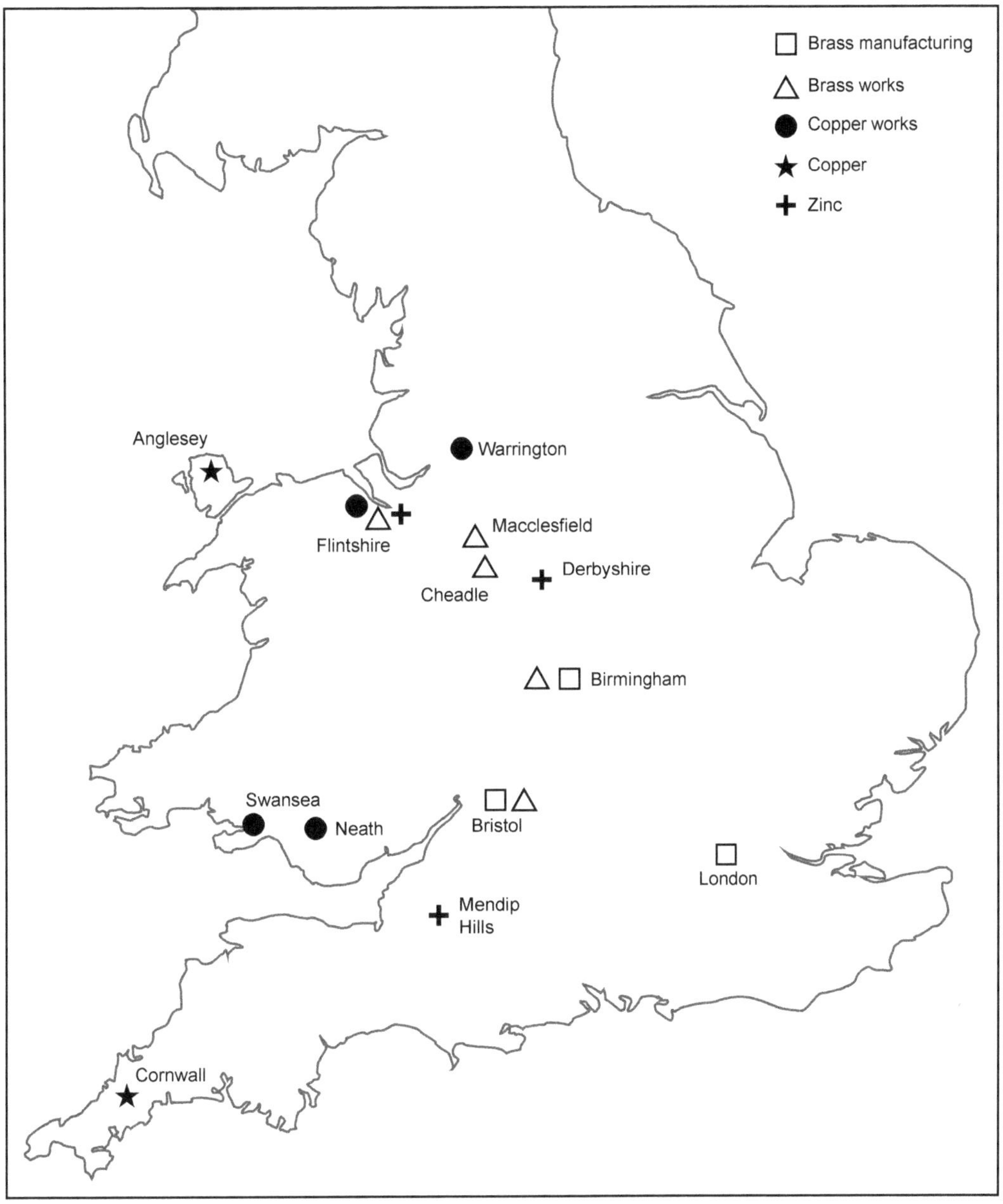

Figure 2.2 The geographic location of the copper and brass industry in England and Wales

reputation as primary manufacturing centres throughout the 19th century (Fig. 2.2).

The 18th-century industry

The development of the brass industry was reliant on a number of geographical factors, not least amongst which were the availability of raw materials. Initially the industry developed in regions where either the raw materials for brass production or a supply of coal or water for power existed such as copper from Cornwall, coal for smelting in Swansea and Neath and water-power for brass battery mills in London and Bristol (Harris 1964, 12–13). The second important product was zinc that was procured from Derbyshire, Flintshire and Somerset, which helps to explain the location of brass producing regions in Bristol, Cheadle, Macclesfield and Holywell (Aitken 1866, 260–261).

The development of the brass industry in Britain meant that five major brass works was established between 1690 and 1703, near to the market for two major consumers, Bristol and London. The copper smelting industry in Swansea in the 17th and 18th centuries initially served these mills. There the copper was either rolled into copper sheet or combined with zinc to produce brass (Hughes 2000, 39). The output of most copper works was destined for brass production and it was not until much later that markets were found for copper production. Later production centred on works at Cheadle, Staffordshire and in North Wales (Morton 1983, 10–11) but the two main suppliers to the Birmingham brass industry in the mid 18th century were Cheadle and Bristol.

From the 1680s Bristol became known for copper smelting, using coal-fired reverbatory furnaces. In the early 18th century the area along the Avon valley from Bristol to Bath developed as a brass production centre under the auspices of four Quaker partners, including Abraham Darby as manager. They encouraged immigrant brass workers to the area from the Low Countries. Furnaces and battery mills were established: the Baptist Mill in 1702; mills at Keynsham on the Avon in 1705 and further mills on the River Chew south of Bristol. Local supplies of coal and water provided power for brass production, rolling into sheets and slitting into wire (Day 1973, 32–40). In the 1730s, William Champion, at his Warmley brassworks, carried out metallic zinc smelting, using calamine for the first time in Europe. He used zinc found locally in the Mendip Hills and copper from Cornwall. This allowed Bristol brass production to achieve predominance in the European trade. The works became one of the largest sites of its type for that period and the use of the Newcomen steam engine to pump water supplies for water-driven manufacturing processes was pioneered there (Day 1973, 73–94). The Bristol works became closely associated with the export trades of brass products to West Africa for exchange purposes with slave traders (Evans and Rydén 2007, 46).

The decline of Bristol brass production was a result of competition from Anglesey copper sources that allowed alternative markets to develop in North Wales and the Midlands from the 1780s. In the 1760s the discovery of the 'Copper Mountain' on Anglesey led slowly to the development of mines by 1768. Smelting on the island was quickly abandoned and the majority of the copper was exported to Ravenhead in Lancashire, for smelting, whilst Holywell in Flintshire was chosen as a location for copper mills (Harris 1964, 40–43). By the later 1780s the supply of copper from Anglesey exceeded that of Cornwall, and exercised a short-lived monopoly on copper during the 1790s (*ibid*, 54–55).

Brass production began in Cheadle in 1734 when Thomas Patten set up the Cheadle Brass Company that adopted the calamine or cementation method of making brass. The site lay in a location not necessarily best suited to brass production. However, using copper from the Patten family copper works in Warrington (established 1717) and local coal and water supplies, the company exploited the valuable new Birmingham and Black Country markets (Morton 1983, 13–14).

The 19th-century industry

Whereas the 18th-century copper and brass industry was supplied by domestic sources from Cornwall, Devon and Anglesey as well as Ireland and the Isle of Man, this would prove to be insufficient for the needs of the19th-century industry. By the 1820s the raw product of copper was not as heavily controlled in the Birmingham market as elsewhere in the UK. The end of the 18th century had been dominated by the Copper King, Thomas Williams of Anglesey and his successor company William and Grenfell in the early part of the 19th century (Harris 1964).

The influence of William and Grenfell and the resultant decline in the price of cake copper led to their collusion, by 1820, with other major producers (Vivian and Sons, Williams-Foster and Sims, Willyams and Nevill) in order to maintain copper prices. New imports from Russia and Peru provided competition as sources of supply. It was in this atmosphere of increased competition that the Association of Copper Smelters came into being. This was essentially a cartel for the control of the price of manufactured copper. However, the influence of this cartel over the Birmingham market was limited as three companies, the Rose Copper Company, the Birmingham Copper Company and the Crown Company, based themselves out of the city, so that attempts to fix a higher price than these producers was impracticable (Toomey 1985, 312–325). After a period of relative stability in the copper production industry the introduction of a new company, the English Copper Company, by Cornish miners in 1844 led to intense rivalry and a reduction in copper profits. This in turn led to the formation of a second copper association. Prices were low throughout this period, but by 1847 shortages in copper supply due to increased demand from the patent [ship] sheathing, railway engine manufacture and the brass industry itself meant export to foreign markets had to be restricted in order to satisfy the domestic market (*ibid*, 331–334).

By the mid 19th century the majority of the brass produced in Britain was supplied by imported copper. World consumption of copper had reached 90,000 tonnes by 1866, of which Britain consumed 60,000 tons. The principal sources were Chile, which exported 48,000 tons, and North America that contributed 5,000 to 7,500 tons. On average the UK mines only produced around 15,000 tons in the 1850s and 1860s (Aitken 1866, 257). Similarly by the mid 19th century, zinc was largely imported. The UK production of ore was only 4,000 tons by the 1860s, insufficient even to supply Birmingham, which alone consumed 11,000 tons. It was generally imported by the Silesian and Vielle Montagne companies (*ibid*, 261). By the late 1850s and 1860s the effect of the market for smelted copper from South America was beginning to be felt in the UK. This resulted in the demise of the second copper association, as independent smelters flourished, having access to new ores from 1857–1866. The war between Spain and Chile was to affect copper prices at this time and prices slumped drastically. After the 1860s the price of copper was more closely controlled by international markets and the direct influence of UK producers was reduced (Toomey 1985, 337–340). The decline in the smelting industry in Swansea was to have a direct effect on the Birmingham manufacturing trade - as smelting declined fabrication increased, bringing it into direct competition with the Birmingham producers.

Zinc, the other major constituent of brass, was also regulated by trade associations. Much was produced in the spelter (zinc) houses of Swansea and this was also controlled by trade associations. The parallels with the copper industry can also be seen as the century progressed, and increasingly spelter was introduced from abroad. Companies such as the German Metallgesellchaft controlled zinc prices up until the First World War (Toomey 1985, 337–340).

BIRMINGHAM'S INVOLVEMENT IN THE BRASS METAL TRADE

Chris Hewitson and Ray Shill

The Birmingham brass industry prior to the 19th century

Birmingham has been a centre for metal manufacture since the medieval period. The presence of four forges for farriers, smiths and at least one armourer in the borough rentals of 1296 and 1344–1345 attest to this (Demidowicz 2008, 18–25). By the end of the 16th century Birmingham represented the principal market for iron in the region. This is suggested by the fact that in 1592 Sir Thomas Middleton sent a clerk to establish 'the usual prices of iron about Birmingham' before establishing his Willoughby Ironworks in North Warwickshire (Pelham 1951, 18). A century later the landscape of urban Birmingham was dominated by small metal workers. A survey of the smiths' hearths undertaken in 1683 (WRO QS 11/c/61), showed a distribution along Digbeth/ Deritend High Street and on Bull Street, New Street and Edgbaston Street. This has been confirmed by the remains of ferrous metalworking from the 16th and 17th centuries found during archaeological work at Park Street (Nicholas 2009, 227–235) and along Digbeth, High Street Deritend and High Street Bordesley (Hewitson 2011, 43–52; Hewitson and Rátkai forthcoming).

Birmingham's early iron trade had flourished due to a variety of factors, including the abundance of local sources of raw material such as iron and coal, available from the South Staffordshire region (King 2003, 117-146) and water-power to fuel the forges and furnaces on the Tame Valley system (Stephens 1964; Dilworth 1976). Yet it continued to flourish because of the presence of skilled metalworkers and of a growing urban middle class that could take advantage of the improved economics of the late 17th century (Hewitson 2011, 43–52). This resulted in diversification of the industrial base of Birmingham through the production of buckles, buttons, toys, guns and the introduction of the non-ferrous industries (Rowlands 1975, 129).

Brass and copper manufacture in Birmingham took part in this diversification process. The first traces of brass and copper manufacture in Birmingham date from this period. Crucible fragments dating from the very end of the 17th century were recovered during excavations at Park Street. These were found on land plots of brass founders and gun makers. The crucible fragments had traces of non-ferrous metals, such as copper, brass, gunmetal, zinc, pewter and 'tutania' (Nicholas 2009, 227–235). Elsewhere, crucible fragments (recovered from excavations at Rea Street) contained copper alloys and zinc (Duncan *et al* forthcoming). Overall this provides tentative archaeological evidence for non-ferrous metalworking dating to the late 17th to 18th centuries. The rise was rapid, and by the mid 18th century Birmingham had risen to become the principal centre of brass and copper manufacture in Britain (Hamilton 1967, 122).

The supply chain of Birmingham's metal manufacturers also changed. The local supplies of raw materials were insufficient to supply the demand of Birmingham's manufacturers and this meant that by the 18th century it became a market for imported iron from abroad. The cessation of the Great Northern War between Sweden and Russia in 1721 saw an increase in iron production, and this supplied the Birmingham markets via the ports of Bristol and Hull (King 2003; Evans and Rydén 2007, 55–56, 186–187). The supply of brass to Birmingham came from the Low Countries in the late 17th century (Hamilton 1967, 134–139). In the 18th century Bristol, Cheadle and later Macclesfield developed as English manufacturing centres, and brass in ingots was imported into Birmingham (Aitken 1866, 240; Day 1973, 101–102; Morton 1983, 13–14). A solitary brass works was established in Birmingham during the middle of the 18th century: Turner's brass works on Coleshill Street was the first established in Birmingham in 1740. It was described by the Swedish Industrial spy RR Angerstein: 'The brassworks...belongs to Mr Turner and consists of nine furnaces with three built together in each of three separate buildings. The furnaces are heated with mineral coal, of which 15 tons is used for each furnace, and melting lasts ten hours' (Berg and Berg 2001, 38). The sheer quantity of raw materials, copper, calamine and coal required for brass production combined with Birmingham's poor transport links meant that large-scale brass production was not economically viable for much of the 18th century (Hamilton 1967, 214–216). It was much more cost effective to import ready processed ingots, which could be then rolled and manufactured into a variety of goods.

This changed towards the end of the 18th century, when the arrival of the canals reduced the cost of transporting raw materials (Wise and Johnson 1950, 181). The production of copper from 1771 to 1780 fell from 3,347 tons to 2,932 tons. Consequently, the price of brass rose from £72 per ton to £84 per ton (Day 1973, 102). This increased the potential profits of the brass manufacturers whilst drastically reducing those of the brass founders. The monopoly of Bristol and Cheadle brass makers over the Birmingham brass founders influenced the decision of the latter to become independent. On the 9th October 1780, a 'serious address to Birmingham merchants and manufacturers of hardware' was published in Aris's Gazette. The writer of the address, possibly Matthew Boulton (Aitken 1866, 243–244), urged the brass founders to build their own brass house:

> ... shall so Respectable a Body of Merchants and Manufacturers become the Dupes of a set of Capricious Monopolists in the articles of brass and

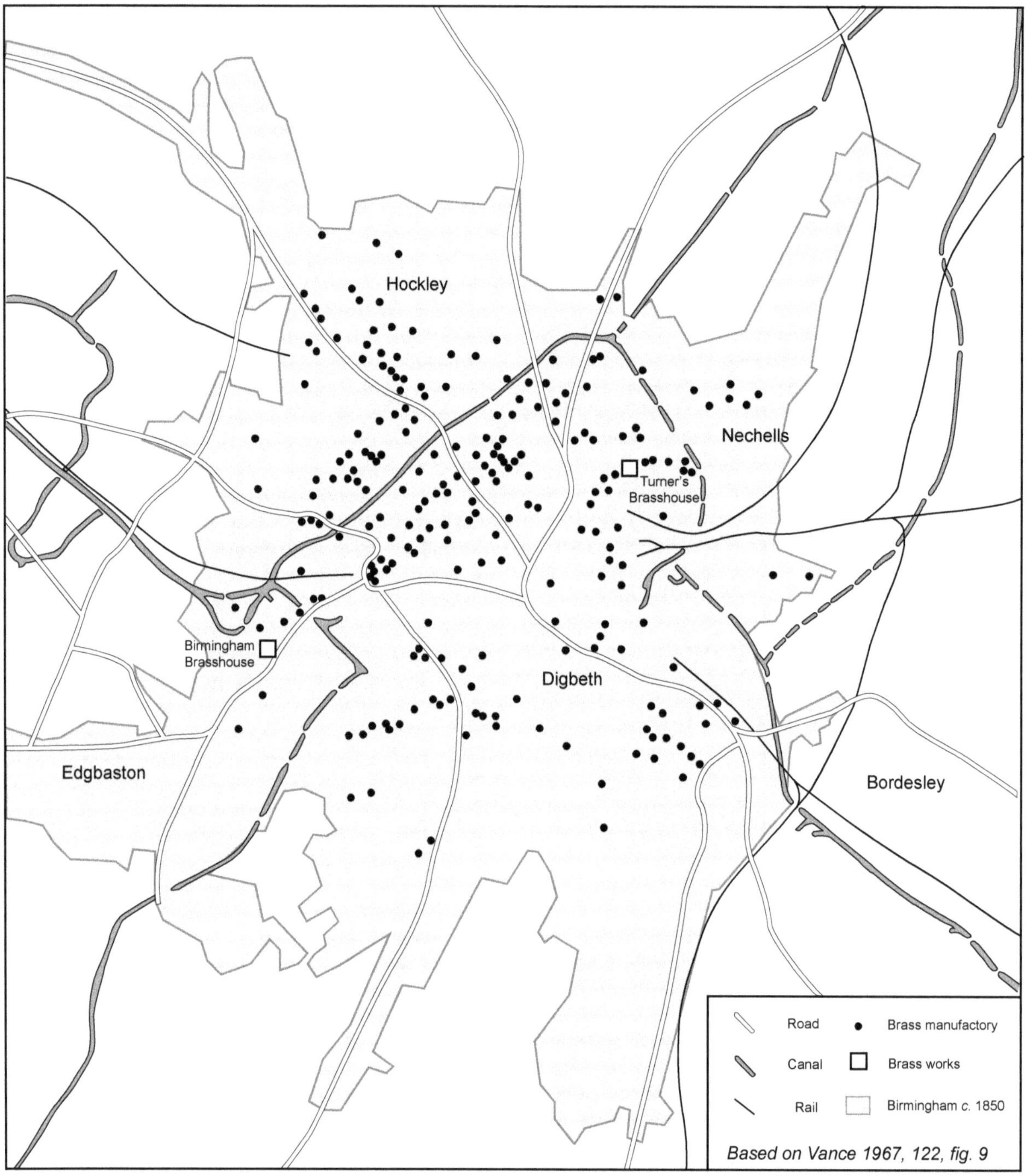

FIGURE 2.3 THE DISTRIBUTION OF THE BRASS INDUSTRY IN BIRMINGHAM

> spelter on which their trade depends? … be no longer governed by Strangers when you have the power to help yourselves at home. …the making of Copper and Brass is familiar and without risqué, what then should prevent the Merchants and manufacturers from making their own metals (ABG 1780).

A fund was raised by subscriptions of £20,000 by local manufacturers and the Birmingham Metal Company was formed on the 2nd February 1781. The headquarters of the new company, The Brass House, was erected in 1781, 'by the canal in Broad Street with … corpulent tunnels or tapering chimneys which reared behind, under each of which were two furnaces' (quoted in Hamilton 1967, 124). This ultimately broke the monopoly of the Bristol and Cheadle companies, although they still maintained a share in the Birmingham markets for a number of years to come.

The majority of brass in the 18th century continued to be used for the production of small products known in the local market collectively as 'toys', for the production of buckles or for the demands of other trades, notably the gun, coining and engineering trades (Wise and Johnson 1950, 181). The scale of production meant that the majority of objects could be produced in small manufactories or within a semi-industrial domestic setting during the 18th-century.

The industrial sector originally lay in areas south of the market square described in a topographical letter of 1755: 'the lower part is filled with the workshops and warehouses of manufacturers, and consists chiefly of old buildings' (quoted in Wise and Johnson 1950, 177). Excavations on Park Street (Patrick and Rátkai 2009) and Digbeth and Deritend (Forster *et al* forthcoming) have confirmed the location of the 18th-century manufacturers. Towards the middle of the 18th century there was a movement to new estates situated north of the town centre, such as the Weaman Estate, associated with gun manufacture (Wise 1949), the New Hall estate, that became known for buckles and toys (Wise and Johnson 1950, 178), and ultimately to the Jewellery Quarter (Cattell *et al* 2002, 8–23. The zones of Birmingham's industrial brass industry are discussed in Chapter 9.

As workshops increased in size they developed from dwellings and were modified to include such features as 'warehouses, shops and other outbuildings' located to the rear of the original terraced house (Wise and Johnson 1950, 178). The process of the industrialisation of domestic properties is well attested in the Jewellery Quarter (Cattell *et al* 2002, 8–23) and examples of shopping for workers are still visible today. The factory system, in part borrowed from the silk industries of the Derwent valley, Derbyshire, was epitomised by the Soho Manufactory to the northwest of the town started by Matthew Boulton in 1764 (Gale 1946; Jones 2008).

The 19th-century Birmingham brass industry

Brass manufacture was a staple trade of Birmingham by the early 19th century, due in no small part to the demand for brass products for engineering and domestic purposes (Wise and Thorpe 1950, 215). The original domestic arrangements of workshops were replaced in the second quarter of the 19th century by larger factories, as firm sizes increased. This was particularly characteristic of the north and northwest areas of the town (Wise and Thorpe 1950, 216) of which the Cambridge Street site is an example (see Fig. 2.3). The larger manufactories included Messenger and Son (Broad Street), T Smith and Sons (Bartholomew Street) and Robert Walter Winfield (Cambridge Street).

The most important development of the first half of the 19th century in brass manufacture was strip casting (Aitken 1866, 267–268). The technique of strip casting (see below) meant that copper and zinc metals could be mixed with relative ease. This meant that brass manufacture and brass founding were no longer separated and Birmingham brass workers could produce their own brass within their factories. The rapid development of steam technology pioneered in the city and adapted for use in stationary machines also meant manufactories were not tied to water-powered locations (*ibid*, 242–243). Improved canal and road transport also overcame the disadvantage of an inland location. By the mid 19th century brassworkers were distributed throughout the city (Vance 1967, 95–127) and brassworking flourished in Birmingham

Like a century before, the beginning of the 19th century saw diversification of the Birmingham brass industry. There were businesses that specialised in brass casting, supplying the trade with rough (unfinished) or finished casting. In addition to the foundry there was the trade of stamping that grew with demand. Stamped brass foundry (rough and finished) work encompassed a range of products, such as mouldings for furniture ornaments, girandoles, picture and looking glass frames, coffin furniture, door handles, door knobs and cornice-pole ends. A third type of manufacture shaped metal items by spinning the metal on a lathe. In the mid 19th century the industry was at its height. From 1831 when it was estimated 1,785 persons were employed and 1864 where 9,500 persons were employed, the industry expanded by a factor of six (Aitken 1866, 361–362). William Aitken's comprehensive review of the Birmingham brass industry (1866, 225–380) still represents one of the best sources for understanding many of the processes at work. He discusses a number of different areas of the industry:

Cabinet, Bell and General Brass Foundry. This discipline made brass work for the cabinet makers and carpenters, and included items such as castors, hinges, bolts, table fastenings, handles, knockers, railings for desks, balustrades for staircases, picture frames, hat rods, screens, poles and fire guards. R W Winfield specialised in general brass foundry, chandeliers, gas fittings, as well as rolling, tube making and wire drawing (Aitken 1866, 274–85).

Rolled Brass, Wire and Sheathing. Rolled brass was a material on which the stamped brass founder depended, without it, soldered brass tube, patent cased tube or cased stair rods could not be made (Aitken 1866, 310–321).

Tube making. Tubes included welded tubes, cased tubes (brass on iron) and seamless brass tubes for locomotive boilers and they were also used to supply the bedstead, stair rod and curtain pole trades. The technology that developed the seamless brass tube found a particular use within locomotive and portable engine boilers. Birmingham and the Black Country had several important tube mills that supplied the railway company workshops or the independent railway and road locomotive builders (Aitken 1866, 321–331).

Stamped Brass Foundry. Stamped brass foundry involved brass being shaped in the press and then passed onto a lathe for machining. Such skills were essential to the making of fittings for the beer, gas and plumbing trades (Aitken 1866, 292–310).

Cock making, Plumbers Brass Foundry and Engineers Brass Foundry. Manufactured items included cocks for plumbing, beer engines and engineering purposes (eg stationary engines, portable engines, road and railway steam locomotives). An important line was the plumber's trade, where cocks, tubes and other parts were made from brass. These became a major industry when improvements to domestic life were introduced such as improved water

TABLE 2.1 THE DISTRIBUTION OF THE TRADES IN THE BRASS METAL INDUSTRIES *C.* 1895

Strip casting, metal rolling and wire drawing	5,500
Chandeliers, gas and electric fittings, art metal, oil lamps, burners, water and steam fittings, bronzes etc	10,500
Cabinet and builders fittings, fire brasses, railway, ship and coach furniture, bedsteads, copper work, bells, coffin furniture, stamping, spinning, pressing etc	14,500
Ornaments, coach harness, bicycle fittings, mathematical and musical instruments, button and military ornaments, coins, medals, and checks, cartridge cases, tinplate furniture, locks, gilt toy work, photographic and scientific apparatus etc	8,000
TOTAL	**38,500**

supply and sanitation that resulted from public health improvements from the 1850s. Brass components were also required for beer engines that pumped beer from the casks and barrels at the many beer shops, inns and public houses across Britain (Aitken 1866, 285–291).

Gas Fittings. During the early 19th century new uses were created for town gas in the home, work and for use as street lighting. Town gas started to be used extensively from the 1820s and brass was used extensively in the making of gasoliers and components of lighting systems (Aitken 1866, 340–352).

Lamp Making. A large variety of lamps were made for household, carriage, mining and railway use. Birmingham had several lamp makers that incorporated brass, plated iron and glass into oil lamps for household, carriage, mining and railway use. By the end of the 19th century acetylene lamps became another local product (Aitken 1866, 331–340).

Naval Brass Foundry/ Military Ornaments. Brass was used for the casing of various instruments and fittings found on board ships and other vessels and also for buttons, military ornaments (eg medals) and whistles (Aitken 1866, 353–356).

Metallic Bedsteads. The skills of working with iron and brass were brought together to make metallic bedsteads. Metallic bedsteads were a popular choice for military barracks, hospitals and workhouses and there was also an important foreign market (Peyton 1866, 624–627).

Nails, Pins, etc. Small items that included nails and pins in iron and copper were a long-established industry in Birmingham and the Black Country. The 19th century saw the mechanisation of these processes and the growth in the industry (Martineau 1866, 613–616; Phipson 1866, 601–605).

Art Metal, Bicycles. Brass was used in the second half of the 19th century in the new trades of art metal, bicycle making, church brass work, stained glass and electric light fittings (Aitken 1866, 536–551).

Those who worked in the brass trade carried out a wide range of tasks that were skilled, semi-skilled or unskilled. They included the strip casters who made the brass, the pattern makers who made that shapes for the brass to be cast, the founders who poured the brass for the rough castings, the finishers who filed the rough edges down, the press workers who produced the stamped brass foundry, the burnishers who polished the finished products and the lacquerers who coated the surfaces and preserved the rich yellow colour of the finished article. By the end of the 19th century the brass industry had developed into a series of distinct individual occupations which were interlinked, often within the same manufactory but also as a series of individual manufactories. W J Davis attempted to produce a summary of the brass trade development in 1892 where he reviewed the history and gave numbers of people employed in the Midlands brass trade. He decided there were four divisions in the trade (Table 2.1; Davis 1892).

What the table demonstrates is the complexity of what was commonly referred to as the Birmingham brass trade as it moved into the 20th century.

Into the 20th century

The expansion of the Birmingham brass industry over the second half of the 19th century continued undiminished into the 20th century. The industrial structure of Birmingham was more complex than the principal 'four trades of guns, jewellery, buttons and brass' (Wise and Thorpe 1950, 216) of the early 19th century. The manufacture of brass and other non-ferrous metals was a trade that was changing with the needs of fashion and other industries. It continued to grow by serving the new industries of the area and few firms specialized exclusively in any particular section of the trade (Stephens 1964, 140–208).

Technological advance fuelled the development of the brass industry. In 1894, Alexander Dick, founder of Delta Metal, expanded the process of extrusion to copper and brass (Drozda and Wick 1983, 11–13). Extrusion had previously been confined to softer metals like tin and lead but could now be used to create objects of a fixed cross-sectional profile as the brass was pushed or drawn through a die. Brass forging came into existence at the start of the 20th century in Britain and in Germany. Extruded brass rod was cut into lengths and heated in an open-ended furnace. The rod section then went to a power press to be shaped by a die (Metal Industry 1922). The First World War saw the use of hot pressings and stampings instead of castings, bringing the advantages of increased speed in manufacture and of a reduction in skilled work and machining (Hiley 1957, 33). The capstan lathe came into general use at the turn of the century, replacing the ordinary brass workers' lathe (Hiley 1957, 63–64).

Keate's (1910) list of the trades (which he admitted was by no means comprehensive) includes many of those in existence in the 19th century: metal merchants, brass casters, designers, modellers and pattern makers in wood, brass or iron for the trade, cabinet brass founders, railway and naval brass founders, engineers and plumbers, brass founders, stampers and spinners, art metal and repoussé workers, bell founders, harness, motor car, chandelier, gas and electric fittings manufacture, buttons, badges, medals, military and naval ornament makers, nails, rivets, bolts and nuts, bedsteads, wire workers and weavers, pin makers and brass chain makers, rolling, wire and tube mills and specialist work.

The West Midlands (Staffordshire, Warwickshire and Worcestershire) increased its dominance in the British brass and copper industries. An estimate suggested 37,700 workers were employed in the industry in Birmingham alone in 1921 (Cook 1936), a figure that suggests little decline had occurred since 1895. The ammunition industry fuelled by the demands of the First World War created a large demand for brass shell cases. A rising population encouraged a demand for domestic items such as cabinet and plumbers' brass foundry and whilst gas lighting was decreasing there was an increase in demand for electrical products. The number of rolling mills in the district benefited from an increased demand for metal-strip and metal-rolling mills remained an important supplier of rolled brass, tubes and wire throughout the 20th century.

Change in the structure of the brass industry occurred in the first two decades of the 20th century. Firms were larger and non-specialized. Patterns of expansion saw the growth of individual trades but also the amalgamation of firms as they grew in scale during the same period. Firms grew by vertical or horizontal integration or simply by entering other trades. Vertical integration (Winfield's provides an example), the process by which firms bought up elements of their supply chain, and horizontal integration where firms bought up rivals in a particular trade, had been occurring throughout the 19th century, but increased in scope into the 20th century (Stephens 1964, 148). No longer were businesses confined to a single site but they occupied multiple sites or had interests in locations well beyond Birmingham. Guest, Keen and Nettlefolds, were an example of both horizontal and vertical integration. It was formed in1902 when the south Wales iron and steel concerns of Guest and Co. and the Dowlais Iron Co., and the Patent Nut and Bolt Co. of Smethwick absorbed Nettlefolds Screw Manufactory (Stephens 1964, 151; it was located on Broad Street, see Chapter 9). The number of employees increased, for example Evered and Company had 1,000 employed in their Birmingham works and 300 in their Smethwick works in 1900 (Stephens 1964, 146). Other major industrialists and firms in the local brass trade included the T R Baylees (Kings Norton Metal Company Ltd) J H Cartland (James Cartland and Sons Ltd), Neville Chamberlain (Elliots Metal Company Ltd), G H Dugard (Heaton and Dugard Ltd) R G Evered (Evered and Co Ltd), Ralph Heaton (The Mint) and Sir Gerald Muntz (Muntz Metal Ltd) (Keate 1910).

However, after 1920 demand for brass decreased. Fashions changed and new metals such as aluminium, phosphor bronze and stainless steel entered the market (Stephens 1964, 200–208). For example in aluminium casting, the Birmingham Aluminium Casting, that initially occupied the Cambridge Street site from 1903–1920, had as many as 2,000 workers in 1927 in their new Smethwick works (Allen 1927, 420). Just before the First World War, the plastics industry began to develop new materials such as thermoplastics as alternatives to metals (Rowell 1939, 35) and they began to replace brass as an all-purpose material in the domestic and foreign markets. Towards the end of the 20th century few firms still engaged in the supply of rolled non-ferrous metals.

CHAPTER 3: THE 18TH CENTURY BASKERVILLE HOUSE AND THE FOUNDATION OF THE CANAL (PHASES 1 AND 2)

Will Mitchell and Ray Shill

AN 18TH CENTURY ESTATE: BASKERVILLE AND EASY HILL HOUSE (PHASE 1: 1745–1810)

Historical background

Ray Shill

The site lay within an area of farmland adjacent to the Harborne Lane (Broad Street). This was outside the medieval core of Birmingham centred on St Martins Church that included the whole of New Street by the end of the 17th century (Foster 2005, 142; Forster and Rátkai 2008). It continued to be largely undeveloped in the post-medieval period. Westley's map of 1731/2 does suggest the area to the south of Broad Street had been developed for brick manufacture and it is depicted as the Old Brickiln Close on Bradford's map of 1750 (Fig. 3.1).

By the early modern period, Birmingham had begun to develop rapidly and with this expansion came the rise

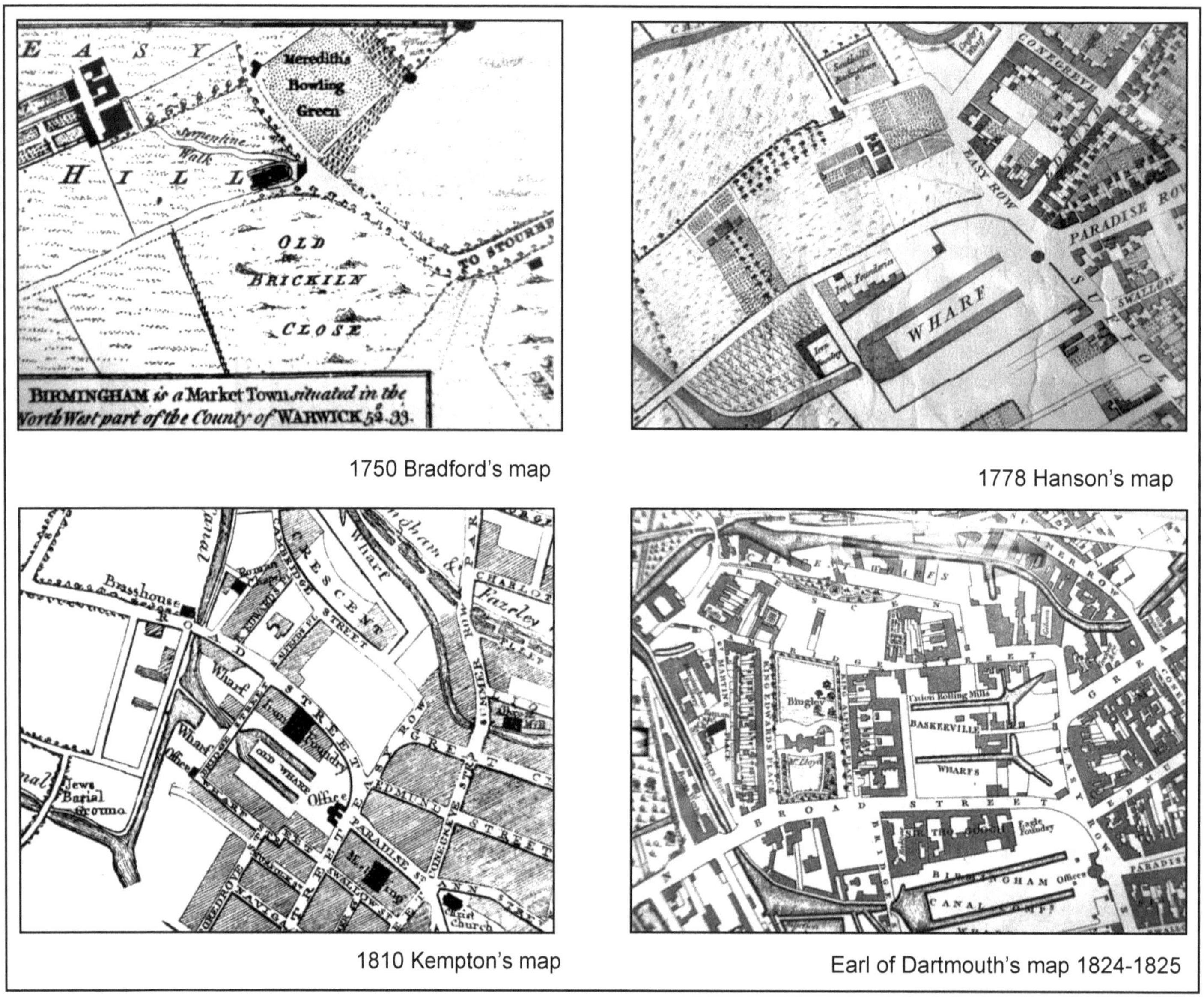

1750 Bradford's map

1778 Hanson's map

1810 Kempton's map

Earl of Dartmouth's map 1824-1825

FIGURE 3.1 THE CARTOGRAPHIC DEVELOPMENT OF THE SITE PRIOR TO THE CAMBRIDGE STREET WORKS, 1750–1824 (MAPS ARE COURTESY OF BIRMINGHAM ARCHIVES AND HERITAGE)

FIGURE 3.2 BASKERVILLE HOUSE (COURTESY OF BIRMINGHAM ARCHIVES AND HERITAGE REF: L42.01 12/475)

of a new middle class who began to build villas and houses removed from the traditional medieval centre of Birmingham. John Baskerville, the renowned printer and typesetter constructed Easy Hill house in 1745 (Benton 1914, 4). The house was first visible on Bradford's map of 1750 (Fig. 3.1) and the building faced east, had its own grounds and a carriage drive joined Easy Row. At the northeast corner was a bowling green north of which was a perimeter road that later became Cambridge Street. First visible on Hanson's map of 1778, the Birmingham Canal (Newhall Branch), which loops around the site on its north, west and south sides, was in place by the later 18th century having been constructed in 1768–1772 (Foster 2005, 142). To the south of Harborne Lane (Broad Street) was the newly laid out Eagle Wharf.

Easy Hill House remained in the Baskerville family until the death of Mrs Baskerville. The house and grounds were advertised for sale by auction, the sale to take place on 19th May 1788 (ABG 1788). In the advertisement Baskerville House was described as '*a handsome Hall with an elegant mahogany staircase and gallery, three parlours, two China Closets, three bed chambers on the first floor and four lodgings in the Attic*'. There were marble and stone chimney pieces throughout the building and 'good' cellars. Out offices included kitchens with servants' rooms over, a butler's and common pantry, brew house, two pumps (one hard and one soft water). There was a four-stalled stable and coach house, a garden with green house and garden house and also a workshop and warehouse. The grounds comprised seven acres, part of which was laid out as shady walks, adorned with shrubbery, fish ponds and a grotto. Much was enclosed within a brick wall. Baskerville's home was later occupied by John Ryland, pin maker. Ryland's son, Samuel, assisted Joseph Priestley to escape a mob in 1791 (15th July; Rose 1960). This band of rioters set fire to Baskerville House causing serious damage and it fell into disrepair after the so called Priestley Riots. The building became a ruin for some twenty years and is depicted as such in the late 18th century (Fig. 3.2). The whole estate was changed forever with its sale to Thomas Gibson (ABG 1811; NA Rail 810).

Archaeological evidence

Will Mitchell

Evidence of occupation relating to the earlier medieval period was not forthcoming and the scarcity of evidence in the historical and archaeological record confirms this area was far from the core of medieval Birmingham. It therefore appears likely that the area was used for agricultural purposes. Only one residual sherd of green glazed medieval pottery was retrieved. This came from a potential fragmentary cobbled surface on the northern side of site (1456; Plate 3.1).

PLATE 3.1 THE 18TH-CENTURY GROUND SURFACE

There was no evidence of structures relating to the Easy Hill (Baskerville) house period of occupation. However, many of the buildings relating to this would have lain to the east, outside the excavation area. The excavation area is likely to have been located upon the extensive back gardens and the open land surrounding this property. Any outbuildings relating to this early phase were removed due to the extensive developments undertaken in the early 19th century.

Several residual sherds of blackware (17th to 18th century) were identified across the site. Many of these came from the lowest levels of overburden deposits beneath the brassworks or from around the cobbled surface (1456; Fig. 3.3). Located at the southernmost end of the excavation area and south of the Baskerville basin, was disturbed buried topsoil (1546; Fig. 3.3). This contained a large quantity of broken garden pots which were associated with this period.

THE DEVELOPMENT OF INFRASTRUCTURE: THE CANAL BASINS (PHASE 2: 1810-1820)

Historical background

Chris Hewitson and Ray Shill

The development of the canals in the late 18th century in Birmingham is well documented and discussed (eg Wise and Johnson 1950; Hadfield 1985; Shill 2002), and its role in contributing to the superiority of Birmingham's brass trade was profound. Unlike the iron industry which had the advantage of local supplies before the 18th century, the brass industry was always reliant on importation of raw materials. By the mid 18th century Birmingham's roads were in terrible repair. In 1726 Parliament appointed a committee examining the state of Birmingham's roads and found them to be 'ruinous and impassable in places' (Hamilton 1967, 214). It was not until the establishment of the first canal from Birmingham to Wolverhampton, linking the town to the Staffordshire and Worcestershire canal via the Staffordshire collieries in 1772, that reliable bulk transport was established. The route (now the Birmingham Old Line) was followed by an extension to the Coventry Canal near Fazeley (Birmingham and Fazeley) in 1783 and links to Warwick and the River Severn at Worcester in the 1790s (Hadfield 1985, 63-73). The provision of communication routes was an important factor in Birmingham's ascendancy in the brass trade. Development around the canals in general was rapid, and by 1840, on a two mile stretch of the Birmingham and Fazeley canal there were 124 wharfs and canal-side industries (Crowe 1994, 74).

The Birmingham Old Line originally terminated on the Newhall estate and was later called the Newhall branch (Hadfield 1985, 63–65). It ran southwest–northeast and skirted the contours of the higher ground on which

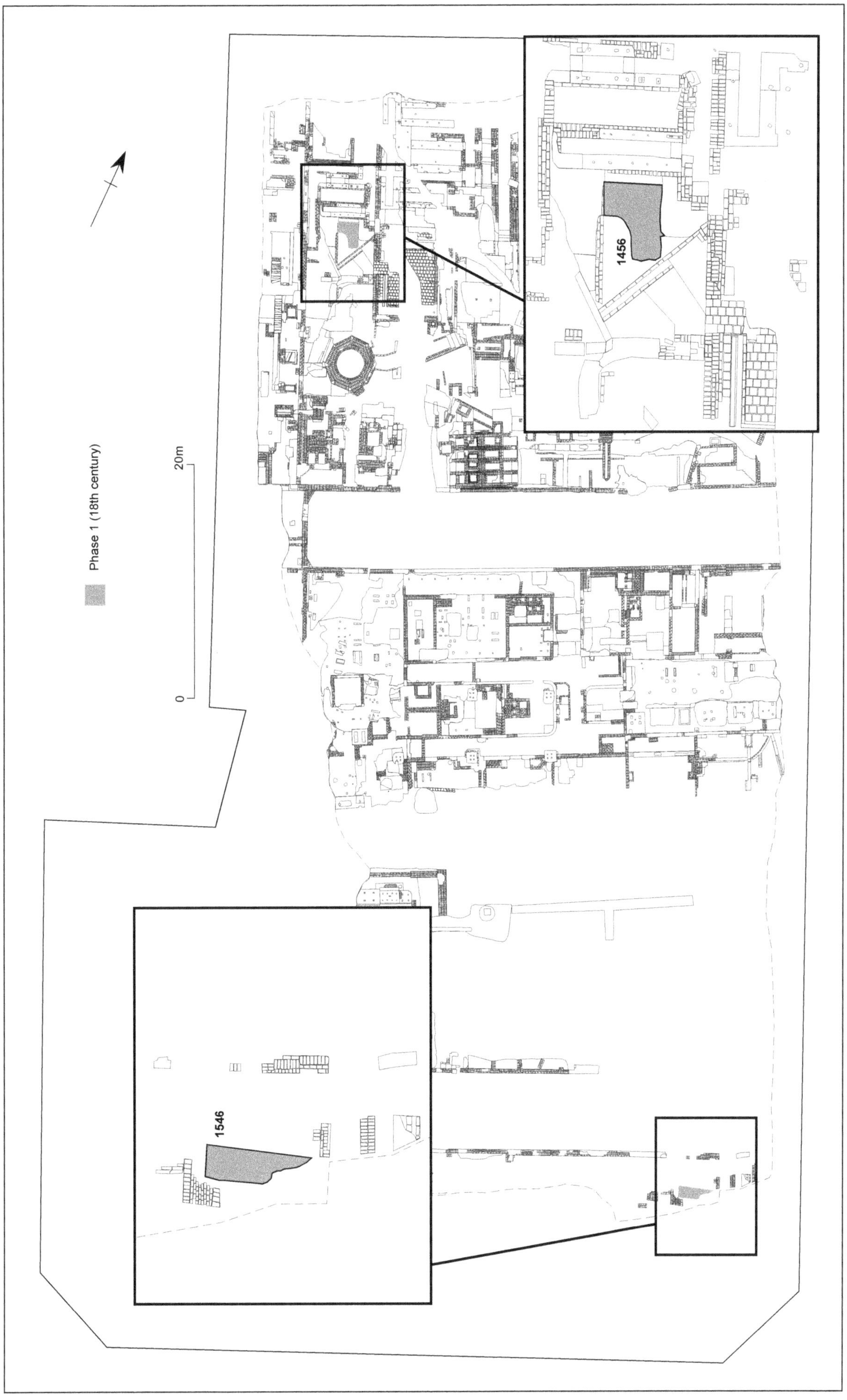

FIGURE 3.3 PHASE 1 EXCAVATIONS ON THE SITE

the Cambridge Street site was located. Thomas Gibson laid out a series of wharves within the site in *c*1810 (see below), joining them via a short link branch from the Newhall branch to the north. They are not illustrated on Kempson's Map of 1810 (Fig. 3.1). The first image of the wharves is visible on the Earl of Dartmouth's Map of 1824–1825 with several buildings initially laid out around them including the Union Rolling Mill. They are referred to as 'Baskerville Wharfs' at this time (Fig. 3.1).

Thomas Gibson was born at Corley near Coventry. He came to Birmingham and after initially working for a merchant called George Russell, he entered into partnership with William Shore ironmonger as an iron and steel merchant (BJ 1840). The partnership was extended in 1803 to include Mr Tomlins. Gibson, Shore and Tomlins were recorded as iron and steel merchants at Suffolk Street and Digbeth (Chapman 1803). The Suffolk Street property was short lived but property in Digbeth included the Town Mill, between Digbeth and Bradford Street (Stephens 1964). It belonged to Thomas Gooch and had formerly been leased to the Lloyd family as a slitting mill and Gibson, Shore and Tomlin's continued to operate it as such. They also owned a warehouse for iron on Great Charles Street, near the Crescent that bordered the Birmingham Canal Navigations, Newhall Branch (Chapman 1808). Tomlin's had ceased to be a partner by 1808 and Shore left the partnership in 1814 (ABG 1814). This left Thomas Gibson and his family to develop the Baskerville estate and run the slitting mill in Bradford Street. He withdrew from the merchant, rolling and slitting trade during 1823 leaving the business to his sons (ABG 1823).

Gibson and Shore acquired the Baskerville property and bowling green at Easy Hill and during 1810 were advertising property to let, including the former Bowling Green (ABG 1810). It was a project of Thomas Gibson to open up the Baskerville Estate for industrial development and he conceived the idea of making canal basins on the estate. The concept involved the construction of a private branch from the Newhall Branch Canal into the estate and the making of a lock to raise boats to the higher level. Gibson's private branch was cut during 1811, and wharves, spoil and earth were advertised for sale in June 1811 (ABG 1811). The Earl of Dartmouth's Map of 1824–1825 shows two parallel basins (Fig. 3.1). The one nearest Cambridge Street became known as Gibson's Basin and the other nearer Easy Hill (later Broad Street) was known as the Baskerville Basin. Gibson's and Baskerville Basins provided a canal link to the main Birmingham Canal Navigations.

Gibson's Lock, a privately owned lock, appears to have been of the same dimensions as those on the main line, but with a greater rise than most. The lock was a single rise of approximately nine feet. There were two bottom mitre gates and a single top gate. The channel from the Newhall Branch passed under Cambridge Street at a bridge and the lock was placed close to the south side of this bridge. There was an engine house with a steam pump that raised water to the upper level. Those that leased land from Thomas Gibson also entered into agreements for the use of the lock and a complex set of rules was established for charging or not charging boats ascending the lock (BCA MS 86). Gibson arranged for a new mill, Gibson's Mill, to be made near the lock. Thomas Gibson died in 1840 and the estate was then administered by his executors. Responsibility for working the lock finally passed to Winfield's Rolling Mills, who periodically arranged for the canal to be cleansed. During May 1907 the Birmingham Canal Navigation Company was commissioned to replace the lock gates and fit new paddle gear (BCA MS1422 60/1/7/1).

Traffic along the arm ceased with the closure of Winfield's Rolling Mills. One of the last operators to use the canal was Leonard Leigh who brought loads of coal up to the mills and took away rubbish to a tip (BCA MS 86). Gibson's Lock was removed with the making of the foundations for Baskerville House. The canal cut to the lock also served the carriers' depots on the corner of Crescent Wharf, which included Thomas Sturland, the Shropshire Union Railway and Canal Carrying Company and then Fellows, Morton and Clayton and Co Ltd.

Archaeological evidence

Will Mitchell

Gibson's Basin

Gibson's Basin (Fig 3.4, Fig. 3.5; Plate 3.2) was aligned northeast–southwest and located north of site between the two sides of the Cambridge Street Works. The two sides of the canal walls were of different construction, suggesting the canal was refurbished during its lifetime. The base was 4.3m below the car park surface at *c* 139.7m AOD, and lined with puddle clay at *c* 4m depth. It was *c* 6.5m wide (21.32ft) and a length of 43m was preserved within the site. The northern wall was incomplete and this limited how much could be interpreted. Successive rebuilds were evident in the northern wall, and a distinctive re-facing was defined by two different types of brick. The inner core, and primary construction in 1810–11, (1206/1209) was constructed of hand-made, orange and dark red bricks (8½ x 4 x 2¾ in) in English bond. The re-facing (1145), which probably occurred during a period of major rebuilding in the 1850s, was built in engineering bricks (9 x 4¼ x 3 in), in English Bond. The bricks used were identical to those used for the fabric of the bedstead works.

The southern wall of the canal (1068) was also constructed entirely of this type of brick. It appears that this wall was entirely rebuilt in Phase 4 (1850s) at the same time as the construction of the Bedstead Works (see Chapter 5 below). There was a lowered loading bay on the northern (Rolling Mills) side (1146). Two reinforced steel girders were set into this, which may indicate the location of a crane. The location of the canal footbridges can be surmised from reference made to them on the Ordnance Survey mapping.

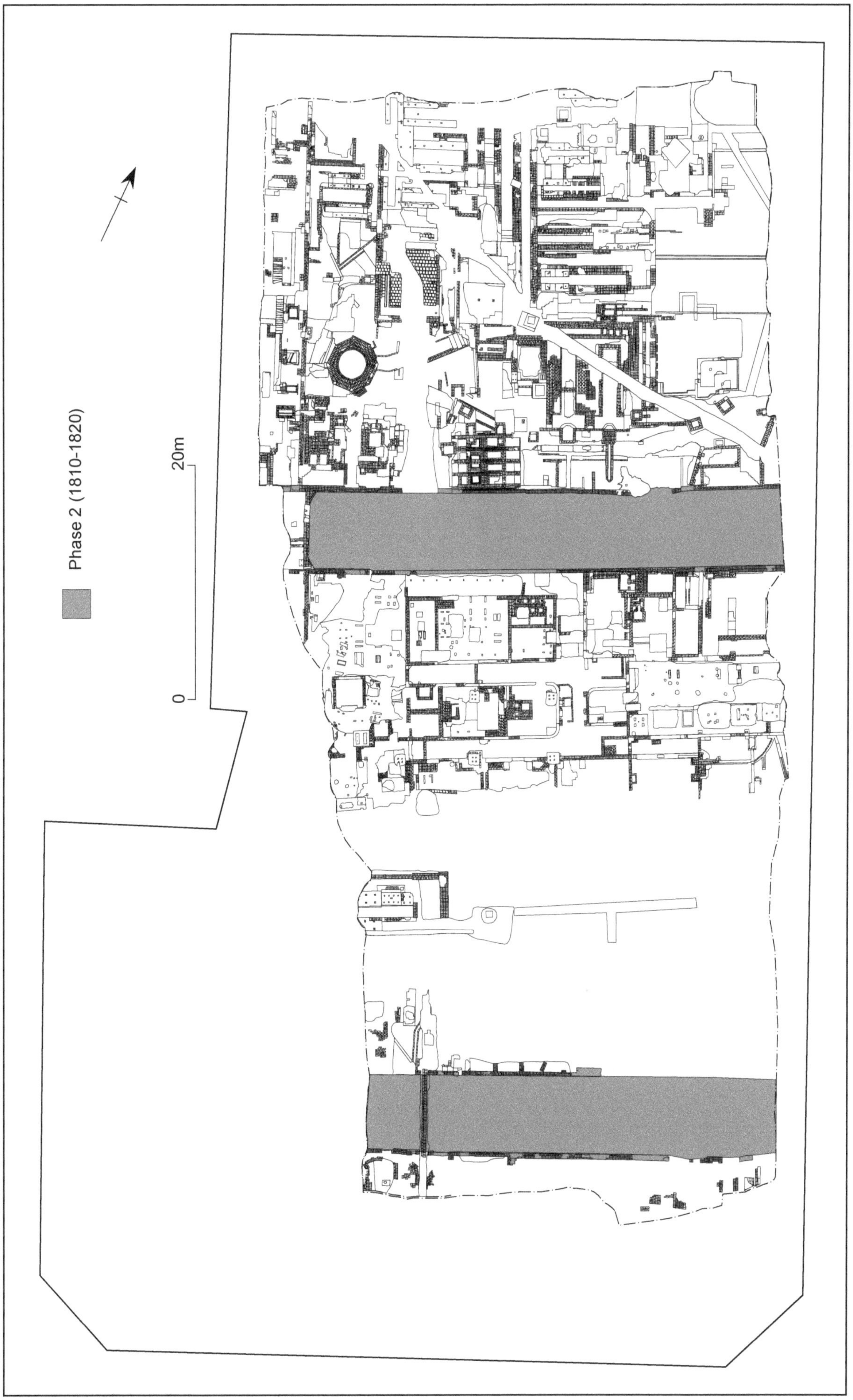

3.4 Phase 2 excavations on the site

Figure 3.5 R W Winfield and Co in 1840s (from Martineau and Smith's Hardware Trade Journal , Courtesy of Birmingham Archives and Heritage ref: LP65.641 49/2023)

These footbridges would have allowed pedestrians access to the two sides of the complex.

The western end of the canal (1499) was shortened between 1889 and 1897. The back of the canal was altered to stop at the western edge of the site, although originally it would have continued. The back wall corresponded with the boundary between the Tube Works and the Rolling Mill. It was constructed in a concave shape to match the front end of a canal boat. A further loading bay was created (1500) where an additional crane was perhaps located. A set of steps (1496) led down onto the loading bay, from the tube drawing mill. The rest of the canal was known to have been backfilled in 1936 during demolition for the proposed civic centre. The difference in height (AOD) between the canal arm identified in the evaluation, and the Newhall canal to which it joined gives some idea of the reason for the additional depth of the canal in this area. The difference in level of the canal towpath on the Newhall Arm of the Birmingham and Fazeley Canal (138.4m AOD as ascertained from the data from the Ordnance Survey 1st Edition) and within the site at Cambridge Street (*c* 144m AOD), suggests a difference in surface level of 5.5m.

Documentary, cartographic, and photographic evidence show that Gibson's Deep Lock, formerly located where Baskerville House now stands, east of the site, was deeper than standard locks in Birmingham. Locks in general raise water levels between 1.8m to 3m, so the additional depth of the canal arm within the site may be due to the water level being significantly lower than that of the adjacent ground level. It is possible that there was a roofed-in wharf

Plate 3.2 Gibson's Basin, east facing (Image by Aerial-Cam)

with cranes to allow shipments to be accessed to the level up to 4m below.

Baskerville Basin Canal

The Baskerville Basin (Fig. 3.4; Plate 3.3) was located *c* 47m to the south of Gibson's Basin. It was aligned northeast–southwest parallel to Gibson's Basin but unlike it, there appeared to have been very little alteration of the wall construction and it represented the original 1810–1811 building fabric. The canal walls (1539, 1540) were constructed of hand-made, red bricks (9½ x 5 x 3 in) in English bond. The canal was 6.6m (21.65ft) in width and a length of 35m was preserved within the site. The walls survived to a depth of 1.2m. The base of the canal was lined with yellow–brown puddle clay (1559). A black material made up of settled industrial material (1560) sat upon the clay base. Above this was the demolition backfill from the period 1922–1925 (1547) that contained building rubble and back-filled waste.

Discussion

Prior to the period of sustained industrialisation which occurred in Birmingham during the early 19th century, this area (the Baskerville Estate) was largely undeveloped. Much of the area was agricultural or private land, ripe for investment. The investment in the transport network made possible by Gibson after the construction of the canals opened up the Baskerville Estate. The area of Gibson's and Baskerville Basin became a profitable location for development and following the introduction of canal wharfs, industrial buildings began to occupy the former Baskerville Estate. The construction of the basins meant materials could be imported and finished goods exported, via the wider Birmingham canal network, to be sent further afield. The road network supplied the workforce and supported the canals in the delivery of goods and materials. Location was the key to the success of the industries that established themselves at the basins.

After the establishment of the wharfs and after withdrawing from the merchant, rolling and slitting trade, Thomas Gibson concentrated his efforts on the further development of his property. A trade depression that followed the end of the wars with France handicapped further industrial development of the Baskerville Estate but matters improved from 1820 when the Union Rolling Mill was established. This marked the beginning of the rapid expansion of the Cambridge Street Works. The change that would occur to the works is depicted beautifully in an image of the 1840s depicting Gibson's Basin, with the rolling mill and boilers located to the north (Fig. 3.5)

PLATE 3.3 BASKERVILLE BASIN, WEST FACING

CHAPTER 4: THE ROLLING MILL, WIRE MILL AND THE TUBE WORKS (PHASES 3, 4, 5 AND 6)

Chris Hewitson, Will Mitchell and Ray Shill

The raw-material section consists of strip casting-shops, metal rolling, wire drawing and tube mills, which are driven by an engine of 150 horsepower supplied by two large boilers. The principal goods manufactured in the department are brass and copper strips and sheets, brass and copper tubes, and brass and copper wire, of which large quantities are turned out (Visit of the Institute of Mechanical Engineers to the Winfield's Works 1897; IMechE 1897, 403–404).

The main area of the site (Area 1) consisted of two parts, those north of Gibson's Basin, the rolling and wire mill and the tube works, and those to the south, the bedstead works (see Chapter 5). This chapter will examine the area north of the canal. As the majority of the buildings were reused and incorporated into later developments few buildings were demolished to foundations level before rebuilding. This created well-preserved structural remains and stratified deposits, which facilitated clarity of phasing, and provided archaeological evidence for additional structures not mentioned in the documentary sources.

The earliest phase of the Cambridge Street Works was the Union Rolling Mill (Phase 3). The Union Rolling Mill was set up by a consortium to produce rolled brass in the 1820s. The earliest structures visible on site related to the engine house and power for the rolling mill and the rolling mill itself. The Union Rolling Mill's history changed radically when it became occupied solely by Robert Winfield in 1829, as he developed the Cambridge Street Works (Phase 4). Winfield invested heavily in the mill, expanding and renovating it. It was massively developed during the 1840s. This work included a purpose-built bedstead works to the south of Gibson's canal and the acquisition of the tube and wire drawing mills. The principal development of the site took place during Phases 3 and 4 between 1820 and 1900. After a change in ownership in 1897 the northern half of the site continued as Winfield's Rolling Ltd with some alteration in 1900–1936 (Phase 5).

TECHNOLOGY – THE DEVELOPMENT OF BRASS PRODUCTION AND PRODUCTS

Chris Hewitson

The production of brass

A number of processes were necessary to produce the brass that was consumed by the manufacturing industry in Birmingham in the 19th century. From the raw materials of copper and zinc a number of technical processes were carried out to produce the finished products; these included the smelting of the copper from its ore (primary processing), and the production of brass from copper and zinc (secondary processing). At this stage the brass was manufactured into products; the processes involved included casting of brass and the production of wrought brass material such as rolled brass sheets, drawn wire or tubes. These were all used in the general brass trade. Final manufacturing involved products passing through a series of hands in the factory system, using a variety of machines including the stamp, press and lathe which aided human skill and labour (Day and Tylecote 1991, 131).

Brass is an alloy of copper and zinc, with a variable composition usually alloyed in a ratio of 80/20 or 70/30. Small quantities of tin or lead might be added to create a metal for specific uses (a particular example being gunmetal). The combination of metals added strength and the result was an alloy that could be polished to make an attractive product.

Smelting of copper

The reverberatory furnace was developed in Bristol from the 1680s as a replacement for the traditional blast furnace method of smelting copper. The fuel was separated from the ore and its heat was carried into the ore chamber by a powerful draught created by the design of the furnace and the use of a tall chimney. The flame from a fire at one end passed directly over the ore and passed out through a flue at the other end. As described at the Hafod Works in South Wales the process was in six stages and involved calcification of the ores (eg roasting; the first three stages) before the fourth stage termed 'running the metal' that produced a 'white metal' of copper sulphate. This was then melted to remove the sulphur and refined to remove further oxygen. In all the process could take several days (Toomey 1985, 277–280).

Early brass production - calamine of calcification

Brass was first made through mixing copper with calamine (zinc carbonate ore) in a furnace. The method owed much in its origin to developments in glass production and relied on a large dome-like furnace, its base perforated with draught holes. Aitken (1866, 265) describes the process probably as seen at the Birmingham brasshouse. Centrally eight crucibles were arranged around a central 'king pot'. The charge of ground calamine, an ore of zinc and smelted copper in granules was placed in the nine crucibles and surrounded by fuel - coal reduced to coke. The furnace was then sealed and left for ten to twelve hours. The brass would be removed and all eight pots were emptied into the king pot, before it was exposed to the heat of the furnace once more. By this method Matthew Boulton estimated that two tons of copper and three tonnes of calamine should produce three tons of brass (*ibid*, 266). The resultant brass was called 'calamine brass'.

Later brass production - strip casting

A revolution in the brass-making trade happened during the 1830s with the increased availability of metallic zinc (spelter), and it became an easier task to mix the zinc with copper by a process that was often referred to as strip casting. Furnaces fuelled by coke heated up crucibles that melted copper tiles or ingots. The zinc was introduced and the mixture stirred with an iron rod or poker (Aitken 1866, 267–268). The process generated a white light and bits of zinc flew off as white flakes. This process was illustrated in an engraving made by A Morrow, which probably depicts strip casting at the Cambridge Street site (Becker 1883; Fig. 4.1).

The Bolton Works at Cheadle give an indication of the types of brass produced by (Aitken 1866, 238). Brass for wire and sheets was cast into plates; the mix generally consisted of two parts copper to one part zinc. More zinc was added to produce wire and ingot brass. They produced four grades of ingot brass;

BB – made of best copper, 2 parts copper to one part zinc.
BC – Inferior copper and slightly more zinc.
AM – Some copper and calamine, tainted with lead.
YY – Made of ash metal and other inferior metal.

The coke used in Birmingham was derived from the gas retort houses and referred to as 'soft coke' (Aitken 1866, 267). The close proximity of the Birmingham Gas Retort House on Gas Street to the southwest must have encouraged the use of 'soft coke' at an early stage in the Cambridge Street Works.

Cast brass

Following its production, brass making underwent two processes in order to make it useful for manufacturing products: cast and wrought brass. As the ingots of brass were made they went onto the founder, the roller or the wire drawer. The founder recast the brass for a variety of uses particularly brass casting, supplying the trade with

Figure 4.1 Image of strip casting at R W Winfield and Co in 1887 (from Martineau and Smith's Hardware Trade Journal , Courtesy of Birmingham Archives and Heritage ref: LP65.641 49/2025)

Figure 4.2 Image of metal rolling at R W Winfield and Co in 1887 (as previous, Courtesy of Birmingham Archives and Heritage ref: LP65.641 49/2027)

Figure 4.3 Image of wire drawing at R W Winfield and Co in 1887 (as previous, Courtesy of Birmingham Archives and Heritage ref: LP65.641 49/2028)

rough (unfinished) or finished castings. Casting brass involved making patterns of the shapes required, often in wood, and using these patterns to form shapes in sand moulds. Molten brass was poured into the moulds with care to avoid bubbles of air becoming trapped in the casting. The product was a rough casting that needed to be filed, machined and polished before it was ready for use or sale (Aitken 1866, 271-4).

Wrought brass, rolled brass, wire and tubes

Wrought brass included the early manufacturing processes of wire drawing and battery. In the latter case brass articles were shaped and hollowed by hammering. Rolling metal at the water mill was a subsequent development. The rolled sheet brass went to the stamped-brass foundry trade. A third type of manufacture, which used a lathe, produced a method of spinning the metal into certain shapes.

The process of producing rolled sheet brass, traditionally known as 'latten', involved a number of stages (rolled sheet brass at Winfield's). Initially brass was melted down in a furnace, using raw materials of copper and zinc (known as spelter), to which a variety of other metals such as tin, lead and cadmium, or alternatively scrap metal were added to create the various compositions of brass. Once melted, the molten metal was poured into ingots and allowed to harden. These were produced not by sand casting but by using cast iron, or slabs of granite coated in oil and charcoal powder. From here the ingot was passed to the 'breaking-down' rolls and repeatedly rolled until it cooled to a point where it began to become brittle. This process is known as 'hot-rolling' (The rolls at Cambridge Street are illustrated in Fig. 4.2). Due to the fact that the brass produced was brittle and liable to break under tension, it was necessary to strengthen it by a process called annealing. This involved the heating of the brass to high temperatures described as 'blood-red heat' prior to rapidly cooling. The brass was then rolled again in a process called 'cold-rolling'. The annealing and cold-rolling was repeated until the brass was at the correct thickness before being coiled (Aitken 1866, 311–312).

To produce wire the flattened rolls of metal were split into several strands before being separated. The process known as slitting was achieved by means of a series of steel discs which operate as revolving cutters. Pairs of rolls were set so that one set of cutting discs overlapped with the other. Sheets of metal were then passed through the rolls to produce several strands of metal. With brass this process was replaced by casting brass rolls that were subsequently rolled thin before drawing. These were then pulled under tension through a succession of draw plates - these are metal plates with holes of diminishing size through which the metal wire is drawn slowly getting thinner in diameter (The process at Cambridge Street is illustrated in Fig. 4.3). In reality the process was mechanised in the 19th century. The wire was set on horizontal drums, that were made to revolve by means of pinions and a long shaft attached to the power source (steam engine). The wire was initially drawn through the hole of the draw plate using pliers (usually mechanical) and attached to a screw on the drum. The drums were then revolved and the wire was drawn through the draw plate. The process was then repeated through smaller holes until the correct width of wire was produced. The process of drawing the wire repeatedly heated and cooled the wire, and through this process it was annealed. The process of placing the wire in the muffles for final annealing would make it brittle. It was therefore struck violently on the work bench in order to expand the outer particles of the wire – which subsequently allowed the inner particles to expand in the muffles (Aitken 1866, 312–313).

The production of tubes was undertaken in two ways, to produce seamed and seamless tubes. Seamed tubes were produced from slit metal (see above). If the tube was large this was passed through a pair of rolls, one concave, the other convex, that made the slit metal concave in profile. This was then passed to the draw bench. Like the wire drawing bench the slit metal is drawn through a tool that altered its shape (Again Fig. 4.4 shows the process at Cambridge Street). One end was attached via tongs to a chain attached to machinery. The strip was then drawn through the draw tool and came out in the form of a cylinder. The edges of this cylinder were then soldered together with a combination of iron and brass solder. Often the tube was redrawn on the bench (Aitken 1866, 323). Seamless brass tubes became important due to problems encountered with steam engines where seams would burst under high pressure with disastrous consequences. The tubes would be formed from short cast brass tubes that would be repeatedly annealed to soften the material and rolled to produce the correct diameter of tube. The rolls designed by Charles Green in 1838, consisted of four concave steel rolls that worked together to create a tube of the desired diameter. The internal diameter of the tube was maintained by a 'spit' or 'manadaril' in the centre of the tube at the correct diameter (*ibid*, 327–328).

The use of water mills

In the 17th and 18th centuries the copper smelting industry of Swansea provided the mills of Bristol with copper which was combined with zinc from the Mendip Hills to produce brass. Copper was also sent to the mills around London and Greenfield Valley in Holywell in northwest Wales (Hughes 2000, 39). In particular refined copper was sent from the White Rock Copper works to the Swinford, Bye and Publow mills on the River Chew, North Somerset five miles south of Bristol to produce battery ware (Day 1973, 65). The Swinford Mill was converted to rolling in 1740, whilst a fourth mill, Wollard, was rolling brass in the 1790s (*ibid*, 204–219). Battery mills functioned by the use of a central wooden drive shaft connected to the water wheel that rotated cogs that in turn operated large wooden hammers. These would have a variety of cast-iron heads that would beat over an anvil and be used to beat copper flat (Hughes 2000, 42). Rolling mills functioned by the use of a water wheel connected to a shaft, which passes

FIGURE 4.4 IMAGE OF TUBE DRAWING AT R W WINFIELD AND CO IN 1887 (AS PREVIOUS, COURTESY OF BIRMINGHAM ARCHIVES AND HERITAGE REF: LP65.641 49/2029)

through the lower set of rolls. A second set of rolls was geared in the opposite direction to drive the copper sheet through the rolls. A slotted roll was used to cut the sheet to strips for wire (*ibid*, 44).

Rolled metal was produced at water-driven mills in Birmingham from at least the 18th century. The economic advantage of water-power had long been realised in Birmingham, with mills established for the edge-tool industry from the late 16th to 17th centuries. The processes of turning bar iron into slit metal was introduced by the Foley's to Stourbridge area in the 1620s (King 2003, 182–184). Rolling mills powered by water were probably present in the midland region from before the Civil War (Court 1933, 242) and was certainly in use by the time the process was described by Dr Plot in 1686 (Plot 1686). Angerstein describes a mill he visited south of Birmingham the mid 18th century, the location of which is open to question but could well be one of the mills on the River Rea, possibly around Edgbaston:

'I walked two English miles south of Birmingham to look at a rolling mill for sheets for boxes and narrow sheets of copper and silver plated tin to be used for the manufacture of buttons. This mill is not shown to strangers, but due to the recommendation of a file-cutter, whom I had taken with me as company, I was let in....Otherwise there was nothing special to be seen at these works except that one water wheel drove four stands' (Berg and Berg 2001, 39).

Wire manufacture by mechanised means was first developed in the 14th century in Germany and was introduced into the country in the 16th century (Aitken 1866, 313). Wire production was reliant on the slitting mill for strip metal. Slitting mills were common throughout the West Midlands from the 1620s but arrived relatively late in Birmingham with the first slitting mill introduced in the town at Lloyd's Slitting Mill, the Old Town Mill in Digbeth, first depicted on Westley's Prospect of 1731 (see Shill, this publication, Chapter 2). Water power was probably also used in the initial production of tubes as Aitken (1866, 322) suggests: tube drawing benches were first in use in Birmingham in 1782 before steam power had been universally adopted. It may also be possible that a mechanised process was being adopted.

The development of steam mills in Birmingham

The pressure on the diminishing and unpredictable water supplies of the Rea valley system put Birmingham at a distinct disadvantage to Bristol (Hamilton 1967, 217). The disadvantage of geography began to be overcome by technological advances. The development of the Boulton and Watt steam engine at the Soho Manufactory in the

1770s allowed industry to develop with a power source that could be flexibly located and where output was not fixed by the flow of rivers or streams. Although a mill in 'Calves Croft' in 1758 was described as having a 'water engine' close to the mill pool, probably a Newcomen Engine (Stephens 1964, 81–139), it was Matthew Boulton and James Watt at the Soho Manufactory who are credited as having recognised the potential of steam power for use with machinery. The process of conversion appears to have been relatively slow. Boulton experimented with the mills to achieve more power, firstly installing a second water wheel, then a horse mill, and finally introducing steam engines, at first only to pump water back into the mill pool to keep the water wheels working (*ibid*). Experiments appear to have been occurring elsewhere in Hockley Brook with the conversion of the mills to steam power throughout the second half of the 18th century. The Aston Furnace had installed a Newcomen engine to operate the bellows by 1783 (*ibid*) before converting to a paper mill worked by steam power in the early 19th century. It was the development of rotative motion, the double-acting principle in 1782 and the parallel motion in 1784 by William Haskell at the Soho Works that allowed Boulton and Watt engines to be used in the rolling of metals (Gale 1968, 16–18, plate 9). The first example outside the Soho Works would appear to be an example in an advertisement dated 1783 for:

'*Charles Twigg and Co., Rollers of Metal, Grinders and Borers of Gun Barrels, at the Steam Mill, Snow Hill. N.B. This mill is erected for the above purposes, and also for the polishing of steel goods, finishing buckles, buckle shapes, and a variety of other articles usually done per foot lathes. The whole is worked by steam engine, and saves manufacturers the trouble of sending several miles into the country, to water mills*' (Aitken 1866, 242–243).

This suggests that steam power had been adopted for rolling by the 1780s, replacing water power for the first time, transmitted along shafts from a central steam engine. A report of the Birmingham Philosophical Institute in 1836 suggests dates for the rapid adoption of the steam engine in Birmingham's manufactories. A date prior to 1790 for rolling brass, copper and other metals tallies with Charles Twigg and Co's advert. The technology was adopted later for drawing wire (1808) and metal tubes (1822; Court 1938, 258).

THE CAMBRIDGE STREET WORKS - A HISTORY

Ray Shill

The Union Rolling Mill (Phase 3: 1820–1840)

During 1824 a piece of vacant land was leased from Thomas Gibson (BCA MS 322/1) which lay between Cambridge Street and the Gibson's Arm. The lessees Daniel Ledsam, Joseph Ledsam, William Potts, Matthew Dixon and Robert Walter Winfield used this property to set up the Union Rolling Mills. All were metal manufacturers who needed rolled brass and copper. Matthew Dixon was a manufacturer of silver and plated wares; Daniel Ledsam made nails, screws, shoe heels and tips and William Potts was involved in the manufacture of brass moulding, stair rod and astragals (Wrightson 1824). Charles Henry Capper had an early association with the Union Rolling Mills whilst manager there, as well as running the adjacent Union (Crescent) Foundry. The Union Rolling Mills become occupied solely by Robert Winfield in 1829 (BCA MS 322/35).

Engine power for the works seems to have been derived from the mill race powered by the Union Mill engine. This engine was described in 1847 sales notices (BG 1847) as being of 95 horsepower beam engine. These notices also mentioned that power was provided to adjacent properties. This sale may have been occasioned by partners leaving the partnership. There were eleven names in the partnership in 1845 (LG 1845), but James Potts had died and Ambrose Warner and William Steel had decided to leave.

During 1848 John Fawkner Winfield, contacted Gilbert Hamilton from James Watt and Co, for the company to supply a new rolling mill (BWC - Letter Book 63). The Watt correspondence indicates the mill was in a poor condition. Hamilton stated that his firm did not make rolling mills but suggested contact may be made with John Bradley and Co of Stourbridge. Undeterred, John persisted with his enquiries with Hamilton and a quotation for a mill was provided in 1850. Watt and Co completely rebuilt the rolling mill and provided new boilers at the Union Mill during 1851 and 1852 (BWC – Letter Book 123) and during this period Winfield's had to source their rolling elsewhere. A description of the arrangement was published in the Electrician in 1822 when the beam engine was used to power electric lights in the town (see below). The engine that provided power for the dynamos was then said to be thirty years old. It was a low pressure engine with a 24in diameter flywheel. These facts correspond with the year (1852) when Boulton and Watt reconstructed the rolling mill (Anon 1882). It was probable that the engine working the mill was replaced with a larger, but evidently second-hand, beam engine of 160 horsepower. This was described in the 1897 sale catalogue as a: '*Condensing beam engine 8ft stoke 48 in cylinder- double cast iron beam 26ft by 4ft crank 26ft long, 5 plunger rods, condenser 6ft by 3ft diameter, cast iron receiving tank 9ft 6in by 6ft by 8ft 3in, force pump*' (BCA MS 322/30). This in turn drove the crank, driving wheel and flywheel. Linked to these through spurs was a series of wheels in the race that drove the rolling mills, wire mills and other machinery (BCA MS 322/30).

Robert Winfield's Cambridge Street works (Phases 3 and 4: 1820–1898)

Robert Walter Winfield began making small items of brassware such as stair rods at a property on the corner of Cambridge Street and Easy Hill during 1820. He first appears in trade directories listed as a brass founder and brass tube maker of Cambridge Street and 84 New Street (Wrightson 1821). He was then, one of 114 separate brass founders listed in the trade in Birmingham. This small operation grew into a much larger business with expansion occurring from 1829 (BCA MS 322/35). The original part of the Winfield's factory grew to cover almost all the land between Cambridge Street and Attwood's Passage. Robert Walter Winfield acquired the lease of the Union Mills (rolling and wire) in September 1853, and the lease of 42 Cambridge Street (the tube mill) from George Crowther in October 1857 (BCA MS322/8; MS322/10). He thus combined the rolling, wire and tube drawing plant into one operation. Parts of the former Baskerville House site also appeared to have been absorbed leaving Winfield's with a continuous line of buildings along Attwood's and Baskerville Passage. The main offices were at 1 Cambridge Street, on the corner with Easy Hill and this was an extensive showroom. These premises were separated from the other Cambridge Street properties by Gibson's Mill and the Crescent Foundry, which remained independent.

Winfield also set up a separate coal and coke business from Baskerville Place. Narrow boats would bring coal from the collieries to the Baskerville Place wharf for sale or use at the Union Rolling Mill or Winfield's works. They also built up a coal merchants business in Browning Street before moving the whole business to Tindal Street in 1852 (Slater 1852). The wharf occupied for the coal trade was acquired for a new bedstead factory opposite the rolling mills. The bedstead factory is discussed in detail in Chapter 5.

Robert Winfield employed some of the best workmen and he supported this trade through patents. Winfield advertised that he was the patentee of the metallic, military, travelling and house bedstead for home use or for export, and used the services of John Simms, at 141 Fleet Street, London, as an agent. He also supplied brass, copper, iron and tin and cased tubing and made cornice pole ends, rings and brackets, plain and ornamental gas fittings, railings light fittings and stair rods (Wrightson and Webb 1843).

The heyday of brass making at Cambridge Street followed Winfield's success at the Great Exhibition at the Crystal Palace in Hyde Park (BCA MS322/186). Further publicity, through exhibitions such as the Paris Exhibition, improved custom, and during its height Winfield's won a series of awards for its metallic products (BCA MS1342/376) including:

Grand Council Medal, London 1851
Prize Medal, London 1862
Gold Medal, Paris 1867
Prize Medal, Vienna 1873
Gold medal for Chandeliers and Gas Fittings
Gold Medal for Metallic Bedsteads, Furniture etc
Gold Medal (Rappel) for Art Metal Work and
Bronze Metal for Tubes, Metal, Wire etc, Paris Exhibition 1878

Winfield's factory comprised brass making, foundry, gas fitting, chandelier, metal rolling, tube drawing, wire drawing and metallic bedstead departments and the publicity of the exhibitions meant it became one of the best known brass manufacturers in the country.

By the late 1840s Robert Walter Winfield had taken his son, John Fawkener Winfield into the business and was trading as R W Winfield and Son. John died in January 1861 and Robert subsequently took Charles Weston and James Atkins, two of his managers, into partnership in 1864 (BDP 1864). This was the situation a year later, when the British Association visited Birmingham. Winfield's was one of a number of local firms that opened their doors to visitors, demonstrating brass founding and metal rolling (BCA MS322/ 192).

Robert Winfield's health failed and he became an invalid and died in December 1869 (BCA MS322/159-166). Management then devolved on the former partners Atkins and Weston. Robert Winfield's two daughters also became responsible for the works. The management process operated through their husbands, Charles Busbridge Snepp and Philip Browne. Both were clergymen, Snepp was the Clerk for Perry Barr, whilst Browne was at Edgbaston (BCA MS 322/9).

During the 1870s Winfield's faced increased competition from other manufacturers, particularly in the metallic bedstead trade with local firms, Hoskins and Sewell, Peyton and Peyton, Taunton Brothers and Samuel Whitfield (Shill 2006b, 34–54) and gained significant shares of the British and foreign markets. Winfield's concentrated on quality and along with their Broad Street neighbours, Messengers, became leaders in a new industry - the art metal trade. Their best-selling lines, however, still included bedsteads and gas fittings and it was from these trades that the main profits were derived (eg PO 1875, 50).

Charles Walker Torr gained his knowledge of the trade through working in the foundry. Torr rose through the ranks to become manager of the works and eventually a partner. His name started to appear on patents from 1874 (eg. Patent No. 2624, 1452, 1924). James Atkins left the partnership in February 1879, leaving the firm in the hands of Philip Browne, Charles Snepp, Henry C Taylor and Charles Torr (LG 1879). With Atkins' departure the last link with the times of Robert Winfield was lost.

Torr paid particular attention to the possibilities of electric lighting and, through him, Winfield's became one of the pioneering firms that made electric light fittings. During 1882 Torr entered into an agreement with Crompton and Co, electrical engineers of Chelmsford. This arrangement became known as the Crompton-Winfield Association. They contracted to supply electric power and lighting to the Birmingham Town Hall for the Music Festival (28th August to 1st September 1882; BDP 1882). It was the first use of public electric lighting in the town. Electrical power was supplied by a dynamo driven by shafting linked to the beam steam engine at the Cambridge Street rolling mill.

This concept encouraged further use of electric lighting powered by dynamo. Other events at Birmingham Town Hall benefited from the electric light and in October 1883 the Crompton-Winfield Association provided lighting for the Leeds Music Festival. A more general scheme of laying mains and lighting Birmingham using the Crompton-Winfield system was suggested in 1883. Agreements were made and parliamentary orders given for an organisation known as the Incandescent Electric Light Company to supply electricity to the Borough of Birmingham. The primitive state of electrical science deterred municipal investment at this time. A new scheme promoted by Arthur Chamberlain and George Hookham later came into being and a Birmingham Electric Lighting Order received royal assent on 12th August 1889. Chamberlain and Hookham's concession was transferred to a new company, the Birmingham Electric Supply Company (Vince and Bunce 1902, 283–287). In 1886 Winfield's sought to break their ties with Crompton and the partnership ended in acrimonious court case.

Another experimental business was stained glass. This was conducted in a part of the premises at 1 Cambridge Street. In 1887 they supplied a window to the Priory Church, Malvern (BDP 1887). Later in 1891 they supplied stained glass windows to the refurbished Grand Hotel in Colmore Row (BDP 1891). Also during this period the London office changed to 47 Holborn Viaduct. Winfield's stained glass department was recognised and won awards including the *Gold Medal, Paris 1878, Gold Medal, Bradford 1882, First Prize, York 1879 and First Prize, Sydney 1880* (Dent 1894).

In February 1887 the works was enrolled as a limited liability company and traded as Winfield's Ltd. Both Charles Snepp and Philip Browne had died and the surviving partners at that time were Julia Ann Snepp (Robert Winfield's daughter), Charles Walter Torr and Henry Charles Taylor. Manufactured products included chandeliers, gas fittings, bedstead tubes, art metal, stained glass, and electric light fittings (NA BT31/3815/23936).

The first year of trading under this name yielded a profit, but 1888 was to prove to be a bad year financially and otherwise. In September 1888 a serious fire broke out in the bedstead works that principally affected the main four storey block. Charles Torr praised the action of the Fire Brigade, without whose assistance more serious damage may have occurred. Amongst the salvage material sold off after the fire was a load of scrap iron-cased tubing. The loss of patterns and stock was an inconvenience, but the most serious outcome was the interruption to a major earner, the metallic bedstead trade. The damage caused was estimated to be about £5000 (BDP 1888a; BDG 1888).

Falling profits led to a change of management and eventually a reduction of the share price of the company. Charles Torr was displaced as managing director in April 1889. The new board of directors included S H Thompson, J Lee, W Wilson and P Gabriel. William Reuben Lane (an accountant and secretary to Winfields) became acting manager of the company. A further financial blow came in January 1890 when an action was brought by the trustees of the Scott Estate of Great Barr to reclaim monies due from a loan of £27,000 made by Sir Arthur Scott in 1883 to finance the electric supply venture. The action was brought against Julia Snepp, Charles Torr, Winfield's Ltd and their solicitors Milward and Co. An important result of this action was the transfer of security from Winfield and Co to Winfield's Ltd (BDP 1890). Charles Torr died in May 1890 and was interred at Erdington Church. Julia Snepp lived a life style beyond her means and in 1891 she faced bankruptcy proceedings. Since the death of her husband in 1880, she had benefited from his estate and was also executrix of Philip Browne's estate. Her bankruptcy left her free from any further attention from the Scott Trustees.

In 1892 their premises in Baskerville Place were leased out to the Birmingham Municipal Technical School. For two and a half years mechanical and drawing classes were held at this location. There was a brass shop, and instruction included training in the use of foot lathes, filing, screwing, pattern making, casting, soldering, tool making and chasing (BDP 1892).

The Scott Trustees appear to have concentrated their efforts on Winfield's Ltd. in order to reclaim their investment. Company orders continued to wane and intellectual property (such as new patents) was in decline (although registered designs were authorised in 1895; Patent no 5686), whilst their competitors were developing new lines. When the Institution of Mechanical Engineers visited in 1897 they found a works where practically everything was done, from casting of the metal to turning out the finished article (IMechE 1897).

However, little investment was made in the plant and the rolling mills deteriorated. James Watt and Co was commissioned to make various repairs in 1897 (BCA MS322/32). The continued ill-fortune of the company led to the appointment of a receiver in May 1897. W R Lane remained as manager and supervised the transfers of these businesses. There may have been hopes to safeguard the business as a going concern but the systematic break-up of the estate followed and different parts of the Cambridge Street Works were sold off separately during 1898 (BCA MS322/ 15-31). Nine departments at Cambridge Street were affected by the sale:

(1) Tube Metal and Wire
(2) Chandeliers and Gas Fittings
(3) Electric Light Fittings
(4) Metallic Bedsteads
(5) Art Metal Work
(6) General Brass Foundry
(7) Wrought Ironwork
(8) Stained and painted Glass Tiles
(9) Church Work

The tube and wire works (Phase 4: 1830–1898)

No. 42 Cambridge Street, west of the rolling mill (possibly within the Union Rolling Mill land) had various occupants initially. In the 1830s the property was probably let by Thomas Cocks, iron founder who subsequently moved to 5½ Broad Street (Wrightson 1833, 1835; Pigot 1842). Part of the land was adapted as a sawmill and leased by Thomas Crowther, timber merchant, box and cask maker (BG 1843). The timber trade conducted here reduced during the 1840s and the Crowther family (Thomas and George) began to concentrate this trade at Wolverhampton. The cost of transporting timber through the Gibson's Lock may have influenced the move.

It became the tube and wire mill in 1845. Originally Allen Everitt had property in Cambridge Street in Lower Kingston Row, with Mr Everitt living at 5 Kingston Row on The Crescent (Rate book Ladywood 1834) for a wire and tube manufactory known as the Kingston Mill (Wrightson 1833, 1839). By 1845 he rented premises between Nos 42 and 44 Cambridge Street (PO 1845). When Allen Everitt moved to Adderley Street in 1846 the Union Rolling Mills advertised room space suitable for wire drawing (ABG 1846) and it is probable that he had rented mill power and space from the Union Rolling Mill Company, almost certainly at No 42 Cambridge Street.

Subsequently in 1845 it was run by Edward Bowyer and George Selby (Rate book Ladywood 1845/70). They traded as the Birmingham Patent Welded Iron Tube Company. Bower left the partnership and for a time (until 1850) William Hodges of Great Bookham was a partner (LG 1850). The name changed to the Birmingham Patent and Brass Tube Company and this company set up a new mill beside the Cape Arm on the borders of Birmingham and Smethwick. Richard Prosser, engineer, was the next occupant of the tube mill. Prosser experimented with new forms of tube making including a patented idea of anti-weld tubes that had certain concepts that formed part of the development of the local weldless tube industry. The Cambridge Street works had a steam powered hydraulic press invented by Mr Prosser. When Prosser died, the tube mill plant was put up for sale (BJ 1854) and Robert Winfield took over the tube works. It continued to function as a department of the Cambridge Street Works up until the late 19th century (BCA MS 322/10).

Multiple tenant occupancy of the works: Winfield Rolling Mills Ltd (Phase 5: 1900–1918)

When the Cambridge Street works ceased trading, the various departments of the works were sold over a period of four months between January and April 1898 (BCA MS322/ 15–31). The cased-tube department, brass foundry and gas-fitting department became multiple tenant occupancy. The brass foundry business was taken over by J and A Tonks. Sperryns and Co took over a major part of the bedstead works, with the remainder being leased to the Player family, railway lamp makers of Broad Street. Paint makers Llewellyn Ryland, acquired the former varnish making works (Cromwell Works), R Richardson (mechanical engineer) leased buildings adjacent to Baskerville Place and Winfield Rolling Mills Ltd was formed to purchase the rolling, wire and tube mills. Bedstead making was lost to their competitors and Winfield's general role as a leader in the various brass trades fell to Smethwick-based competitors, Evered.

Winfield Rolling Mills Ltd

Winfield Rolling Mills Ltd was formed specifically to purchase the rolling, wire and tube mills in January 1898. The first chairman was Francis Mitchell (steel pen maker) and he was assisted by directors George and William Dugard (partners in the firm of Dugard Brothers; BCA MS 322/15–23). C J Bamford was appointed manager of the rolling mills and he steadfastly worked to improve the rolling, tube and wire trades (BCA MS 322/ 46–155). The bulk of their custom was in the Birmingham area, with lesser amounts of orders going to 'country' sales. Brass wire was used in the screw trades and Winfield's customers were in this line. The rolling mills became a successful business supplying tube, sheets and wire. Sales to Leeds were entered as a separate item and it is likely that these were sales of tubes to private railway and road locomotive builders (BCA MS 322/ 45B). They also capitalised on letting of spare shopping to smaller companies such as Butler and Co, W and J George and W Markes and Co who agreed a tenancy of a former lacquer room at the rolling mills during 1898 where they had access to engine power to drive their lathes.

Winfield's Rolling Mills Ltd made various improvements to the plant and alterations to their premises. From 1899–1902 William Dugard persevered with a system of firing the muffles with gas from a gas producer. They had difficulty obtaining suitable coal but by 1902 had decided to invest in an economiser and from 1903 savings in coal consumption were achieved (BCA MS 322/ 46–155). A major improvement was the construction of a new casting shop (BBP 22091). This property occupied a strip of land that was formerly the end of the canal basin nearest Baskerville Place. Other improvements included a new covered annealing shop (BBP 22163) and a roof over the muffles adjacent to the boilers (BBP 23159). The annealing shop occupied the wharf land between the engine house and the canal arm, a piece of land 42ft wide by 45ft 6in long. The curved roof was covered by galvanised sheeting and the metal stanchions supporting the roof sunk in concrete.

Despite Winfield and Co's early incursion into the electricity trade, the rolling mills continued to be lighted by gas in the early 20th century, whilst the steam engine provided the power for the cutters, drawing benches, rolling mills and the wire mills. Between 1911 and 1912 the rolling mills embarked on a partial conversion of the plant to electric power, which included the taking down and rebuilding of a group of muffles. Part of the rolling mill plant was replaced with new rolls driven by electric motors. The works were also lighted by electricity and all power was supplied by Birmingham Corporation (BCA MS 1422 60/1/7/1).

Winfield's Rolling Mills prospered as suppliers of material to a busy working city such as Birmingham. During the First World War they were taken over by the Government department the Ministry of Munitions. In 1917 they signed over the old boiler house south of Gibson's Arm to the Birmingham Aluminium Castings Company (see Chapter 5; BCA MS 322/46–155). The other two boilers were retained to supply the Watt beam engine. After the war Winfield's Rolling Mills invested in new wire drawing plant and some new electrically driven plant for the rolling mill. The Bamford family also gained a stronger control of the business. C J Bamford became managing director and his sons C C Bamford and F J Bamford gained seats on the board (BCA MS 322/15-23).

The tube works

With the demise of the Cambridge Street Works the tube works went through a variety of guises. Winfield's former gas fittings, brass foundry and cased tube departments were leased to a variety of tenants who occupied the three and four storey premises located on the site. A particularly unsuccessful venture at the site was carried on by John Edgar and Arthur George Tonks. The agreement had been made between the Tonks and the Winfield's Ltd receiver in April 1898. A new company was formed in May 1898 that had the lengthy title J and A Tonks (Late Winfield's Brass Foundry) Ltd (NA BT31/7986/57388). They carried on the trade of brass founders, cased tube and stair rod makers until July 1900 when a receiver was appointed. The plant was offered for sale between the 21st and 23rd of May 1901 and included two gas engines of 16 hp and 115hp, lathes, milling machines, punching presses, polishing and cleaning machines (BDP 1901).

The tube and brass foundry was later leased by W and J George, chemical apparatus makers, when the address became 138 Cambridge Street. A section near the main entrance facing Great Charles Street was occupied by Matthews and Timings, fender makers. Their works separated the gas fitting works in 1918 from that subsequently leased by W and J George (BBP 29213).

Closure of the works and the move to the Icknield Port Loop (Phase 6: 1918–1936)

At the end of the First World War an idea had been formed to obtain land for a new civic centre, a memorial to those who lost their lives during the First World War and a Peace Garden. By 1919 Birmingham Corporation showed interest in acquiring the freehold of the Easy Hill estate from the Ryland family trustees. Sir David Brooks, chairman of the Birmingham Corporation (General Purposes Committee) played an integral role in the negotiations for the land. Initially, in 1913, plans were put in place to widen Broad Street from Baskerville Place to Easy Row. This did not take effect until 1920, when the leases of the properties fell due and the purchase of the freehold for the Easy Hill Estate could be arranged at a cost of £100,000 (BC 1920). Birmingham Corporation also pursued the purchase of the leaseholds belonging to Birmingham Aluminium Castings Ltd and Guest, Keen and Nettlefolds. The persistence of Sir David Brooks finally led to agreement. The final purchase of the leases belonging to the Birmingham Aluminium Castings property was made in March 1922 for £25,000 plus legal costs (BC 1922a; BC 1922b; see Chapter 5 for details of the move of the aluminium works).

All the buildings facing Broad Street, Easy Row and Cambridge Street around as far as the Lock were vacated and pulled down. Even W and J George vacated their premises. Demolition left the rolling mills and the Gibson's Basin untouched and the last part of the former Winfield's factory to survive. The Baskerville Basin was sealed off and filled in (Shill 2006a). Work went ahead to make the Peace Garden and the Hall of Memory leaving the remainder of the demolition site a wasteland until the Civic Centre Scheme was instituted. A photograph of 7 June 1926 shows the newly constructed Hall of Memory and gardens in place of the canal basin (BCA MS 322/181–185). In the background can be seen the surviving Winfield's buildings including the rolling mill and chimney. A watercolour for 1929 also shows this scene (BCA MS 322/181–185). The Ordnance Survey map of 1922 shows the area built up with industrial buildings, and on the Ordnance Survey map of 1936 (Fig. 2.1) these buildings had been replaced by an open land and the Hall of Memory.

The 1920s saw the decline in the fortune of the Winfield's Rolling Mill. A move initially contemplated during the 1920s was made an urgent reality in the 1930s by the extension of the Civic Scheme. A site at Icknield Port became available when Vivian and Sons of Swansea closed down their Birmingham rolling mills. Winfield's moved in and rebuilt the premises with an electrically driven mill. The sale of the leasehold premises to Birmingham Corporation, for £41,000, was agreed in 1935 (BC 1936) with representatives of the Civic Centre Sub-Committee.

The sale of their property was crucial to the new scheme, which extended beyond Cambridge Street to include the Crescent Canal Wharves. The final details of the sale were agreed during January 1936, and in February 1936 the local papers printed a story saying that the old Watt beam engine was to be scrapped. The sale was completed during May 1936. Negotiations allowed possession of the canal at an earlier date so that excavation for the Civic Centre could start. Work had commenced by September 1936, when the tall chimney was demolished (Birmingham Mail 1936). The canal, the lock and the rolling mills were swept away for the new Civic Centre (Shill 2006a). Yet only Baskerville House was completed before the Second World War halted any further work (Shill 2002, 88).

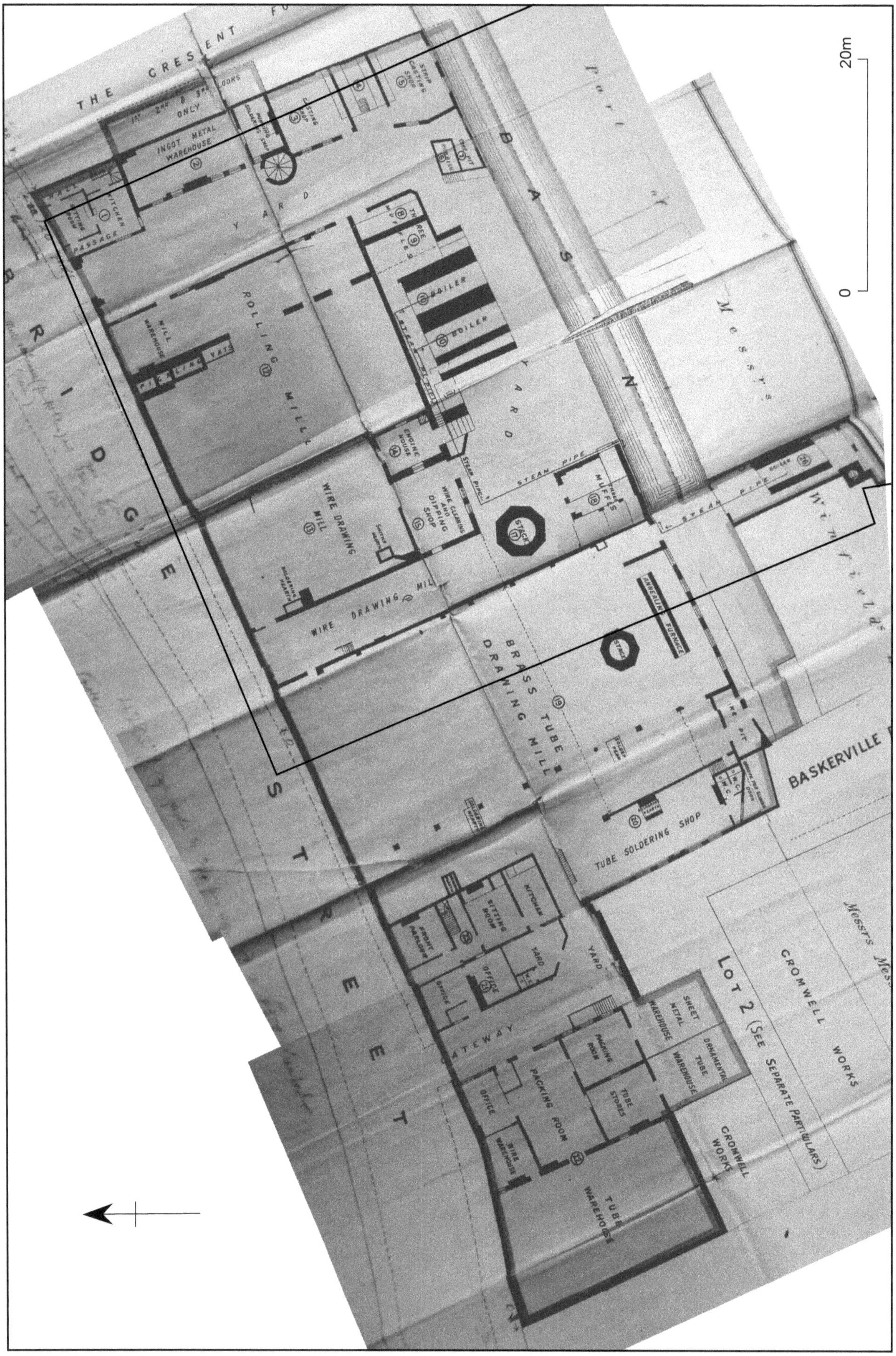

FIGURE 4.5 PLAN OF THE SALE OF THE CAMBRIDGE STREET WORKS DATED 1897 (COURTESY OF BIRMINGHAM ARCHIVES AND HERITAGE REF: MS 322/30)

After1945 Birmingham Corporation planned new schemes to build the Civic Centre and new Philharmonic Hall on part of the vacant site. This proposal was dropped and the wasteland that had been Winfield's tube mills was later occupied by the new Repertory Theatre. The rolling mills and part of the old bedstead works became a car park. The foundations of the old works stayed covered and protected by a car park until excavations prior to the building of the new Library of Birmingham.

THE CAMBRIDGE STREET ROLLING MILL, WIRE MILL AND TUBE WORKS – THE EXCAVATIONS

Will Mitchell and Chris Hewitson

The archaeological preservation within Winfield's Cambridge Street Works was exceptional given its location and the general state of preservation of similar remains in the area. Our understanding of the site was improved by the survival of a plan from the sale of the works in 1897.

The plan of the sale of the works provided the functions of each of the buildings found on site (Fig. 4.5; BCA MS 322/29). From this it could be deduced that the majority of the northern side of the site beyond the Gibson's Wharf towards Cambridge Street was involved with heavy industrial brass work. Rolling, drawing, annealing, pickling, cleaning, dipping and casting were some of the techniques employed to produce the finished article. This area was served by a centrally located steam engine and a series of boilers. The 'strip' casting shops and brass ingot store were located along the eastern side of the site. The more refined work (assembly of finished items and small casting) was undertaken south of the canal and was not described in the sales plan. A second boiler for a steam engine was located south of the canal, and this presumably served the jigs and machinery within the bedstead works. This area will be discussed in greater detail in Chapter 5.

Archaeologically there are three phases within this section, Phase 3 relating to the Union Rolling Mill, Phase 4 relating to the Cambridge Street Works and Phase 5 to when the works was sold and worked again as a rolling mill alone. Discussion of the works has been separated into their functional parts. These have been described as:

- Power generation and transmission
- The rolling, wire drawing and tube drawing mill
- The furnace and casting shops
- The muffles and the process of annealing
- Cleaning, the pickling vats and dipping shops

Elements of all these processes were discovered during excavation of the site (Plate 4.1).

Power generation and transmission

Power for the Union Rolling Mills (later Winfield's Cambridge Street works), including the rolling, tube, wire

PLATE 4.1 THE ROLLING MILL, AREA 1, EAST FACING (IMAGE BY AERIAL-CAM)

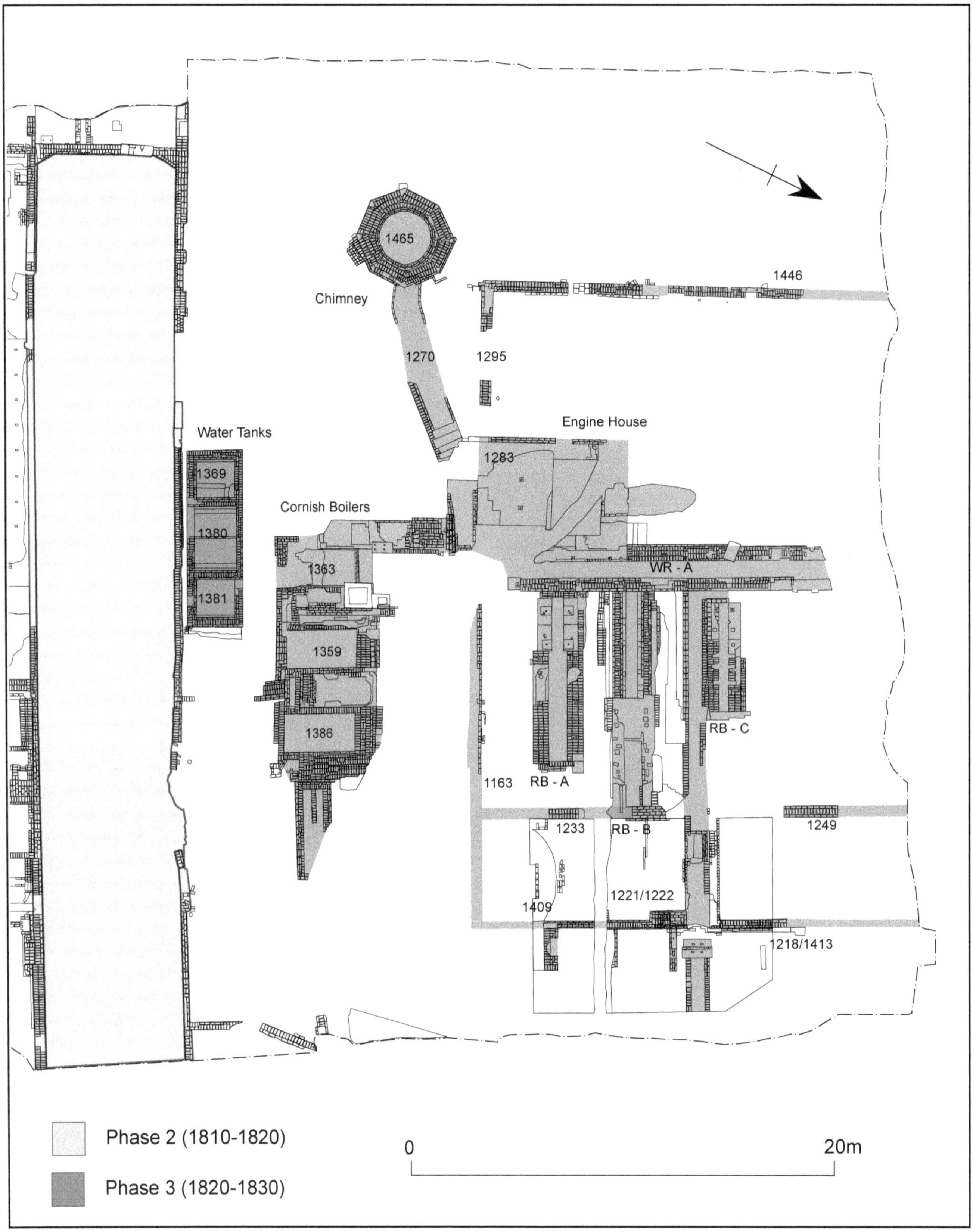

Figure 4.6 Phase 3, the Union Rolling Mill

mills and warehouses, was generated by a beam engine located within a purpose-built engine house in the rolling mill. The engine (which saw a succession of technical upgrades) was a condensing beam engine, which drove flywheel connected to a series of spur wheels within the 'Wheel Race(s)'.

The beam engine worked by providing steam under pressure to a cylinder; this forced the cylinder to compress and decompress in turn which drove a large mounted beam. The 'up and down' vertical motion of the beam was transferred to circular motion by means of a crankshaft connected to a large flywheel. The flywheel was in turn connected to a series of spur wheels (inter-connected gears) within the wheel races. These drove a series of drive-shafts that ran perpendicular to the wheel race and powered the rolling mills, wire drawing mills and tube mills. The whole of the power for the rolling mill was thus provided by a single engine.

Steam for the engine cylinder(s) was provided by large boilers, the exhaust fumes of which were drawn through a series of flues to a large purpose built chimney. The arrangement of water pumping was not clear in the archaeological record, but it is known that water was pumped into the boilers via the main engine and later, by the boiler feed pump. The engine and boilers were subject to a variety of upgrades throughout their lifetime and although all traces of the machinery had been removed, possible evidence of restructuring was represented in the fabric of the building foundations. It is also very likely that the engine doubled as a pump to supply the boilers with water. The arrangement of the pumping machinery also saw technical upgrades.

Summary of historical evidence

Phase 3 (Fig. 4.6)

1824 — Union Rolling mills constructed (BCA MS 322/1).

Phase 4 (Fig. 4.7)

1847 — The first mention of the steam engine was in 1847 when 'Three boilers and a 95 horsepower steam engine' were advertised for sale (BG 1847).

1851/52 — Watt and Co. replaced the old boilers and their foundations and provided a new condenser (BWC – Letter Book 123).

1882 — Mention of the steam engine is made in 'The Electrician' where it is described as a 'low pressure condensing engine, having a flywheel 24 ft in diameter, running 48 revolutions per minute….A belt from the flywheel is carried to a camshaft ….which drive the dynamo machines' (Anon 1882).

1897 — A comprehensive description of the engine and its components was made for the purposes of the sale of works (BCA MS 322/30).

The engine was described as a '160hp Condensing beam engine- 8ft stroke 48in cylinder-double cast iron beam 26ft by 4 ft crank 26ft long, 5 plunger rods, condenser 6ft by 3 ft diameter, cast iron receiving tank 9ft by 6ft by 8ft in, force pump'.

The boiler house had two double flue firing boilers by E Danks, Oldbury. These were fed by 'two large storage tanks over centre to supply boilers fed by canal water-pumping being done by mill engine'.

The rolling mill had seven pairs of rolls (intermediate, breaking down, finishing and preparing rolls), driven by spur pinions/wheels attached to mill-race wheels.

The mill-wheel race had 14 cast iron spur wheels and one wrought iron crank to engine spear rod.

The yard had a double cylinder donkey pump.

From these sources it can be deduced that the engine of 1897 was of improved efficiency and that horsepower had greatly increased from that of 1847. The engine was also involved in pumping at some point in its history as there is historical reference to pumping ('37ft of cast iron flanged water pipe to well pump under engine' - (BCA MS 322/30), which is confirmed in the archaeological evidence. The archaeological evidence for pumping was in the form of large pits beneath, and at the front of the engine, of which the former was later covered.

Phase 5 (Fig.4.7)

1899–1902 — System of firing the muffles with gas from a gas producer introduced (in 1902 an economiser was installed and in 1903 savings in coal consumption were achieved).

1911–1912 — Partial conversion of plant to electric power (including the taking down and rebuilding of a group of muffles. Part of the rolling mill plant was replaced by new rolls driven by electric motors at this time.

Phase 6

1936 — In 1936 the engine was due to be scrapped, prior to this it was described as follows: 'The beam engine at the rolling mills was built by James Watt. After 100 years carries on supplying original work of driving power for rolling, tube, wire mills and the warehouse. The driving

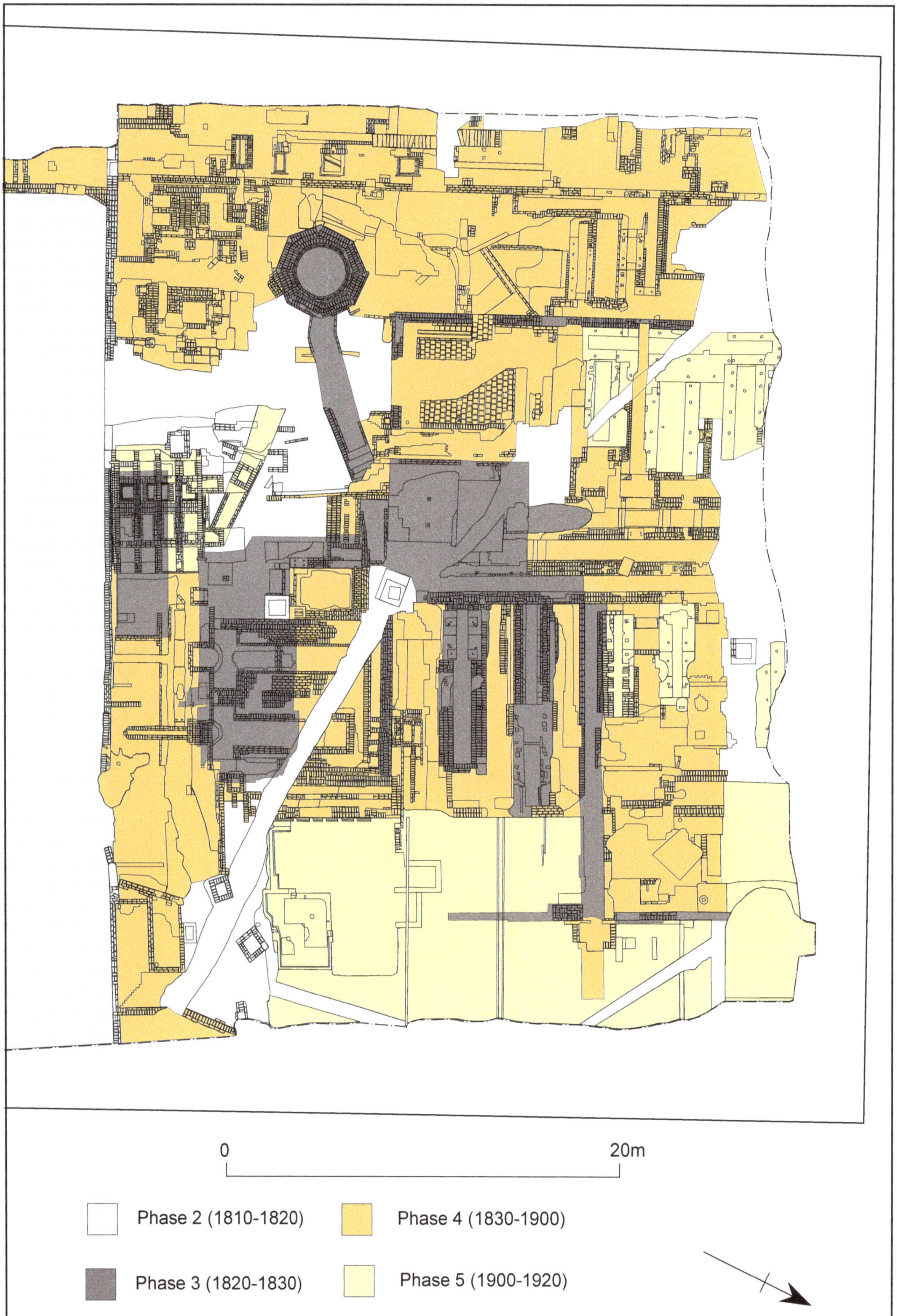

FIGURE 4.7 PHASE 4 AND 5, THE ROLLING, WIRE AND TUBE WORKS

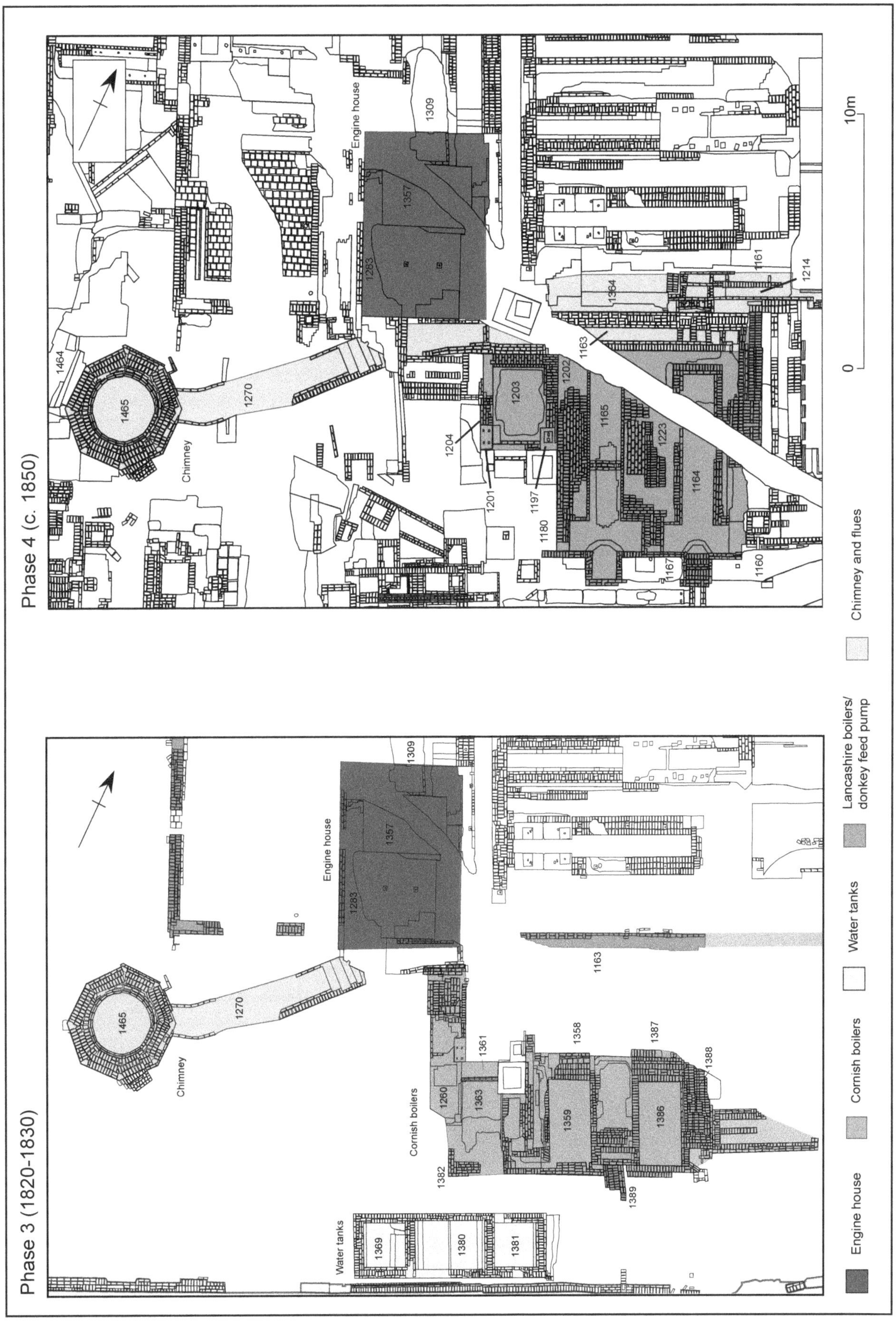

4.8 THE DEVELOPMENT OF THE STEAM ENGINE

> wheel (26ft in diameter) makes 19 revolutions per minute- the even larger flywheel (28ft in diameter and weighing 30 tons) revolves 70 times per minute. The engine has been repaired 3 times and just after being built, had a new condenser put in. he engine originally had a wooden beam, which was replaced by a cast iron beam. It now has a twin steel one.' (BG 1936; BCA MS 322/175–180; Fig. 4.8, Plate 4.2, 4.3).

The engine house (Phases 3, 4 and 5)

Nothing of the steam engine survived, having been scrapped in 1936. However, the foundations of the building, (constructed in 1824 and including the machine base) within which it was housed, did survive. The steam engine is known from historical records to have been a condensing beam engine, which was built by James Watt and upgraded throughout the history of the works. The early 19th century was the period when this method of powering the rolling mills became economically and technologically practical. Steam-powered copper mills were constructed around the same period in Swansea (Hafod Copper Works 1819), along similar layouts (engine house, rolling mills with water for boilers provided by the canals). There are also examples known from Bristol (Hughes 2000, 48).

Archaeological evidence

The engine house structure was located approximately 8m to the northwest of the boilers and boiler feed pump. The steam pipes connecting these boilers and the engine would have been located above ground and their rough locations are present on the 1897 sale of works plan (BCA MS 322/30). The engine house was a purpose-built building 7m in length x 5m in width, the design of which was dictated by the form of the steam engine contained within. The steam engine of the Union Mills, being of a beam engine type, would have partially relied on the superstructure of the brick engine house to support the beam.

The engine (including the beam supports cylinders and condenser) was situated upon the engine base. The engine would have been aligned north to south with the cylinders, condenser and boilers on the south (canal side) and the flywheel and driving wheel located in the Mill wheel race on the northeastern side of the engine, facing the rolling mill.

Nothing of the beam engine had survived, having been scrapped in 1936. However, the foundations of the brick structure within which the engine was sited had been preserved. The engine house structure had substantial brick walls 0.75m–1.2m thick constructed of handmade red bricks ($9 \times 4\frac{1}{4} \times 2\frac{3}{4}$in) in English Garden Wall Bond (1283). The southern side of the engine house was the location of

Plate 4.2 The layout of the steam engine and boilers, east facing (Image by Aerial-Cam)

PLATE 4.3 LARGE PIT BELOW THE ENGINE HOUSE, SOUTH FACING

the machine base. This base was approximately 2.85m x 3.7m and up to 1.6m thick in places. It was constructed of a mixture of brick, crushed brick and cement. There were two large upright holding down pins, each over 5m in length towards the centre of the base, these would have held the vertical beam supports. The southern wall of the engine house was difficult to define, the foundations having been truncated by later alteration and demolition, it is clear though that if the beam was 26ft (7.92m) in length (BCA MS 322/30), the engine house must have been at least this to accommodate the beam. A gap in the northern elevation of the engine house structure, illustrated on the plan of 1897, may be where the beam came through the building to be attached to the flywheel in the wheel race. A substantial lump (0.95m thick) of concrete, situated to the north of this opening (1309) represents a further section of the machine support structure. The main southern elevation must have been located further south, probably aligned parallel to the main rolling mill elevation (1163).

To the north of the engine base there was a large rectangular pit 2.3m in length x 3.6m in width, excavated to a depth of 4.4m (Plate 4.3). This had been backfilled with a dark brown sandy-silt and demolition rubble fill (1357) during the demolition of 1936. A second pit was located beneath the engine base. This pit was rectangular, 1.6 m in length x 3.6m in width, it was also excavated to a depth of 4.4m. The two pits were separated by a shared wall approximately 0.23m thick, and at the base of the pit there was a door for access between them. It is likely that these pits were water storage areas for the boilers or condensers.

The scale of these power generation elements and the investment that they represented meant that they were unlikely to have moved location, or dramatically changed much during the lifetime of the works. Structurally the foundations that have been identified are likely to be the foundations introduced when the Union Rolling Mills were constructed. Elements of the engine itself were subject to periodic change.

The boiler house - phases 3 and 4

Two types of boilers were employed at the Cambridge Street works. The first was a Cornish-style boiler. This was the earliest form of fire-tube boiler, which had a long horizontal cylinder with a single large flue containing the fire. The fire itself was on an iron grating placed across this flue, with a shallow ash-pan beneath to collect the non-combustible residue. As the furnace relied on natural draught (air flow), a tall chimney was required at the far end of the flue to encourage a good supply of air (oxygen) to the fire. For efficiency, the boiler was commonly encased beneath by a brick-built chamber (Hills 1989, 128–132). These were later replaced in the 1850s by two Lancashire boilers. The Lancashire boiler is similar to the Cornish, but had two large flues containing the fires (Hayes 1983, 23–24; Hills 1989, 133–134).

The boilers were first described as 'three boilers and a 95 horsepower steam engine' and were advertised for sale in the Birmingham Gazette and Journal (BG 1847). Four years later in 1851/52, Boulton and Watt were hired to supply the mill with new boilers. Their letter book mentions that 'one of the new boilers will be finished this week, and the other in the following one'. This work included the removal of the 'pieces of the old boiler' and bricklayers 'clearing out all the defective parts of the old foundations' (BWC Letter Book 123). In the sale catalogue of 1897 the two boilers were described as 'Double flue firing boilers by Edwin Danks of Oldbury 23ft long by 9ft diameter'. Associated with these were their brick settings (27ft x 25ft) Details relating to the sale plan note 'Two boiler settings for Lancashire boilers and protection shed over fire holes'. These are also seen in the accompanying sale plan (BCA MS 322/30). It appears therefore, that between 1851/52 and 1897, there was a further replacement of the boilers.

Plate 4.4 The water tanks, east facing

The water tanks (Phase 3; 1369, 1380, and 1381; Fig. 4.8; Plate 4.4)

8.5m length x 2.8m width x 2.1 depth (maximum).

This structure was likely to be contemporary with the Union Rolling Mill phase of building, although it is not present in any of the mapping from the period, including the detailed 1897 plan. It was a large rectangular structure made up of three chambers and was constructed of brick with iron reinforcements. The structure was made of hand-made red brick (9 x 4 ¼ x 3 in), the interior of the chambers were lined with yellow firebricks. The central chamber (1380) was rectangular and was the largest of the three. Two square chambers (1369 and 1381) located either side of this were connected to the central one via an arch through their side wall. Residues adhering to the edge of the structure (1369) were examined by metallography and this suggested that they had been formed by the reaction of coal as a fuel. The current interpretation that they were water tanks is derived from their form, but the residues may suggest that later structures built over the top (see the central muffles below) were coal fuelled. There were vent holes along the top edge of these chambers, also connected to the central chamber. The bricks were severely heat affected, and there was vitrification and slag accumulation on the interior of each chamber. It is probable that the structure comprised three water storage tanks associated with the first three Cornish style boilers. There is reference to the boilers being fed by canal water with the pumping being done by the mill engine and the well pump under the engine (BCA MS 322/30). These may have acted as storage facilities for water that was extracted from the canal prior to use in the boilers.

The cut for the structure produced a fill (1411) that produced smithing slags that were probably some of the earliest on the site. Metallographic analysis suggested these were possibly associated with ironworking (see McDonnell, this publication, Chapter 7). The difficulty with these samples is that they could easily have derived from the adjacent site of the Crescent Foundry (an iron foundry). They do however suggest some ironworking was occurring on the site in Phase 3 or early Phase 4, dependent on chronology.

Cornish boilers (Phase 3, 1824; Fig. 4.8, Plate 4.5)

The foundations of the primary phase of boilers were made up of three rectangular bases situated within thick external and central dividing walls. The bases themselves (1359, 1363 and 1386) were of roughly the same dimensions (1.6 x 3.1m) and they were equally spaced, the central dividing walls being 2m apart. Overall the bases were incorporated into a rectangular structure 12.2m (40.02ft) in length x 5.2m (17.06ft) in width. They were located in what might be termed the boiler house, which as a structure was orientated northeast to southwest.

The external walls (1260, 1358, 1361, 1382, 1387 and 1388) were 1–1.5m thick and had been built in stages, probably around the in situ boilers. Within wall 1388 were two pieces of pale coloured slag, which XRF analysis

PLATE 4.5 THE CORNISH BOILER BASES, PHASE 3, NORTH FACING

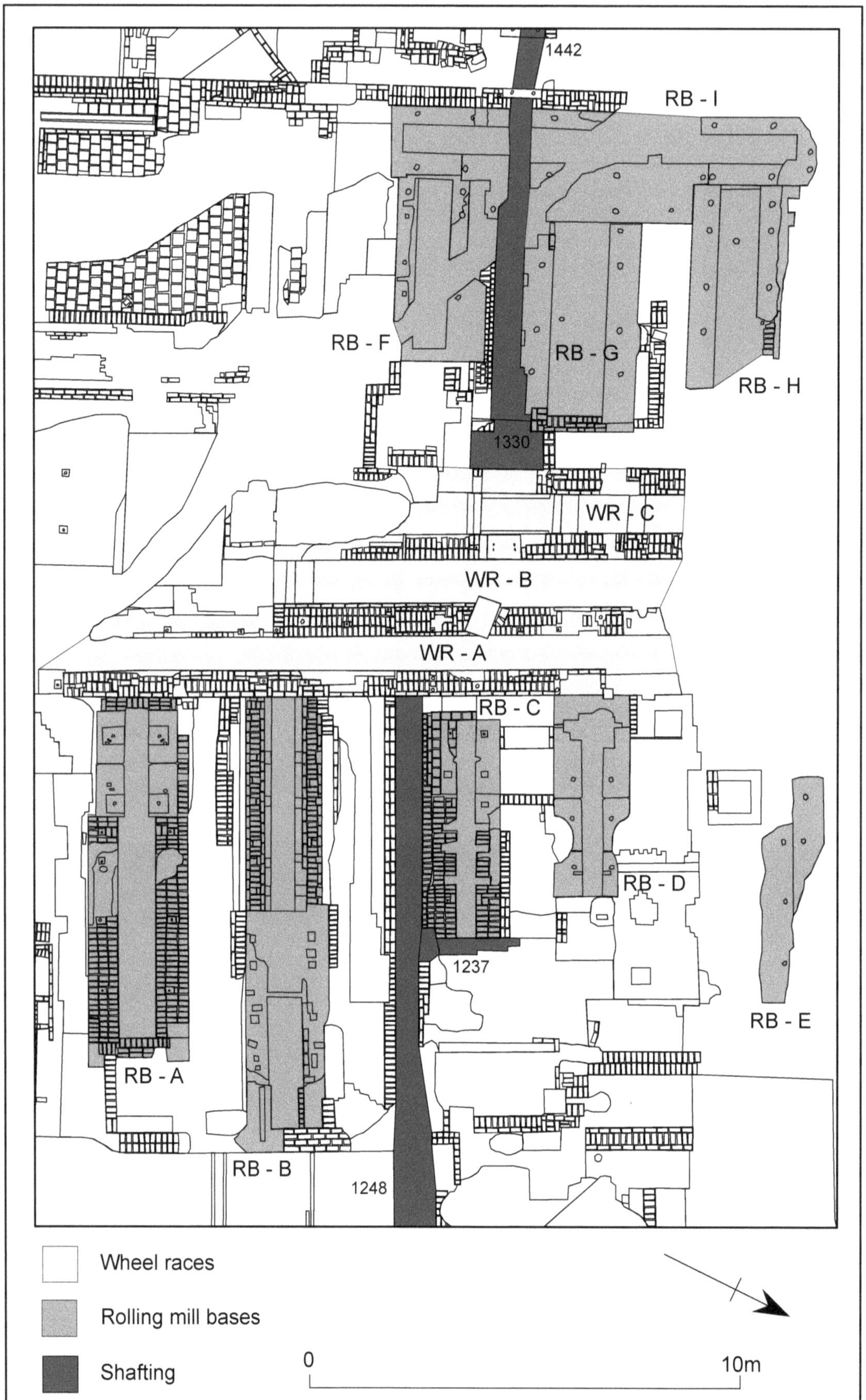

Figure 4.9 The wheel races, transmission and rolling mill bases

suggests were very high in zinc but low in copper and may therefore have been residual residues from strip casting in the vicinity. The brick settings were integral to the functioning of the boilers, as flues were required to be located beneath. The bases were likely to have supported cylindrical Cornish-type boilers. This type of boiler is illustrated in contemporary pictures. The southern side of the boilers was the location of the ash pits, one of which survived (1389).

The bricks used in the construction of the boiler bases were hand-made red brick (9 x 4¼ x 3 in) in English bond. The floor surface was constructed of the same type of bricks. The westernmost base (1363) survived to a significant height and was the most complete. Plinth bricks had been used in its construction around the upper edge.

During 1851/52 the boiler house was upgraded and the three boilers were replaced with two. The westernmost boiler base (1363) became filled with general brick rubble before being covered over by a brick floor surface (1180). The other two boiler bases became incorporated into the reconstructions.

Lancashire boilers (Phase 4; 1851/52; Fig. 4.9; Plate 4.6, 4.7, 4.8)

The boilers were upgraded during 1851/52, and this meant the total reorganisation of the bases. The upgraded boilers were of a Lancashire type, these were a later development (patented 1844) than the Cornish boilers. This type of boiler was more efficient and therefore would have increased steam pressure to the steam engine, which would also have seen an upgrading at this period.

The foundations of the secondary phase of boilers were made up of two rectangular bases situated within thick external and central dividing walls. Their construction had partly re-used the two easternmost bases of the primary phase of boilers. The bases themselves (1164, 1165) were of roughly the same dimensions (1.3m x 7m or 4.26ft x 22.96ft. The discrepancy between the foundation dimensions and the documented boiler dimensions is due to the fact that these foundations represent the supporting structure for the boilers only), the central dividing wall (1223) was 2.5m in width.

Overall the bases were incorporated into a roughly square structure approximately structure 9m (29.52ft) in length x 8.5m (27.88ft) in width. The external walls (1202, 1163) were 1.3 to 2m thick as a result of incorporating the walls from the earlier phase. The brick settings were constructed of yellow firebrick. These were constructed in a sloping fashion at the southern end of the base. The southern side of the boilers was the location of the ash pits. The bricks used in the construction of the boiler bases were a mixture of machine-cut, yellow firebricks and ordinary machine-cut red/grey brick (9 x 4 ½ x 3 in) in mixed bond. The floor surface was constructed of the same type of bricks.

PLATE 4.6 THE LANCASHIRE BOILER BASES, PHASE 4, NORTHWEST FACING (IMAGE BY AERIAL-CAM)

Plate 4.7 The Lancashire boiler bases, south facing

Plate 4.8 The Lancashire boiler bases, drum base, west facing

Built up to these boiler bases and contemporary with their construction, there was a series of yard surfaces (1160, 1167 and 1180). These were all constructed of engineering brick and there were several alterations evident. On the northern side of the bases there were a series of flues (1163 and 1346). Again, these flues contained alterations and rebuilding, suggesting periodic upgrading. This upgrading may have been at times when it was anticipated there would have been increased combustion products entering the flue system (due to larger boilers etc.) or at a time when the flues needed to be repaired because of their general ware.

The steam pipes feeding into the steam engine from the boilers would have probably been of the flanged type and all above ground. The locations of some of these pipes are present on the 1897 sale of works plan (BCA MS 322/30), but no in situ evidence remained.

The Boiler Feed Pump (Phases 3 and 4; Fig. 4.8; Plate 4.2)

The donkey pump was a supplementary small steam-pumping engine that was for the sole purpose of feeding the boilers with water and was probably a later addition. This would have been a simple addition that meant more power could have been obtained from the boilers by adding water more quickly. It was housed on the western side of the boilers in a small rectangular structure. It was described in 1897 as a 'double cylinder donkey pump with cast iron flywheel,' which had a 'wood shed over boiler feed pump' (BCA MS 322/30).

Archaeological evidence

The structure itself took the form of a rectangular pit structure 4.5m in length x 3.1m in width and 0.85m in depth (1204) constructed of hand-made red brick (8½ x 4 x 2¾ in) in English Garden Wall bond, with an internal dividing wall (1271, not illustrated), presumably for the purpose of separating the pit to house the double cylinders. It also contained a firebrick-lined base (1203). There were two iron settings (1201 and 1197), which were the likely locations for the tying down points for the feed pump structure.

The southern wall of the boiler feed pump (1204) abutted the northern wall of the most easterly boiler base dated to Phase 1 (1363). This suggests that the donkey pump was introduced during the lifetime of the original boilers (1824–1851/52), and as the structure was present on the 1897 plan, its use must have continued into the boilers' second phase.

The Chimney and Flue(s) (Phases 3 and 4; Fig. 4.8; Plate 4.9)

The chimney and flues removed waste fumes from the earliest furnaces, boilers, the donkey pump and areas including the muffles. The flues were subterranean tunnels around 1m in width that ran along the northern side of the boilers and from the southern furnaces. The combustion products from the boilers were carried along flues to be expelled up the chimney. The chimney also provided

PLATE 4.9 THE CHIMNEYSTACK BASE, WEST FACING

draught to these boilers. The chimney was described in the 1897 sale catalogue as the 'mill chimneystack approximately 164ft in height and 14ft x 14ft base' (BCA MS 322/30).

Archaeological evidence

The foundations of the main chimneystack were identified (1465). The chimney was located outside of the southern corner of the Union Rolling Mill. It was octagonal and had a large firebrick flue (1270) running into the base of the eastern elevation. The chimney was identified in the documentary sources, and the dimensions described in 1897 are identical to the foundations exposed. The combustion products from the boilers were carried along the flue to be expelled up the chimney. The chimney also provided draft to these boilers.

The base itself (1465) was 4.3m x 4.3m (14ft x 14ft) and excavated to depth of 0.8m (internal). The main octagonal construction was made up of hand-made, 9 x 4 x 2½in red bricks set in a stretcher course, with the central circular space being made up of dark red, soot stained brickwork (vitrified firebricks). A fragmentary brick floor surface (1464) surrounded the chimney base.

A flue (1270), built into the chimney's eastern foundation was constructed of a mixture of red bricks (of the same type as the chimney) and was covered by a vaulted roof constructed of wedge-shaped firebricks. This flue ran in an east to west direction from the northern side of the boiler bases. There had been much alteration to the flue system probably due in part to the reorganisation of the boilers. At various locations along the length of the flue, the roof had been rebuilt using iron plates.

Any direct relationship between the boilers and these subterranean brick flues was unclear. However it is certain that the flue would have connected directly to the boilers on the northern side of the boiler house. Additional evidence of flue rebuilding was noted immediately north of the boiler house. At this point the flue was split into two separate parts (1161 and 1214). This separation may be related to the upgrading of boilers in 1851/52 when the foundations were known to have been rebuilt.

The Wheel Race(s) (Phases 3 and 4, Fig. 4.9; Plate 4.10, 4.11)

The wheel-race structures were large linear brick built trenches which would have housed the flywheel/ driving

PLATE 4.10 LAYOUT OF THE ROLLING MILL BASES, EAST FACING (IMAGE BY AERIAL-CAM)

PLATE 4.11 WHEEL RACE A, NORTH FACING

wheel and spur wheels, the purpose of which was to transfer the power generated by the steam engine to the rolling and wire drawing machines. The power would have been transferred to these machines through the use of directly driven spur pinions connected to these spur wheels.

Located within the wheel-race(s) were fourteen cast iron spur wheels along with their associated shafting, plummer blocks and brasses, cast iron sill plates, holding down pins and brick built foundations. The flywheel/ driving wheel would also have been located in this location. The flywheel of 1882 (24 ft in diameter, running 48 revolutions per minute) appears to have been upgraded before 1936 when it was 28ft in diameter and weighing 30 tons and revolved 70 times per minute. The driving wheel was mentioned once in 1936 to be 26ft in diameter and made 19 revolutions per minute.

Archaeological evidence (Fig. 4.9)

The location of a single linear trench is marked by two parallel dashed lines on the 1897 sale plan (BCA MS 322/30). It is marked as running in a northerly direction from the north corner of the engine house. The archaeological evidence has exposed three trenches, each lying parallel to one another, and each located immediately north of the engine house, in the centre of the rolling mill/ wire drawing mill buildings. The trench marked on the plan is the furthest east of these three trenches.

The location of the flywheel and spur wheels can be identified by the presence of their holding down pins and cast iron sill plates. These were visible along the original top edge of the mill wheel race foundations. These '... cast iron sill plates, holding down pins and brick built foundations' were mentioned in the 1897 sale catalogue (BCA MS 322/30). A good approximation can therefore be made as to the exact positions of the flywheel/driving wheel crank shaft and spur wheel axles. Broadly, they are aligned with the rolling and wire drawing machine trenches.

For simplicity of explanation each trench has been assigned a letter (A, B and C). Each trench shares a wall with its neighbour and was modified throughout its lifetime. Inevitably, with the constant movement of heavy machinery, the foundations and machinery would have required upgrading. The easternmost race (A) is the race identified on the 1897 plan and probably housed the spur wheels for the rolling mill. The central race (B) was the probable location of the main flywheel and driving wheel(s) and the westernmost race (C) probably housed the spur wheels for the wire drawing mill.

Voids were present beneath the base of each of these trenches. It is possible that these were connected to a deep culvert, which carried away any liquids which would have accumulated at their base (eg after cleaning, grease spillages etc). A definitive answer to this question was not possible as this remained unexcavated.

Wheel Race A (WR - A)

Approximately 15m of this mill race was identified. It was the easternmost of the wheel races and began at the southeast corner of the engine house and continued in a northerly direction beneath the northern edge of the excavation area. There appeared to be a succession of major modifications of these mill race foundations. The southern end of the mill race nearest the engine appeared to be of an earlier (and probably primary) construction. This was clearly identifiable in the eastern wall which was constructed of hand-made red brick (8¾ x 4¼ x 2½ in) in English bond (1252). This wall had been rebuilt in a northerly direction using machine–cut engineering brick (9 x 4 x 3¼ in) in English bond (1354). A further rebuild, of the same construction materials, was identified at the far northern end (1356). The western (and parallel) wall of the wheel race (1345) was constructed of these later engineering bricks. Only part of this wheel race went to a depth of 2.2m, the majority was preserved to a depth of 1.5m. It is likely that there was a larger wheel (perhaps the driving wheel) at this deeper location. The level of the base of this deepest section is comparable to that of wheel-race B and it is aligned with the bowl shaped base section, interpreted as the location of the flywheel.

Walls: 1252 - eastern elevation. (Phase 3). 1354 and 1356 - eastern elevation. 1345 - western elevation. (Phase 4). Length 15m (exposed), width 0.8m, depth 2.2m (maximum).

Wheel Race B (WR - B)

Approximately 9m of this mill race was identified. It was the central wheel race of the three trenches and began at the northern corner of the engine house and continued in a northerly direction beneath the northern edge of the excavation area. The main walls (1345 and 1305) were constructed of machine-cut engineering brick (9 x 4 x 3¼–3½in) in English bond. At the southern end of the trench, nearest the engine house, the base was bowl shaped in its construction to accommodate a large wheel, perhaps the flywheel.

Walls: 1345 - eastern elevation. 1305 - western elevation. (Phase 4). Length 9m (exposed, width 1m, depth 2.7m (maximum).

Wheel Race C (WR - C)

Approximately 5.5m of this mill race was identified. It was the westernmost wheel race of the three trenches and again it continued in a northerly direction beneath the northern edge of the excavation area. The walls were of different constructions, suggesting alteration at a later period. The eastern wall of the race (1305) was constructed of machine-cut engineering brick (9 x 4 x 3¼–3½in) in English bond. However, the western wall of the race (1304) was constructed of hand-made red brick (8½ x 4¼ x 2¾in) in English Garden Wall bond. This wall appeared to be the earlier of the two. The base contained a shallow stepped construction, presumably to contain a wheel.

Walls: 1305 - eastern elevation. 1304 - western elevation. (Phase 4). Length 5.5m (exposed), width 0.9 m, depth 1.3m (maximum).

The rolling mill and mill machinery (Phases 3, 4 and 5)

The Rolling Mill formed the main part of the works. It was the area in which the brass was rolled into flat sheets. The flat sheets would then go on to be turned into different brass items including tubes, or stamped brass items which would be stamped using fly presses or drop stamps.

The area originally occupied by the Union Rolling Mill was described in detail in the 1897 sales catalogue: 'Square of buildings greater part of which his one storey (approximately 96ft x 77ft, a floor space c. 7500 sq ft) comprising metal rolling mills, warehouse, wire drawing mills, engine house, wire cleaning shop with carpenters shop over part of the building and two large storage tanks over centre to supply boilers fed by canal water - pumping being done by mill engine' (BCA MS 322/30). Details of the structures contained within are also described. Two pairs of breaking-down rolls, one pair of intermediate rolls, three pairs of finishing rolls and one pair of preparing rolls were mentioned in the 1897 inventory along with their component parts (BCA MS 322/30).

Each pair of rolls was set within cast iron housings and driven by pairs of spur pinions. They were fixed on oak or cast iron beds with holding-down pins and brick foundations. Comprehensive dimensions of these articles are given in the sales catalogue. The breaking-down and getting-down roll, and pinion housings, were 'out of line and level' according to Watt and Co's report and it was recommended that 'these ought to be put right at the earliest possible moment, so as to lessen the power that passes through the train of wheels' (BCA MS 322/32).

In 1911–1912 there was a partial conversion of the plant to electric power. Despite Winfield and Co's early incursion into the electric trade, the rolling mills continued to be lit by gas, whilst the steam engine provided the power for the cutters, drawing benches, rolling mills and the wire mills. Now part of the plant was replaced with new rolls driven by electric motors. The works were now lit by electricity and all power was supplied by Birmingham Corporation (BCA MS 322/34).

Archaeological evidence (Fig. 4.6, Fig. 4.9; Plate 4.10, 4.12, 4.13)

Several phases of construction can be identified within the main structure of the rolling mill building and the structures contained within. The primary build (Phase 3; Fig. 4.6) was the construction of the mill building itself. Additional building and reconstruction was noted within the rolling mill. The rolling machine bases, wheel race(s), floor surfaces etc all received later treatment. Almost the entire floor plan of the rolling mill was exposed in the excavations. This was the area of the original Union Rolling Mill which later became incorporated into the larger Winfield's complex. The northern elevation remained unexcavated beneath Cambridge Street, which must have been widened at some point in its history. The internal floor space measured an area 20m (65.6ft) in length (maximum) by 23.5m (77.08ft) in width. If the length of the boiler building (29ft) is included as part of the length of the rolling mill this brings the total length to 94.6ft.

The dimensions of the rolling mill, as recorded in 1897 (and not including the mill warehouse), were 96ft x 77ft, this being almost identical to the excavated dimensions (94.6ft x 77.08ft). The dimensions of the Union Rolling Mill would therefore have been approximately 96ft (length) x 81ft (width - this includes the area marked on the 1897 plan as the mill warehouse).

The walls of the main rolling mill building (formerly the Union Rolling Mill; Phase 3; Fig. 4.6) were four courses thick (0.5m) and constructed of hand-made red brick (9 x 4 ½ x 3 in). These walls (1163 and 1295 - south wall, 1218/1413, 1221/1222 and 1409 - east wall and 1446 -

PLATE 4.12 ROLLING MACHINE BASES A AND B, WEST FACING

PLATE 4.13 ROLLING MACHINE BASES F, G, H AND I, SURVEYED BY MARY DUNCAN, NORTHEAST FACING

west wall), appeared to have received very little alteration, except in the later period, (early 20th century, Phase 5) when the southeastern corner was extensively altered. This occurred at the same time as the eastern muffles were destroyed for the insertion of the square concrete structure.

The mill warehouse

A large internal wall (1233/ 1249), interpreted as the western wall of the mill warehouse, was also preserved. This was constructed of the same building materials, in a stretcher bond, and the foundations had survived to a depth of 1m. The south-eastern corner of this wall (1249) contained a number of gaps the largest of which may be the location of an arch, as seen in the 19th-century pictorial representation of the rolling mill (BCA MS322/197, BCA MS 322/181-185; Shill 2006b, 36; PO 1871, 14).

The rolling mill bases

Located to the east and west of the mill race(s) were the rolling mill bases. The locations of these were not marked on any of the historical plans but details of the machinery were described in the 1897 sales catalogue (see above). A number of different types of rolling machine were described. However, all that remained were the large brick built foundations upon which these machines were mounted. Each of these foundations was built to a considerable depth, presumably to accommodate the spur pinions which attached the main spur wheels located in the wheel race. Above these foundations would have been mounted a series of different types of rolls. These rolls worked in pairs and metal ingots were passed between them by teams of workers in order to gradually reduce the thickness, so that it eventually became sheet metal.

There were two areas of rolling mill bases within the main mill building. The first was located to the east of the wheel-race(s). This area contained five rolling pits, three of which (A, B and C) were likely to have been related to the primary phase of rolling mill construction (Phase 3), based upon the materials used. Within the central pits of Rolling Base A, the fills 1347 contained large quantities of metal off-cuts associated with non-ferrous brassworking.

The second area of rolling machine bases was west of the wheel race(s) in the area marked on the 1897 sale of works plan as the Wire Drawing Mill. This area contained four pits, similar in construction to the structures known to be rolling machine bases. These probably represent later rolling machine bases. The foundations of these rolling machine bases overlay several earlier brick structures (1300, 1306, 1308, 1330 and 1338, see below), which are likely to be the truncated remains of these wire drawing bench foundations. The rolling machine bases located in this area (F, G, H and I) were of a later construction (concrete and/or stone and cement like that of machine pits D and E), perhaps being built around 1900 when parts of the works were taken into new ownership and the rolling mill plant was rebuilt.

The historical information suggests that there were seven sets of rolls in use in the later 19th century (BCA MS 322/30). An attempt was made to excavate each of these rolling machine pits. The width between some of the walls, however, meant that machine excavation was not possible. The rolling pit dimensions and construction materials are individually described below. It is important to note that each of these lengths of brickwork may have supported several sets of rolls and that the machines may not have been used contemporarily with each other. The construction of each set of rolling machine bases is different. The spacing between rolling pits A and B and also C and D was the same (1.3m) suggesting this was the optimum working space need to pass the rolled metal between rolls.

The rolling machine bases east of the wheel races

Rolling Base A (RB - A; 1251): Machine-cut, 9½ x 4½ x 3½ in red bricks set in an English bond. The rolling pit was a stone construction at the western end and brick construction at the eastern end. Two distinctive sections. Six pairs of holding down pins still in situ. Length 8m, width 0.55m–0.9m, depth 1.8m (x22 courses), Phase 3/4, Plate 4.12.

Rolling Base B (RB - B; 1250): Machine-cut, 9 x 4½ x 3in red bricks set in an English bond. Constructed entirely of brick. Had large square cement and brick base in the centre with arch running beneath. Fifteen pairs of holding-down pins (or holding-down pin positions). Length 11m, width 0.55m–0.8m, depth 2m (x 22 courses), Phase 3/4, Plate 4.12.

Rolling Base C (RB - C; 1344a): Machine-cut 8½- 9 x 4½ x 3in bricks. Constructed of brick with a cement surface. Fourteen pairs of holding down pins (or holding-down pin positions). Length 5.3m, width 0.45m, (unexcavated depth), Phase 4.

Rolling Base D (RB - D; 1344b): Large stone and cement construction. Two semi-circular niches opposite on another at eastern end. Three pairs of holding-down pins. Brick floor working area between bases 1344a and b. Length 5m, width 0.5m, (unexcavated depth), Phase 5.

Rolling Base E (RB - E; 1342): Large stone and cement construction. Four pairs of holding-down pins (visible). Length 5.6m, only partially excavated the rest remained beneath the northern excavation edge, Phase 5.

The rolling machine bases west of the wheel races

Rolling Base F (RB - F; 1299): Concrete/crushed brick and cement construction. Four pairs of holding-down pins. Length 5m, width 0.75m, depth 0.3m (not fully excavated), Phase 5, Plate 4.13.

Rolling Base G (RB - G; 1300): Concrete/crushed brick and cement construction. Four pairs of holding-down pins. Length 6m, width 1.2m, depth 1.7m, Phase 5, Plate 4.13.

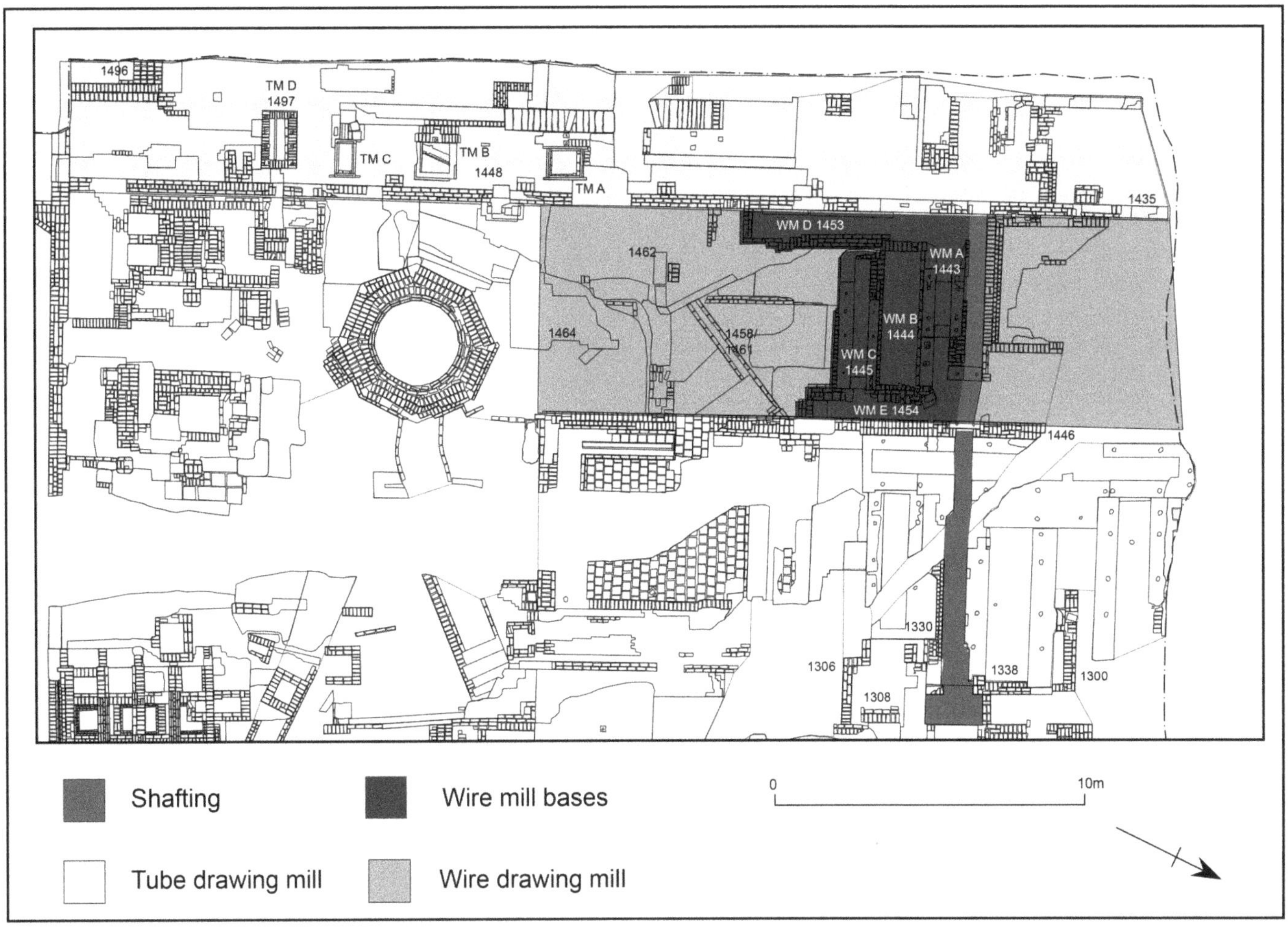

FIGURE 4.10 THE WIRE MILL

Rolling Base H (RB - H; 1301): Description: Concrete/ crushed brick and cement construction. Four pairs of holding-down pins. Length 5.3m, width 0.9m, depth 0.9m, Phase 5, Plate 4.13.

Rolling Base I (RB - I; 1302): Orientated north to south at western end of pits F-H. This may be an additional drive shaft. Concrete/crushed brick and cement construction. Six+ pairs of holding-down pins. Length 9.5m, width 0.65m, depth 0.3m (not fully excavated), Phase 5, Plate 4.13.

The drive shafts

An additional linear trench (1237/1248) ran from wheel race A in an easterly direction to the outer elevation of the rolling mill (Phase 3). This trench ran between rolling pits B and C and was constructed of hand-made red brick (9 x 4½ x 2½in) in Stretcher bond. This trench was 17m length, 0.6m width, 0.85 depth (not fully excavated) and was interpreted as an additional drive shaft, which further powered the driving mechanism of the rolling machines. It appeared to continue towards a small machine base 1406, Phase 4, sealed beneath later concrete surfaces. It contained a number of deposits that included brass sheet, wire and slag deposits from non-ferrous metal working.

An additional linear trench (1330/1442) ran from wheel race C in a westerly direction into the wire drawing mill (see below; Phase 4). This trench ran between rolling pits F and G and was constructed of machine-cut engineering brick (9 x 4½ x 3in). This trench was 15m length, 0.6m width, 0.8 depth (not fully excavated) and was interpreted as an additional drive shaft, which further powered the driving mechanism of the rolling machines.

The wire drawing mill and machinery (Phase 4; Fig. 4.10)

The process of wire drawing involved the pulling of brass under tension until it was stretched very thinly. Initially brass would be prepared in cylindrical form, either by direct casting into this form, or by splitting the sheets into narrow strips by passing rollers with through semi-circular grooves in a slitting mill. The sheets would then be stretched on wire drawing mills, ie drawn by large wheels through plates with varying sizes of holes, until they were the right width.

According to the 1897 Sales Catalogue the wire drawing mill was known to have a 'jigger shop on the first floor' (BCA MS 322/30). The wire drawing benches contained within were driven by shafting as in 1897 the 'line of shafting that drives these mills' was in a 'dilapidated state' (BCA MS 322/32). From the sale of works plan it is clear that in 1897 there were two areas of wire drawing mill. The first area was located to the west of the main mill race(s) and within the original part of union mills. A secondary rectangular wire drawing mill occupied the space between the original western elevation of the Union Rolling Mills and the brass tube-drawing mill.

Archaeological evidence (Fig. 4.10)

The wire drawing mill described below is the mill which occupied the space between the original Union Rolling Mills' western elevation and the brass tube drawing mill. It had a first floor which housed the 'jigger shop'. The other wire drawing mill is briefly described above as part of the rolling mill, having been truncated by the insertion of rolling machine bases in the early 20th century. The severely truncated wire drawing mill was 6m (19.7 feet) in width by 17m (55½ft - exposed) in length. No original floor surfaces remained, having been truncated by later developments. South of the wire mill were the chimney foundations. The pictorial representations of the chimney show that it was outside the wire mill. The eastern wall of the wire drawing mill was made up of the western wall of the Union Rolling Mill (1446), and the western elevation was made up of the eastern wall of the brass tube drawing mill (1435). There was no southern elevation and the northern elevation lay outside the excavation area. Several earlier brick structures (1300, 1306, 1308, 1330 and 1338) are likely to be the truncated remains of these wire drawing bench foundations. Much of what remained within the wire drawing mill were features that would have been beneath the floor levels, such as drainage (1458/1461) and culverts (1462/1464). However, within the northern end of the wire drawing mill were the foundations of several large machine pits (1443, 1444, 1445, 1453 and 1454). Two of these machine pits were similar, having been constructed of timber (1443 and 1445). The layout and construction of these machine pits was different to that of the rolling machine bases and it was possible these were breaking-down machines for initial rolling of the brass wire.

The wire drawing machinery

Machine Pit A (WM A; 1443): Aligned east–west and constructed upon the northern wall of pit 1444. Timber frame construction and joined with mortice and tenon joints and iron tie bars. Four pairs of holding-down pins. Length 3.5m, width 0.45m, depth 0.4m.

Machine Pit B (WM B; 1444): Aligned east–west. Machine- cut 9 x 4 x 3in bricks. Set in English bond. X2+ pairs of holding-down pins (or holding-down pin positions). Length 5m, width 1.05m, depth 1.35m, (Plate 4.14).

PLATE 4.14 WIRE-DRAWING MACHINE BASES A, B AND C, EAST FACING

Machine Pit C (WM C; 1445): Aligned east–west. Machine-cut 9 x 4 x 3 in bricks. Set in English bond, with a timer-frame construction, joined with mortice and tenon joints above. Five pairs of holding-down pins. Length 4.5m, width 0.45m, depth 0.2m.

Machine Pit D (WM D; 1453): Aligned north–south at the western end of pits 1443–1445. Constructed of machine-cut 9 x 4 x 3in bricks. Length 6m, width 0.5m, depth unexcavated.

Machine Pit E (WM E: 1454): Aligned north–south at the eastern end of pits 1443–1445. Constructed of machine-cut 9 x 4 x 3in bricks. Length 3m, width 0.5m, depth unexcavated.

The brass tube drawing mill and machinery (Phase 4; Fig. 4.10, Plate 4.15, 4.16)

Brass tubes were created in one of two ways. The first involved the rolling of the tubes from sheet brass and then the soldering together of the two edges. These would be unsuitable for high-pressure tubes such as those required for steam engines. Instead they would be produced by casting tubes in brass around 20–30in long and then drawing them to lengths of 15ft or more. This was done in a similar manner to wire drawing and the tubes would be forced over a cylinder (known as a triblet) of the required size.

The brass tube drawing mill was described in the sales catalogue of 1897 as a 'one storey building approximately 121ft x 74 feet- superficial area 8906 feet forming brass tube drawing mill (separate chimney about 80ft high and 11 x 11ft at the base), contains three soldering hearths annealing furnaces and coke pit' (BCA MS 322/30).

Archaeological evidence (Fig. 4.10, Plate 4.15)

The brass tube drawing mill was located west of the wire drawing mill. It covered the entire expanse of space between Cambridge Street and Gibson's Arm Canal. Power to the works was supplied from the engine located in the main rolling mill. It contained its own chimney, annealing furnace and soldering hearths (these were however, located outside of excavation area). Approximately 32m long (104.96ft) x 4m wide (13.12ft) was exposed within the excavation area - the rest had been truncated during the construction of the new REP Theatre in 1971. The main wall (1435) was constructed of machine-cut red brick (9 x

Plate 4.15 The tube-drawing mill, south facing

Plate 4.16 Tube machine Base D, 1497, east facing

4½ x 2½in). Along the eastern elevation of its length were a series of strengthening buttresses.

The engineering brick floor surface (1448) of the mill contained three square and rectangular brick built pits (A, B and C) each with timber base-plates around their edge and holding-down rods on their western side. These may have been types of hearth or casting pits. Two of the pits (B and C) were filled with clean dark red sand, which may suggest a casting function for these pits (or perhaps soldering hearths which are not marked on the plan). Timber rails laid into the brick floor in front of these structures would have had a shock absorbing effect when items were moved across the floor between them.

The tube drawing machinery

Tube Machine A (TM A). Square, length 0.9m, width 0.9m, depth 0.8m (not fully excavated).

Tube Machine B (TM B). Square, length 0.9m, width 0.9m, depth 0.8m.

Tube Machine C (TM C). Rectangular, length 0.5m, width 0.95m, depth 0.7m.

A further machine base (Tube Machine pit D; TM D; 1497, Plate 4.16) was identified south of these pits. This base was constructed of engineering bricks and was 1.77m in length x 1.04m in width. There were four pairs of holding-down pins set within it. There was access to the canal loading bay via a set of steps (1496) at the southern end of the brass tube drawing mill.

Casting and annealing hearth structures (Phase 3, 4, 5)

The strip casting shop (Phase 4; Fig. 4.11)

A single wall relating to the strip casting shop was exposed in the excavation area. This wall (1142) had a distinctive change of angle, which mirrored the wall identified in the 1897 plan. The interior of the workshop remained unexcavated (being beneath the excavation edge). The wall itself was constructed of machine-cut, 9 x 4 x 3in red bricks set in an English bond. Overall, approximately 6m x 1.5m of the structure was exposed. Within the building slags were recovered (1329, SF100) from the internal fills indicative of hot-working of non-ferrous brass (see Chapter 6 and McDonnell, Chapter 7). The slags also suggested that coal may have been used as a fuel and overall this would be consistent with its location as a strip casting shop.

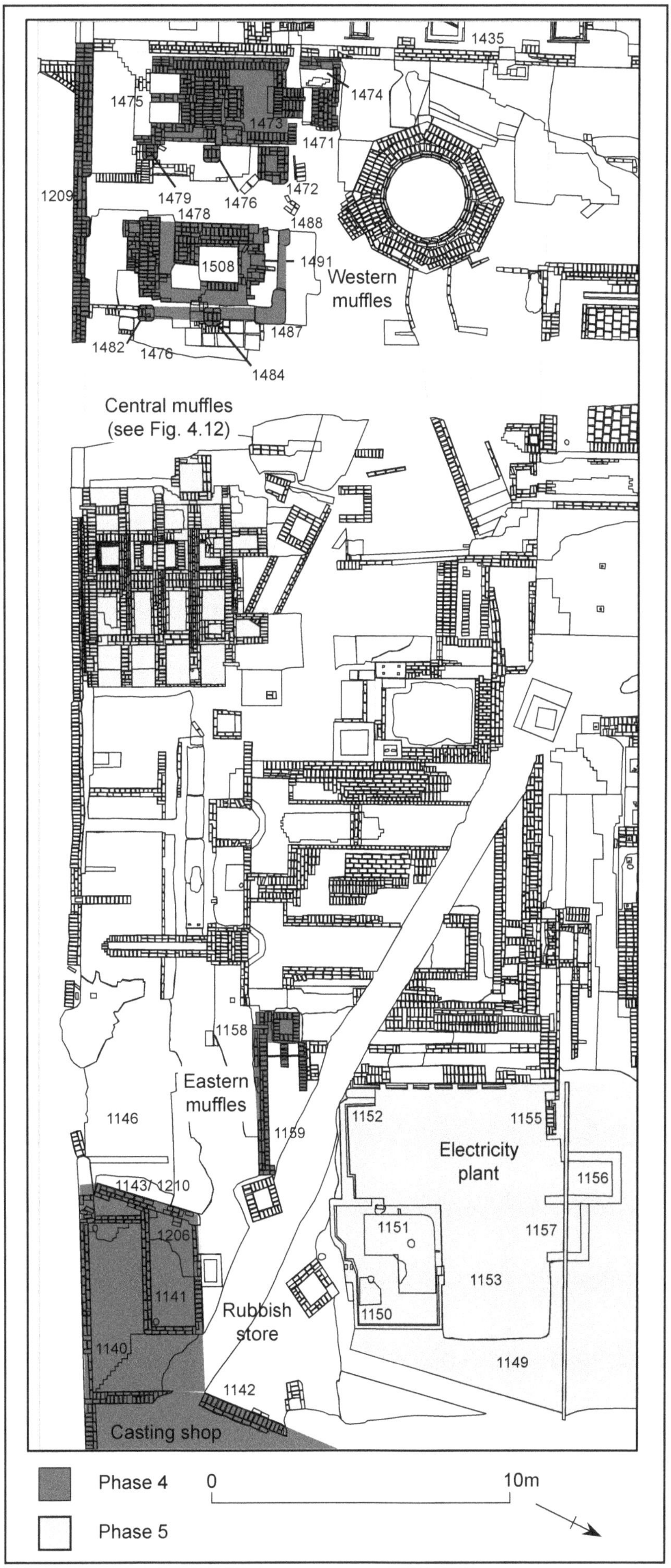

FIGURE 4.11 CASTING, ANNEALING AND THE ELECTRICITY PLANT

The 'muffle' buildings (annealing) (Phase 4 and 5)

In order to make the brass flexible it was necessary to undertake a process known as annealing. This involved heating the brass to a high temperature before allowing it to cool. This changed the structure of the brass and made it less brittle and therefore less likely to break when it was either rolled or drawn.

The locations of two separate muffle buildings were marked on the 1897 sales plan (BCA MS 322/30), they are described below as the eastern and western muffle building. These muffle buildings contained annealing furnaces. More is known of the western muffle building as there is detailed documentary evidence, supplemented by well-preserved archaeological evidence. The eastern muffle building was severely truncated by later developments (the plating/tinning structure - early 20th century) and only part of the foundations survived.

The sales catalogue of 1897 mentions three 'brick and iron annealing furnace(s)' within the muffle house. The muffle house itself was a three-storey building, with the 'three muffles on the ground floor', the first floor was the 'lacquer still room', and the second floor was 'formerly used as a plumbers shop and glass cutters shop' (BCA MS 322/30). A further three 'brick built annealing furnace (s)' are mentioned. These are also described as the 'three rolling mill muffles' These two sets of annealing furnaces are the two separate 'three muffles' mentioned on the sale plan.

William Dugard (Company Director at the time) persevered with a system of firing the muffles with gas from a gas producer. They had difficulty in obtaining suitable coal but by 1902 they had invested in an economiser and by 1903 saving in coal consumption was achieved. During this time there was a partial conversion of the plant to electric power, which included the taking down and rebuilding of a group of muffles (BCA MS 322/34).

Archaeological evidence

The western muffle building (Phase 4; Fig. 4.11, Plate 4.17)

The two surviving 19th-century illustrations identify the western muffle building as a three-storey building with a pitched roof, arched windows and arched accesses on the ground floor. There is an additional extension built on to

PLATE 4.17 THE WESTERN MUFFLES, NORTHEAST FACING

the eastern elevation at the ground-floor level. This also had a sloping roof and open arched access.

The dimensions of the building were 9m x 6.5m. The base of its southern elevation was the northern wall of Gibson's Arm Canal (1209) and its western elevation was the eastern elevation of the brass tube drawing mill (1435). The northern and eastern elevations were open-sided, using pillars to support the elevation above (1472, 1482, 1484, 1487 and 1488). The engineering brick pillar bases were to support the arches. There were further pillar bases within the building (1479 and 1538).The construction method of using pillars would have provided easy access and open ventilation to the annealing furnaces. Along the southern site of the building was an access corridor with an engineering brick floor (1476). This surface ran between the bases of two distinct structures contained within the building. The processes undertaken within these structures must have involved fire as the bricks were heat affected and there were small square hearth features connected to flues contained within. They are likely to be the annealing hearths mentioned in the sources.

The easternmost of these structures (1478) was made up of a thick wall (1.85m) around a brick lined pit (1513) filled with layers of heat-affected sand (1508, 1515-1518). On the northern side of the structure were two firebrick-lined, square hearth features (1491). It was 4.5m in length by 2.5m in width.

The westernmost of these structures was 4.5m in length by 2.7m in width and was constructed of yellow firebricks (1475) which were heat affected due to prolonged contact with heat. On the eastern side of the structure was a firebrick-lined, square hearth feature (1473) and on the northern side was a square feature (1474) which led into a flue (1471). A gate attached to this would have regulated the air flow into the furnace. There had been rebuilding of this structure and pits on the southern side and in the centre of the structure had been filled in and covered with bricks.

The eastern muffle building (Phase 4; Fig. 4.11)

The area within which the eastern muffle building would have been located was occupied by the plating/tinning structure, which was constructed in the early 20th century. All that remained of the muffle structure was a firebrick square feature (1158), which was probably part of the firing system and a part of a wall orientated east to west (1159).

The central muffles (Phase 5, Fig. 4.11; Fig. 4.12, Plate 4.18)

The water tanks for the boilers (see Phase 3 above), was later overlain by a two phase rectangular structure, which was located adjacent to the northern side of Gibson's Arm Canal (1209). These features were not present on any of the mapping evidence before 1922. A building occupying the area appears on the Ordnance Survey map between 1905 and 1922 and is visible on the watercolour of 1929 and the photograph from 1936. On these it is an open sided, corrugated-iron, and barrel-roof building. It is very likely that this was the location of the rebuilt muffles mentioned in the archives from 1911–12 and the archaeological evidence confirms this fact. A group of muffles (1182, 1186–89 and 1227) found within the yard area, north of Gibson's Canal were likely to have been the muffles rebuilt in 1911 to 1912. Archaeologically, the layout was very similar in nature to the known annealing hearth base located just to the west, within the muffle

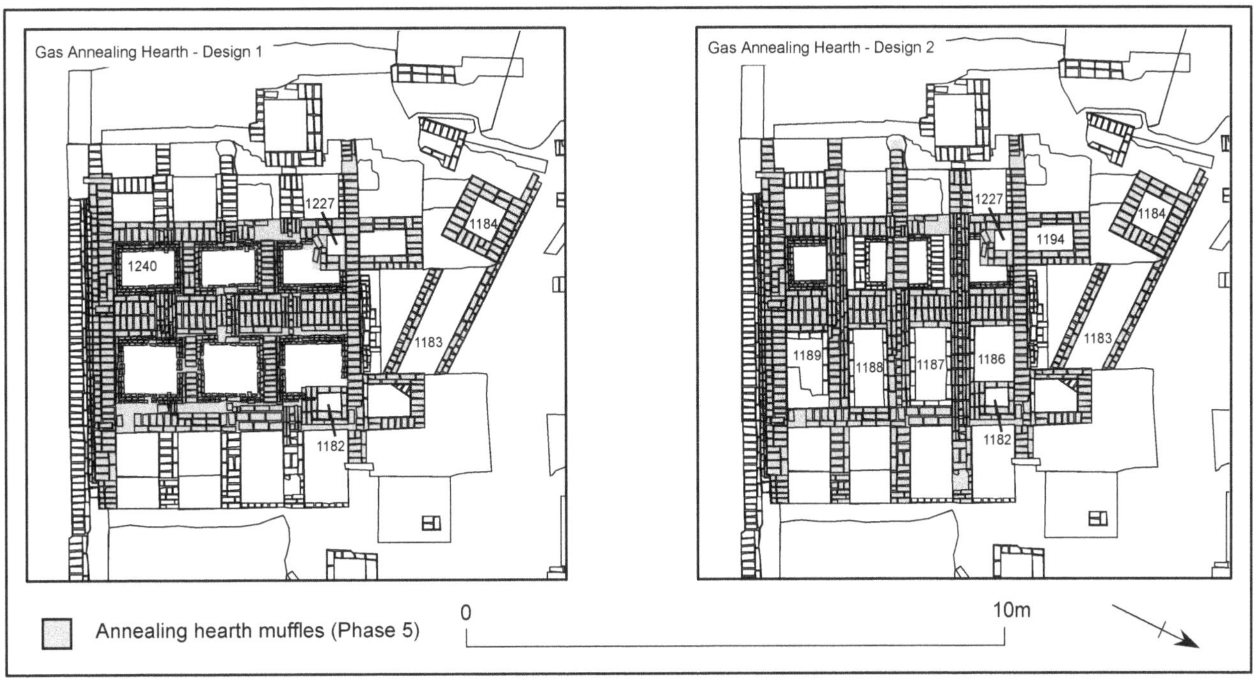

FIGURE 4.12 THE GAS-FIRED ANNEALING HEARTH

PLATE 4.18 THE CENTRAL MUFFLES, INITIAL PHASE 5 LAYOUT, NORTHEAST FACING

house, confirming its interpretation as a muffle (see above the western muffles). The curved roof mentioned in the sources, and covering this part of the yard, is present in both the 1929 watercolour and the photograph of 1936 (BCA MS 322/181–185).

Gas annealing hearth – Design 1 (1240; Phase 5)

Feature (1240) appeared to be the first design of the muffles, made up of six stepped brick pits connected to fire pits (1182 and 1227) and flues (1182, 1183, 1184 and 1194). It was 6.3m in length by 4.5m in width. The structure was constructed upon a slab of concrete/brick aggregate, which had capped the structure beneath (1369, 1380, and 1381). The presence of residues adhering to the interior of 1369 suggested that it was associated with coal as fuel - this does not correlate with the later design - yet this interpretation was based on sulphur peaks in the metallographic analysis that may as easily correlate with a gas fired muffle set-up. The bricks used were machine-cut red brick (8¾ x 4¼ x 2¾in). The pits on the eastern side of the feature were 1.3m x 1.2m and 0.46m and the pits on the western side were 0.9 x 1.2m x 0.46m. The pits were filled with deposits of clean burnt sand. This suggests that these may have been casting beds. On the northern side of the structure were two firebrick-lined, square hearth features (1182 and 1227), similar in form to the crucible hearths identified within the bedstead works. These were filled with black slag and ash material (1279) that contained slag fragments (see McDonnell, Chapter 7).

Gas annealing hearth – Design 2 (1186–1189; Phase 5 - the 'rebuilt muffles' 1911–12)

The final phase of the structure involved the reorganisation and bricking over of the pits, and its restructuring into four equal linear sections. Each of these linear sections was 0.7m in width and covered the whole length of the structure (6.3m). The two firebrick-lined, square hearth features (1182 and 1227) remained in use as did the flue. They were 6.3m in length by 4.5m in width. The layout was very similar in nature to the known annealing hearth base located just to the west, within the muffle house. Taking this evidence into account it is likely that, in its final phase at least, this structure was the 'rebuilt muffles' (annealing hearth) mentioned in 1911–12.

The electricity plant (Phase 5; Fig. 4.11)

Within the southeastern corner of the rolling mill was a large square structure, which was constructed at a lower level than the original floor surface. The structure itself was made up of a group of concrete walls (1150 and 1152) metal I-bars and concrete floors (1151 and 1153). The concrete used was pre-war in origin. It was pinkish grey/white with aggregate inclusions. The walls survived to a depth of 1.3m below floor level in some places. In the southeastern corner of the structure was a square tank with sloping floor and inlet/outlet pipes in two of its corners. In the northeastern corner of the structure was a hearth/ fire hole (1155) structure that connected directly to a flue, which would have ultimately fed into the main flue and chimney. Adjacent to this were two concrete chutes (1156 and 1157) that may have been coal chutes from the floor of the adjacent building (1149).

Around the same time as the construction of this structure part of the rolling mill was rebuilt and concreted over. This concrete surface (1149) extended from the eastern wall of the mill warehouse to the northern elevation of the structure (1152). The conversion of the plant to electric power, in 1912, resulted in the rebuilding of the eastern group of muffles. They were replaced by this large concrete structure which housed the electricity plant. The likelihood is that this replaced the steam engine in the early part of the 20th century. It does not appear on the 1897 sales plan (BCA MS 322/30) and is located in the position marked as 'three muffles'. It therefore must have been constructed after this date.

Cleaning the brass – pickling and dipping (Phase 4)

Pickling was another process in which the brass was dipped in sulphuric acid. This was done before resuming the drawing process, the wire being immersed in some acid liquor or pickle, to remove the slight coating of oxide, which would corrode the drawing plates. This process would restore the lustre of the brass.

The Pickling Vats (Phase 3 and 4; Fig. 4.13)

The foundations of the 'pickling vats' (1232 and 1234 - floor, 1235 - western wall) were preserved. These were located towards the northern corner of the rolling mill against the western wall of the mill warehouse (1233). They covered an area of 6m x 1.5m. The bases of these pickling vats were constructed of a mixture of red bricks and crushed brick and mortar. Drainage was set into these floors to carry away the waste liquids.

Along with annealing, the process of pickling would have been undertaken intermittently during the rolling process to soften and improve elasticity. Particular rolled metals needed to be heated in annealing furnace, this made them scale. They then needed to be lowered into warm diluted sulphuric acid, then rinsed with cold water before being rolled again.

This area was described in the 1897 sales catalogue as the 'wire cleaning shop with carpenters shop over part of the building' (BCA MS 322/30).

Archaeological evidence

Immediately west of the engine house were the remains of an area, identified on the 1897 plan as the Wire Cleaning and Dipping Shop. The area was approximately 5.8m in width by 8m in length and was made up of two main walls (1446 and 1295), within which was a distinctive internal flooring composed of square engineering floor tiles and brick (1285). Surrounding this flooring were the remains of a drainage and guttering system (1294 and 1296), connected to a culvert (1462/1464).

The southern (1295) and western (1446) walls of the structure made up the southern corner of the original Union rolling mill. These walls were constructed of hand-made, 9 x 4½ x 2¾in red brick. There was no evidence of the northern elevation present on the 1897 mapping and it appears that this end of the structure had been altered. A later concrete flooring (1297) was present in this area. The doorway within the western elevation present on the 1897 plan was identified by the use of floor tiles.

A culvert (1462/1464), connected to the drainage channels and attached to the western elevation of the room, ran beneath the wire drawing mill and muffles before joining the northwestern corner of the canal.

The processes undertaken within the room (wire cleaning and dipping) would have required the use of cleaning and dipping solutions (such as acids and lacquers) followed by water for rinsing. The construction of the flooring, with flat floor tiles surrounded by drainage channels, would have made it easier to clean down. The culvert would have carried away the waste products created during these processes.

Welfare and storage

The coke pit and rubbish store (Phase 4; Fig. 4.11)

A small building located on the eastern side of the site and adjacent to the northern side of Gibson's Arm Canal was interpreted as the coke pit and rubbish store. The sale plan of 1897 identifies this structure as 'Coke stores, refuse pit and WC's forming a three storey building' (BCA MS 322/30).

The southern wall of the structure was the northern wall of Gibson's Arm Canal (1206). The structure itself was split into two by a central wall (1141). The southern half of the structure was the 'coke pit'. Access was obtained through a large gap in the eastern side. A brick floor (1140) was located within, and although truncated, must have originally attached to the main yard surface (1146) located next to the canal. Beneath the rubble backfill and against the back wall (1143/1210) at the eastern end of this room, were the in situ remains of a large quantity of coke, confirming its documented use. This coke pit was situated to facilitate the easy unloading of the raw material from the barges. The northern half of the structure was the rubbish store, again part of the internal floor surface (1206) remained. Ceramic drainpipes and drainage confirm the likelihood that this building doubled as a water closet.

Distribution of artefacts, metal waste and residues

The majority of the material derived from the rolling mill site was either not associated with structural features,

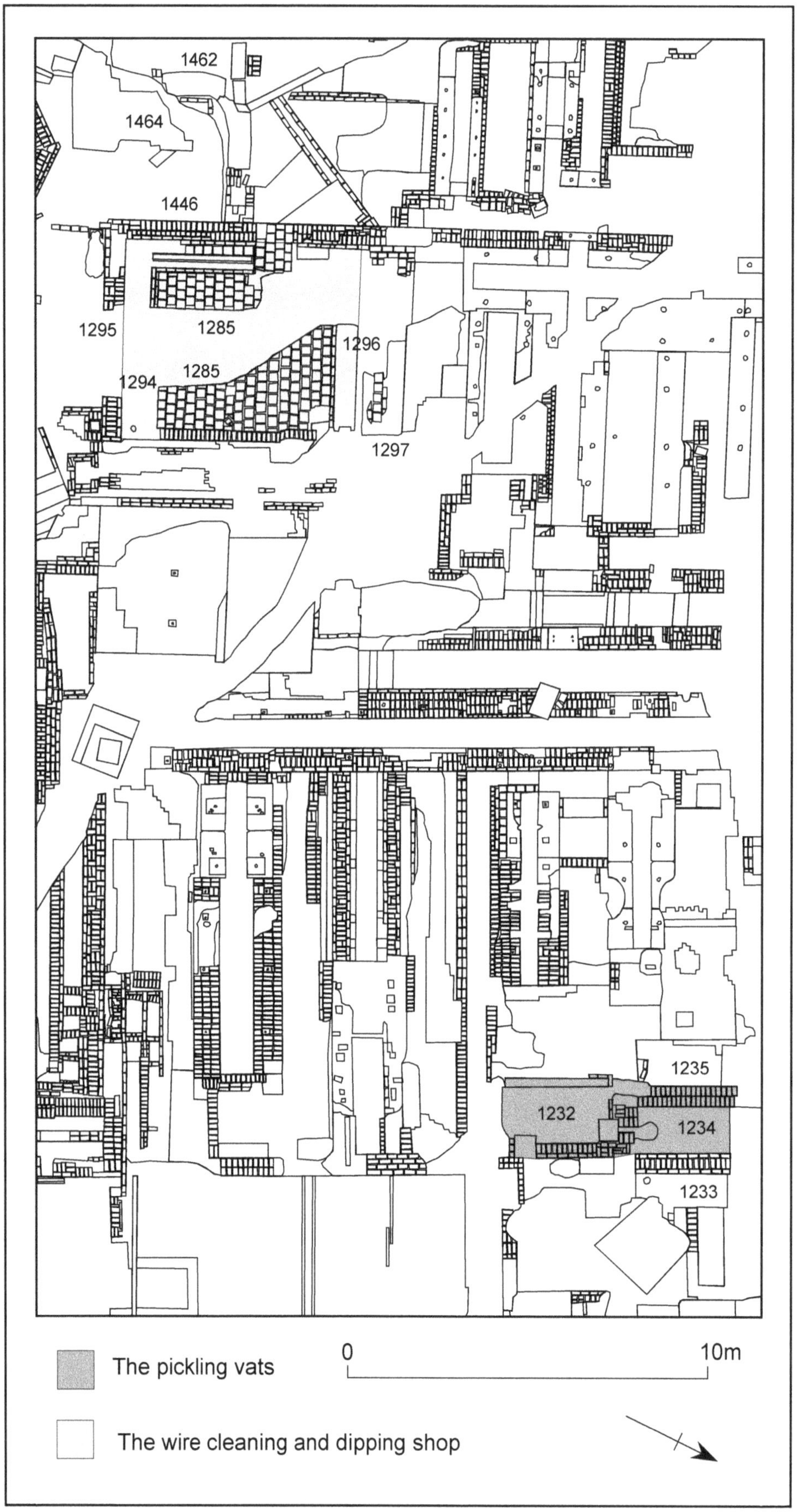

FIGURE 4.13 PICKLING VATS AND WIRE CLEANING SHOP

poorly stratified or unstratified. As such its value for explaining specific functions of machines was not necessarily as great. However, there was a significant spread of material across the site that could be related to production, either from the very end of Phase 4 or Phase 5 (late 19th and early 20th centuries) when production of rolled brass and wire was still occurring.

Large artefacts

The larger machine parts came from a number of overburden layers and cannot be directly associated with structural remains (see Chapter 6). They probably relate to machinery but it is difficult to know precisely what function they served. Some of the elements, including a single ribbed cast-brass machine arm that may have come from a piston arm, and two cast iron arms that may represent elements of a parallel motion, could be associated with the steam engine. It is however, difficult to directly attribute their function. A pulley, counter-weight and jib arm were found in the overburden layers in the vicinity of the rolling mill and appear to be associated with lifting mechanisms for moving metal around the rolling mill.

A large wooden beam (see Chapter 6) was recovered from Wheel Race A. It is likely to have been a cross-beam for the roof structure. This would tend to suggest that the pitched roof of the rolling mill was supported by wooden trusses as opposed to a cast-iron superstructure. The examples from Sheepcote Street (see Chapter 8) also had wooden trusses. This was in contrast to other hot-working industries such as the retort houses of the gas industry that had cast-iron roofs, an example of which can be found at the retort house on Gas Street (Linnane 1998).

Strip brass and wire (Fig. 4.14)

A clear distribution of waste metal was apparent for the wire drawing and metal rolling production. From these, well-stratified material was recovered. Both sheet brass and wire were recovered from the area adjacent to the central muffles (SF49; SF56; SF58), adjacent to the eastern muffles (SF67) and in the area of the electricity plant and strip casting shops (SF63 and SF71). It was also recovered from sealed contexts associated with the flues south of the rolling mill, Rolling Base A (1347) and a small machine base (1406) to the east of the rolling mill.

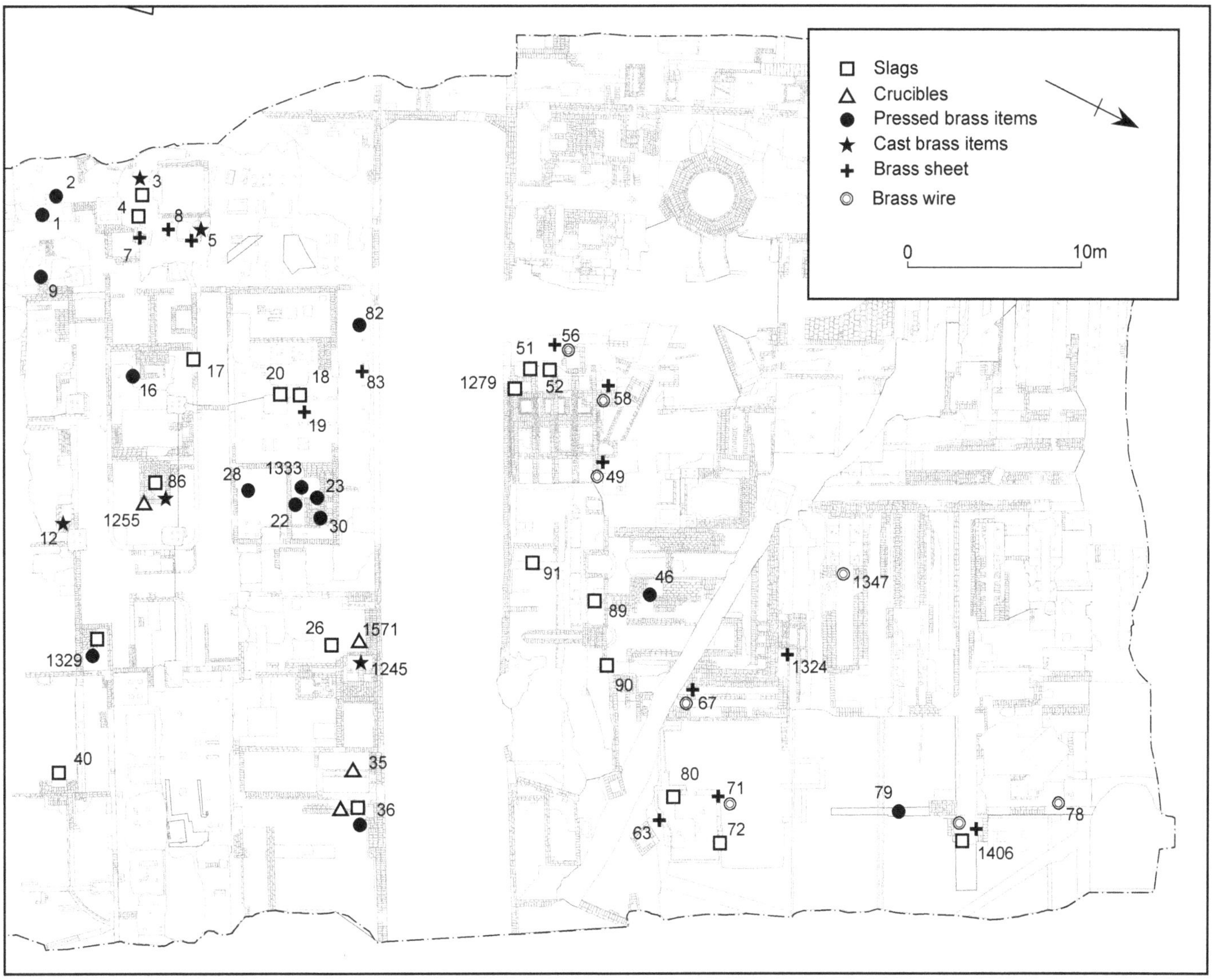

FIGURE 4.14 DISTRIBUTION OF SCRAP AND WASTE COPPER ALLOY ITEMS

The metal waste and residues associated with machine base 1406, came from sealed contexts and were clearly derived from Phase 4. They contained some wire with exceptionally low levels of zinc (12% Zn). In comparison the normal distribution for the site of 34–37% Zn. It also had a single piece of strip that had begun to be worked to wire. The XRF data, but less so the SEM data, showed a difference between the two ends that supported the idea that brass with high zinc content was used and then drawing reduced the zinc content as it was heated and annealed naturally by the drawing process. A further piece of wire (Wire 5) displayed heavily distorted features in the grains of its metal structure due to drawing process.

Two off-cuts of worked sheet metal from 1406, a disc (Disc 9) and an offcut (offcut 3), suggested that they had been subject to cold working and annealing or hot working. In Disc 9 the manufacturing process involved heating to above 250°C and quenching but cold working was also suggested. Offcut 3 lost very little zinc during the working process. This may have been due to the presence of finely dispersed lead droplets in the brass.

Data from Phase 5 was provided by some of the unstratified material from the end of the works lifespan. A single bar ingot from SF58 showed the early elements of hot-rolling had occurred as the metallographic analyses demonstrated that the bar had been subject to some hot working and was not in the as-cast condition. In addition wire from SF58 was also cast annealed and hot worked.

Overall analysis suggested that the wire drawing and rolling process was, as expected, the result of brassworking. It demonstrated that wire of high zinc content was used, and that there was zinc loss during the drawing process. In comparison the use of lead particles in the rolled brass allowed it to be malleably worked without zinc loss. The processes of cold-rolling, hot-rolling and annealing were certainly occurring.

Slags and Slag Residues (Fig. 4.14)

A number of slag residues were distributed throughout the rolling mill, but there was a distribution towards the casting shops and annealing structures. Of these some were in stratified contexts from Phase 3 (1411, SF91; 1388, SF90) and Phase 4 (SF103, 1406), others were in stratified contexts that were probably Phase 4 (1396, SF89) and others were from Phase 5 contexts or unstratified (1329, SF100; SF58). Of these the majority were copper-alloy working slags with a copper content of more than 50%. The exceptions was 1411, SFN 91 that had a copper contents below 50% and a very low non-ferrous alloying content. These may have resulted from ironworking either on site or imported from the adjacent site of the Crescent Foundry. In addition 1388, SF90 was almost entirely zinc and may have been added as part of the strip casting process.

Overall the slags supported the cartographic evidence that strip casting was occurring in the eastern part of the site (not excavated). The slag that had low copper content was associated with the earliest lifespan of the works and may suggest localised ironworking occurring, not necessarily derived from the site.

The demise of the works (Phase 6)

Archaeological evidence for the demolition of the works was obviously seen in the debris covering the site, and within the backfill of the canals (1547). Amongst the debris was artefact evidence of the demolition. This included a wrecking ball, pairs of the workmen's shoes, old beer bottles from the 1930s and most intriguingly of all, newspaper articles with the date and discussion of the Great Depression.

DISCUSSION OF THE ROLLING MILL

The development of the site

The structural evidence encountered on the northern side of the site consisted of the foundations of the superstructure of the rolling mill dating from the 1820s. The earliest Phase 3 structures consisted of a set of three Cornish boilers, three associated water-tanks, the engine base (for a Boulton and Watt Engine), and the chimney base and flues that can be collectively regarded as the mechanism for power generation. In the early 1950s the steam engine was remodelled (Phase 4). This involved the wholesale replacement of the Cornish boilers with two Lancashire boilers, and the equivalent alteration to the bases. Critical technological changes are likely to have occurred in this time period however.

Our current understanding of the mechanism of the steam engine is speculative. This is due to the limitations of archaeological excavation in which only the superstructure of brick supports (and no evidence of the machinery) was recovered. Supplementary information from the historic records may suggest a clearer picture. The engine base had two elements: on the southern side was a brick platform with two tying-down pins that continued into a large pit located below; on the northern side was an open pit of identical size, but *c* 4.5m deep. The overall size of this structure hints that it may have enclosed the beam engine, as the size of the second beam was 7.5m (26ft) in length and very similar to the size of the engine house. The 1897 Sale Plan (BCA MS 322/30) depicts the engine house with two openings at north and south ends, possibly to accommodate this beam.

There were two sides to the engine house, with the beam accommodated on a central wall or support. On the southern side of the engine house the cylinders were tied in place by the two tie pins. Steam pipes are depicted entering the engine house from the boiler and these would have fed into the cylinders. Below the brick surface was a room that may have accommodated the condenser and condensing cistern. The condenser was designed to effect the condensation of the steam after its escape from the cylinder, by admitting a quantity of cold water out of the condensing cistern, a water tank, through an injection cock. The northern half of the engine house would have accommodated the flywheel in the pit. The mechanism that transferred reciprocal power to rotary motion is not currently understood. Although the crank used in later models of steam engines could have accommodated the transfer of power, early Watt engines used a planet and sun arrangement of gears as Watt did not have permission to use the crank patent (Herbert 1849).

Parallels for the engine exist from the earliest period of the development of rotative engines. The oldest working engine (not rotative) in the world is the Smethwick Engine, brought into service in May 1779 and now at Thinktank, Millennium Point in Birmingham. The best extant parallel is probably the contemporary example of the Boulton and Watt Engine of 1817, originally erected for M W Grazebrook's ironworks in Netherton, but subsequently removed and erected on the Dartmouth Circus in Birmingham. Although a blowing engine it does demonstrate the basic layout of beam engines at this time (Buchanan and Watkins 1976, 154).

The vertical steam engine originally installed in the Union Rolling Mill in the 1820s was replaced with a similar one in the 1850s. Yet elsewhere, horizontal steam engines began to be adopted. The Landore Works, Swansea had a horizontal steam engine by the 1860s (Gale 1976, 162) and the Musgrave Rolling Mill at the Hafod Copperworks had one also photographed in 1910 (Hughes 2000, 48, fig. 68). The efficiency of a horizontal steam engine outweighed that of the vertical steam engine – yet such an engine continued to be used on site at the Cambridge Street Works until they moved in 1936 – a throwback to a bygone age.

In the 1820s (Phase 3) elements of the transmission, the wheel races and rolling mill bases of the early rolling mill lay to the north and east of the engine house. Changes were visible in the scale of the transmission. Probably coinciding with the alteration to the engine in the 1950s (Phase 4) wheel race A was extended north and two further wheel races added (B and C). Specific construction phases of the rolling machine bases were more difficult due to the likelihood that they had stayed in use and were refurbished throughout the lifetime of the works but a chronology of rolls existed on site. As the rolls were replaced the bases of the former rolls were left *in situ*. This technological development can be seen by the variance in the material of the bases which goes from brick to concrete, with the concrete structures being constructed later. The initial reorganisation of the rolling machinery may have coincided with the refurbishment of the 1950s. The evidence for this is difficult to see but it may be that there is a two-phase chronology in the brick bases. Rolling mill bases A–C all relate to Phase 3 with some Phase 4 alteration. More Phase 4 rolling bases may have been removed by the insertion of later rolling bases. By 1900 further alteration to the rolling mill had occurred, with the construction of a series of rolling machine bases in concrete (rolling bases D–I, Phase 5). This probably suggests a steady improvement in rolling technology and it is almost certain that some of the later rolls were powered by electricity as suggested by the insertion of the electricity plant in Phase 5 (also constructed in concrete).

Parallels for the layout of the rolling mill are few, and the work undertaken on the site probably represents the first fully excavated example of a rolling mill. Contemporary copper mills were operated in Swansea by John Vivian of the Hafod Copperworks in 1819 (Toomey 1985, 361–363). The new purpose-built mill was constructed on the western riverside section of the Hafod Copperworks. In 1842, due to direct competition from the Morfa Works, John Vivian expanded the Hafod Works, adding a 60 horsepower steam engine. It was again re-built in 1860–1862 and the engine house survives extant. It consisted of a four-bay structure in Pennant sandstone with a pitched roof, and square stack located adjacent (Hughes 2000, 48–49).

A similar copper rolling mill was established on the adjoining site by the Morfa Copperworks of the Williams family in 1828. These two works (with the Hafod mill above) represent the most comparable examples of the buildings of the copper rolling industry to the Union Rolling Mill and the design must have been similar at this time. A plan survives of the original 1828 Morfa copper mills. They depict a single building, of similar scale to the Union Rolling Mills, with the steam engine housed in the corner. The sides of the building are open arcades. The surviving buildings of the Morfa copper mills date to 1840 after the 1828 copper mills burnt down and were replaced. They now form a store for the Swansea Museum. They are long linear buildings built in Pennant sandstone rubble on a much greater scale than the buildings of the Union Rolling Mill. The sides were again open on the eastern elevation to allow ventilation (Hughes 2000, 48–49). Images of works from the 19th century suggest that the sets of rolls were generally located in a significant space. An image of the works at Deptford suggests that the rolls were located in a line with sufficient space to manoeuvre hot metal before it was passed through the rolls (McLain 1976, 222). Yet the contemporary image of the Winfield's Rolling Mill (Fig. 4.2) suggests a confined space. The Blists Hill Ironworks at the Ironbridge Gorge Museum, Coalbrookdale still has a working iron rolling mill that would have operated on almost identical principles to the Union Rolling Mill. The works still operate an occasional basis with iron being rolled in 2010. The open space is a feature of these ironworks and suggests that not every rolling mill base can have been operating at the same time.

The descriptions of the rolls as 'pairs' (BCA MS 322/30) in the sale of 1897 suggests that for the majority of their working life the rolling mill operated a 'two-high mill' and no more. The installation of a 'three-high mill' had happened by the 1860s when Lauth introduced the three-high principle to his works in Birmingham in 1862. In this the third roller was placed above the other two to enable metal to be passed back again without reversing the engines. The first four-high mill in Great Britain was installed at the Henry Wiggin plant in 1930 (Stephens 1964, 140–208). This probably suggests that the improvements recommended in the report by James Watt in 1897 (BCA MS 322/32) may have included recommendations to improve the height of the rolls to three-high.

The change to electricity as a power source appears to have occurred in the early 20th century when the electricity plant was installed. By 1922 electricity was being used to power entire rolling mills as was the case for Hadfield's East Hecla Works, Sheffield, which was a 3,200 horsepower mill (Gale 1976, 164). The first electric furnace for melting copper in Birmingham was of the resistance type and was installed at Kynoch's works in 1920, followed by an electric induction furnace for brass melting in 1921 and for annealing copper alloys in 1926 (Stephens 1964, 140–208). This suggests that the electricity plant was utilised for the rolling mills initially but it is possible that the gas fired muffles were also converted to electricity later. The use of electricity was an adaptation of the works to improve efficiency but was a sign of the change in the industry that would lead to it moving to a new site.

Wire drawing and tube drawing machine bases were preserved to the west of the rolling mill. These machines were smaller in scale. The design of wire drawing machinery generally needs a much smaller amount of space than the rolling machinery. The wire would be drawn on a series of drums that would stretch from the roof to the floor (Fig. 4.3). The tube drawing machinery appeared to be laid out against the north–south wall of the works. The image of the works (Fig. 4.4) suggests that it would have adopted a long linear location and this may support some of these machine bases being associated with the drawing of tubes.

Evidence relating to intermediate processes - hot work such as annealing and pickling - was exposed. The muffle houses, used in the annealing of the brass were identified. These muffle houses showed similarities in their construction, but had clearly been developed, and rebuilt during their lifetime. A chronology of muffle designs follows from the coke designs of the western muffle house (and probably the eastern muffle house) to the gas powered design of the central muffles. The eastern and western muffles were coal powered. In the case of the eastern muffles the necessity for high temperatures meant they were situated against the boiler houses in order to retain heat. The constant rebuilding of the central muffles suggests that they were experimental in design and hence were rebuilt quickly in a short space of time in order to improve efficiency. The advantage of a gas-annealing system was highlighted in an address to the Birmingham Society of Chemical Industry by HW Brownsdon in 1917. It allowed brass to be annealed in a clean fashion as it did not expose it to oxidisation and contact with 'injurious furnace gases' and thus did not require repeated pickling (MacMillan 1917, 984). It would appear that the constant rebuilding of the muffles at Cambridge Street was designed to increase the economy of the works and may in part explain why there was such a small area for 'pickling' the wire.

The layout of the site shows the significance and necessity that power played in the function of the rolling mill. Not only was the steam engine centrally located, it was also central to the functioning of the site. All the machines that it served distributed away from it to the west, north and east. The site appears to have been restricted and this meant that space was at a premium; the rolling mill bases were distributed in a confined tightly packed space; the wire mill was constructed in the space between the boundary with the tube mill and the rolling mill. This may be one reason why the muffles that were likewise in constant use, were poorly positioned. Although the eastern muffles were located in an area that could be defined as for 'hot working' in a triangle between the casting shop and rolling mill, the western muffles were poorly positioned. The engine house, boilers and structures such as the pickling vats and the wire mill prevented direct access from the rolling mill. This may suggest that the western muffles were used for annealing wire, and items produced in the bedstead works (see Chapter 5) as opposed to rolled brass.

The significance of the Cambridge Street rolling mill

The development of Birmingham as a key brass centre cannot be said to be due to any geographic advantage. All three elements of the initial manufacturing process, primary (copper smelting), secondary (brass refining) and tertiary (rolling, wire and tube production) could be accomplished elsewhere at a lower cost. Copper smelting developed in Swansea due to key economic factors: three times as much coal was needed to copper ore and it is economically more viable to locate the industry close to the available coal source (Hughes 2000, 1). Bristol, Cheadle, Macclesfield and Flintshire relied on locally available zinc deposits to sustain market advantage and began to lose market share when local raw materials began to run out. The production of wrought brass (ie rolled brass and wire) required water-power initially and was located next to streams and rivers. Bristol initially held sway in the 18th century over this element of the process before steam power was used. The manufacturing process required none of these elements, and more important was a skilled or semi-skilled labour force. In the 16th and 17th centuries Birmingham and the Black County had developed as a metal-production centre based predominantly on iron, and easily expanded to produce brass articles in the 18th century. In the 18th century it had an even share with Bristol in market production in brass. Yet by the end of the century economic change in sources of the metal allowed Birmingham to capitalise.

Two key developments - canals and steam - placed Birmingham on an even footing with its competitor centres of Swansea, Bristol and London. Canals allowed low cost transport of raw materials for the first time. The initial elements of the canal network in the West Midlands were developed between the 1770s and 1790s and this corresponded with the development of the Birmingham

Brass House. The brass house was a response to concerns about the increase in costs of brass supply but it survived economically because improved transport allowed raw materials to reach Birmingham at low cost. The second development was steam power. Before the advent of steam, water-power had been adopted to power rolling and wire drawing mills. In this respect Birmingham, although not lacking streams and rivers, was at an economic disadvantage to Bristol or indeed Cheadle or Flintshire. The rivers and streams of the Tame Valley system were relatively small and power was limited. The limitations of the Birmingham water-power supply were overcome by the development of steam power that became first practically used for rolling and wire drawing in the 1780s. By the turn of the 19th century many former water mills had converted to steam power and it was becoming to be used in the Birmingham industries. The development of steam power in Birmingham gave them an initial economic advantage as they began to roll brass and produce wire.

In examining the significance of this development it should be seen against a backdrop of the industry in Birmingham. Aitken writes less than 50 years later:

Without the rolling mill the Birmingham brass trade would be small indeed. Rolled brass is the semi-raw material on which the brass-founder entirely depends; without it, ordinary soldered brass tube, patent cased tube, or cased stair rods could not be made. It largely enters into the productions of cabinet brassfounder, of the door, shutter knob, and cupboard turn-maker, of the drawn and pressed brass hinge maker, and bras covered cornice pole maker; in truth it would be extremely difficult to state in which department of the brass trade it was not used (Aitken 1866, 311).

It is almost certain that James Watt had developed steam-powered rolling mills as early as the 1790s at the Soho Manufactory as suggested by his correspondence with John Morris of the Landore Copperworks and Matthew Boulton as to the viability of using steam-power for rolling mills (Hughes 2000, 48). Evidence also suggests that a number of mills developed steam powered rolling facilities in the late 18th and early 19th centuries. These include mills at Nechells in Hockley, Fazeley Street in Bordesley, Lifford in Kings Norton and Dogpool in Stirchley. Yet, critically the Union Rolling Mill was the first purpose built steam-powered rolling mill in Birmingham. Prior to this rolling mechanisms had been established within former water mills. This was a significant step that separated the rolling mill from the use of water and provided a template for other rolling mills that developed afterwards.

In comparison the Swansea industry was reliant on water-power for much longer. The first potential use of steam power can be attributed to the Upper Forest Mill of Lockwood, Morris and Co. Copper where in the 1790s, John Morris of the company asked Matthew Boulton whether steam power could be used for rolling mills and Boulton replies in the affirmative. However, this was never accomplished (Hughes 2000, 49–50). It was certainly being used by the late 1810s when the Hafod Works was established in 1819 and began to roll copper. Within a few years a second mill was established at the Morfa Copperworks of the Williams family in 1828 (*ibid*, 48–49).

Although the influence of the Association of Copper Smelters, based in Swansea, was limited over the Birmingham market (see Chapter 2) the competitive atmosphere in the ferrous metal trade in the 1820s had allowed middle-men, who owned rolling mills to buy cake copper and undersell the association (Toomey 1985, 312–325). It is possible that the Union Rolling Mill was set up to secure Birmingham manufactured brass supplies at a reasonable price at a time when it faced competition from both middle-men and the powerful monopoly of the Association of Copper Smelters. In addition copper was now being rolled in Swansea at the Hafod Works. It is in this environment that the Union Rolling Mills was established. A purpose-built mill was designed to produce affordable rolled metal using the latest in steam power. It provided a template for future works.

As will be seen, its influence can be seen in the design of the rolling mill at Sheepcote Street (see Chapter 8). The details of the working of this mill are somewhat obscure but it is clear that it adhered to the template provided by the Union Rolling Mill. The steam-engine was situated close to the canal for ease of transport of coal. From this it can be deduced that wheel pits ran along the western end of a series of four large open-plan workshops within which rolling machinery (and probably some tube machinery) was housed. The likelihood is that power was transferred via a flywheel contained within a pit and shafting transferred perpendicularly from this. In the final plan separate locations for casting are situated away from the main rolling plant but close enough to allow the movement of material for hot rolling.

CHAPTER 5: THE BEDSTEAD WORKS (PHASES 3, 4 AND 5)

Chris Hewitson, Will Mitchell and Ray Shill

The bedstead department is one of the earliest in which metallic bedsteads were made. The late Mr. Winfield commenced making brass and iron bedsteads, and by elegance of design and many other attractive features succeeded in so far developing this branch that many thousands of bedsteads had been turned out here before other makers took up their manufacture. (IMechE 1897, 403–404).

The southern part of Area 1 was located to the south of Gibson's Arm and was occupied by the bedstead works. The bedstead works functioned from the 1850s until the sale of the Cambridge Street works in 1897–1898. The works underwent a number of changes of ownership after the sale of the works in 1897–1898 (Phase 5) but was owned by Sperryn and Co until they moved in 1910. The bedstead works was amalgamated into the Birmingham Aluminium Casting Company during the First World War and was eventually demolished in the 1920s.

The eastern half of the bedstead works was destroyed during the construction of the new Baskerville House in 1938. Therefore the structures identified on the site related to workshops from the western half of the bedstead works. The nature of these structures and historic depictions of the site suggest they were low one or two-storey workshops, which appear to have been brass-casting shops.

A HISTORY OF THE BIRMINGHAM METALLIC BEDSTEAD TRADE

Ray Shill and Chris Hewitson

Two factors were to influence the expansion of the metallic bedstead trade. Both brass and iron were used in metallic bedsteads and thus Birmingham became a centre for manufacture. Advances in the puddling process of wrought iron and the development of strip casting of brass (see above Chapter 4) improved the supply of raw materials and meant that mass production of metallic bedsteads became a possibility. This new trade, which rose to prominence during the 19th century, was to revolutionise bedstead making, taking it away from the carpenter and the smith and into the realm of commercial manufacture.

Few branches of manufacture have been more rapidly developed within a comparatively short space of time than the trade of making metallic bedsteads. ... The improvements in the manufacture of iron, the use of ingenious machinery in making the various parts of bedsteads and the number of skilled workmen engaged in the trade, have resulted in the production of articles of furniture cheaper, more useful, because more readily moved and fitted, cleaner and more elegant than similar articles made of wood (The New Illustrated Directory Entitled Men and Things of Modern England; Billing 1858).

Metallic folding beds, cots and chairs also came to be a staple product of this industry. Furnishing ironmongers stocked a range of different types and beds could be ornamented or plain.

Origins of the trade

The foundations for the mass production of bedstead were derived from a few basic inventions, which took time to evolve into efficient and practical methods.

The folding bedstead used many of the same principals as the metallic bedstead and was developed in London at the start of the 19th century. The earliest manufacturer was T Butler of Catherine Street (Times 1803). By 1810 his factory had been acquired by Pryer, Steains and Mackenzie who developed the patent brass-screw bedsteads (Morning Post 1810). Other patented metallic bedsteads were manufactured by John Thomas Thompson, travelling equipage maker at his factory at 126 Long Acre, London from 1812 (Patent 3560). It is likely Thompson adopted the drawn and soldered iron tube, as developed in Wednesbury by Cornelius Whitehouse in 1825 (Patent 5109). In 1826 he advertised a metallic bedstead that had brass drawn over an iron tube and his portable 'Wellington' bedstead both of which adopted a patent for the use of metallic tubes in bedsteads.

In parallel bedsteads were being made in Birmingham. The commercial development of the 'cased tube' a concept based around a ferrous metal tube encased in brass can be attributed to a patent granted to Benjamin Cook and Thomas Attwood of Birmingham, (Patent 3460). The concept was not an original idea but the invention had been attributed to Sir Edward Thomason (Aitken 1866, 322–323). Cook and Attwood's patent was no doubt promoted to enhance the business of both patentees. Thomas Attwood specialised in the making of calico rollers (Patent 4798) at Baskerville House Mill, whilst Cook produced a range of brass house furnishings that included bedsteads at a brass works in Weaman Street. Cook submitted a second patent in 1812 (Patent 3609) that incorporated the cased rod in the manufacture of various items of furniture including bedsteads.

Robert Winfield adopted a variation on the cased tube – the continuous hollow pillar design. The Cambridge Street

Works produced beds with legs and pillars made of tube. Robert Winfield had previously worked for Cook (see Chapter 4) before setting up in own business and some of his inventions may have come through the experience working for Cook. Winfield's first bedstead patent was presented in 1827 (Patent 5573), another specific to bedsteads was presented in 1831 (Patent 6206) and others followed. It seems likely that the brass bedstead industry in Birmingham developed from a number of influences but Robert Winfield was able to adapt these ideas to make his bedsteads more efficiently and cheaper.

Iron tubes, whether cased in brass or not, became an essential component for the headers and footers. Angle iron and iron rods were also required for the bed structure. Angle and strip iron made up the laths that joined the headers and footers, and when joined together formed the structure of the bed. Brass was used for ornamentation and in the exterior layer of cased tubes (Aitken 1866, 322–323). Other improvements included methods of joining and fixing the frame together. An essential component in the Winfield bed, as with other contemporary makers, was the dove-tail joint patented originally by John Bennett (1813; Patent 3677) that enabled beds to be constructed more effectively. Further improvements were made by Dr William Church (Patent 9187). Malleable iron castings were used for the assembly of head and foot rails. Malleable iron casting was first used in Sheffield, but soon became a specialist production in Birmingham.

Birmingham manufacturers came to excel in making of metallic besteads, which became more popular after the Great Exhibition of 1851.

Bedsteads of brass are still made in great numbers and in these the skill of the modeller is called into operation to produce the required degree of ornament. The finest bedstead hitherto made of this description was that produced by Messrs. Winfield and Son, Cambridge Street Works, for the Great Exhibition of 1851, at a cost of nearly £500. Brass bedsteads can be made from about £6 and upwards (The New Illustrated Directory Entitled Men and Things of Modern England; Billing 1858).

Whilst factories existed elsewhere in Britain, the greatest concentration was located in Birmingham and across the Warwickshire/south Staffordshire border into Smethwick. There was a tremendous rivalry between firms. Patents and registered designs for style and means of construction were regularly sought and enforced. The glut of patents and designs ensured a variety of product, but sometimes restricted makers as to what each could produce. Lawyers and the courts were frequently the only real winners in the civil actions brought to preserve a company's rights (Furniture Gazette 1875).

Basic elements to this industry were the trades of iron foundry, brass foundry, tube drawing and smiths' work. Metallic bedsteads had frames made from tubes and strips of metal with castings attached to form the joints. A variety of skills were used for the making of metallic bedsteads. Many were carried on by men and youths, but certain tasks were performed by girls or women (see Peyton and Peyton, Peyton 1866, 624–627; S B Whitfield, Furniture Gazette 1875): -

- Angle iron maker, iron puncher and straightener
- Brass turner
- Brazed tube maker
- Burnisher
- Chipper
- Colour mixer
- Edging - *youths generally employed*
- Finisher
- Floater
- Foundry clerk
- French-end caster
- Japanner or black work - *women employed*
- Lacquerer - girl
- Malleable (Iron) caster
- Oil varnisher
- Ornamentor, decorator and painter- *girls and women*
- Penciller - *girls*
- Pillar caster
- Polishers
- Pourer for foundry castings
- Rail bender
- Rail caster, brass rail maker
- Repairer - *youths generally employed in this duty*
- Rifer - *youths generally employed in this duty*
- Sacking blacker
- Scroll bender
- Stock and chill fitter
- Smith or blacksmith
- Transferrer - *girls and women*
- Tube charger
- Turner (brass)
- Wrapper and packer - *women, includes strawing-up*

Those who made metallic beds did so as a core business or part of a series of peripheral or related trades. Robert Winfield was typical of this pattern and made bedsteads as one of wide range of other products. Another example was S Bott and Co, of the Marrian Works, Small Heath who made a range of products including bedsteads, fireguards and fancy ironwork before they filed for bankruptcy in October 1877 (BDP 1877). By the mid 18th century major bedstead manufacturers were distributed throughout the city with a particular distribution along the Birmingham Canal towards Smethwick (Fig. 5.1).

Decline in trade

The most productive years for the Birmingham metallic bed trade were from 1850–1890. After this period increasing competition from European and American manufacturers led to reduced sales abroad. An initial indication of the decline was the bankruptcy of Edward Cook in 1883. Edward traded as Benjamin Cook and Son at 261–262 Bradford Street and was the successor to the

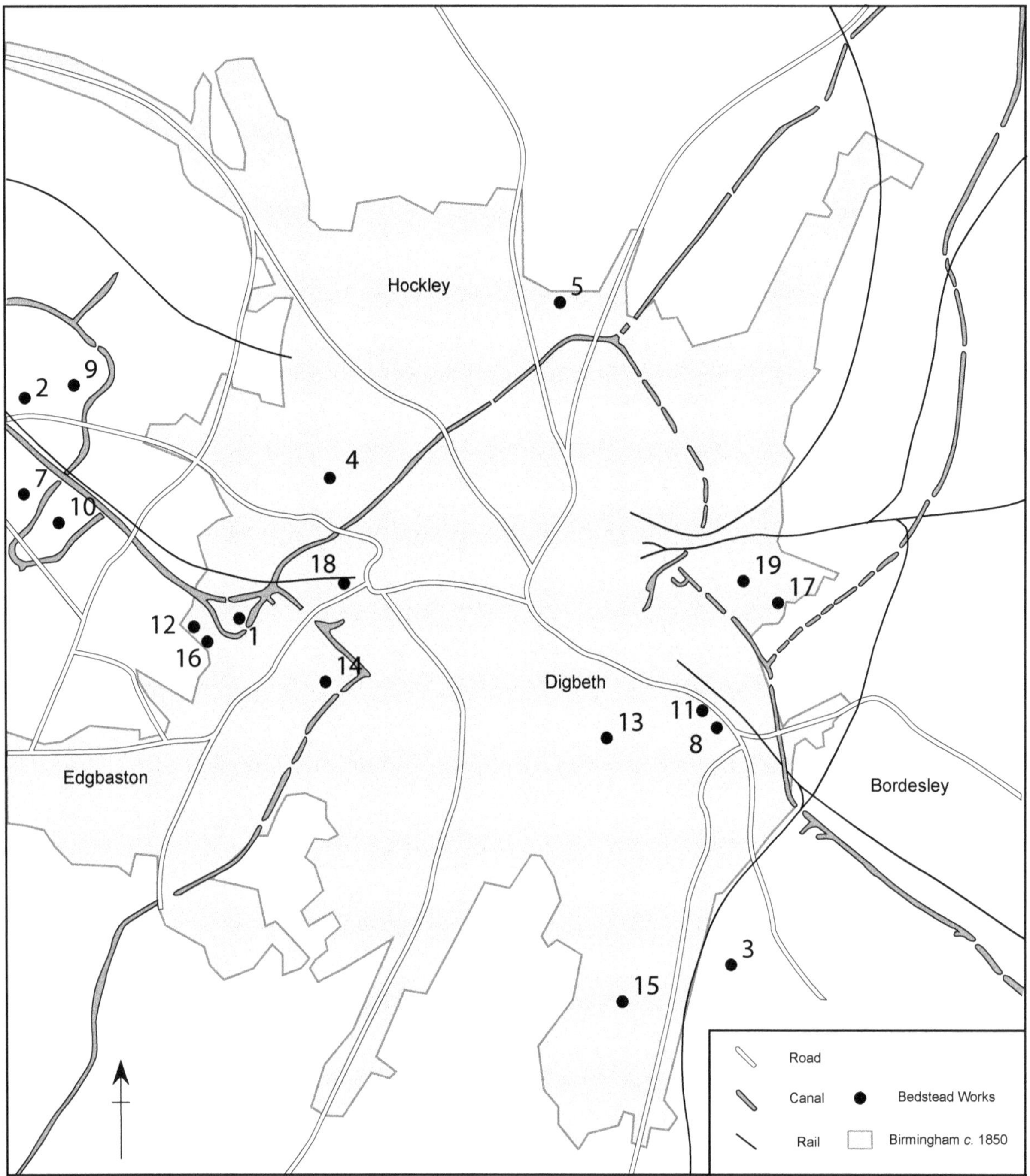

1. Brierley and Sons, Sheepcote Street
2. Charles Bryant, Bryant Street
3. Charles Cartwright, Hertford Street
4. Robert Crosbie, Charlotte Street
5. Evereds, Surrey Works, Smethwick (not ill.)
6. Fitter Brothers, Pritchett Street
7. Frazer Brothers, Wiggin Street
8. Hoskins & Sewell, High Street, Bordesley
9. Hoyland & Smith, Western Road
10. W and W.A Hulse, Rotten Park Street
11. Peyton & Peyton, High Street, Bordesley
12. Philips and Son, Sherborne Street
13. Albert Philips, Rea Street
14. Phipson and Warden, Granville Street
15. T Smith, Athole Street
16. J and J Taunton, Sherborne Road
17. J Troman, Bolton Street
18. R. W Winfield, Cambridge Street
19. S.B Whitfied, Watery Lane

FIGURE 5.1 THE DISTRIBUTION OF BEDSTEAD WORKS

business started by Benjamin Cook (LG 1883). Cook's trade had been somewhat eclipsed by the major bedstead firms such as Hoskin's, Peyton's and Winfield's.

The receivership of Peyton and Peyton, a year later, was a more significant event. Along with Hoskins and Sewell they were the most successful bedstead firm and enjoyed considerable sales abroad. The Peyton and Peyton partnership comprised Edward Peyton and Henry Eagles and apart from the main Birmingham works they had premises at Liverpool, Manchester, London, Glasgow, Dublin, Paris, New York and Melbourne (Australia) (LG 1884). Peyton's financial difficulties came to a head in 1884 when following receivership proceedings, the Dublin premises were offered for sale and disposed of (Freemans Journal 1884). Edward Peyton was able to arrange a suitable settlement with his creditors and the bulk of the business was saved (BDP 1884). Peyton then went on to face another crisis when the union called strike action over pay in 1890 (Hopkins 2002, 112–129). Most bedstead factories were affected and sales were damaged in consequence. Several firms subsequently left the trade or went on to other product lines. Yet metallic bedsteads remained a core Birmingham business to 1920. Demands for commercially made wooden bedsteads started then to increase. Some makers continued to make metal beds, but this was for specialist markets such as for hospital use or as ships berths (Shill 2006b, 48–54).

The bedstead manufacturing processes

A bedstead factory comprised several departments. A detailed description of the process was given in a published account of S B Whitfield's factory, which specialised in making iron beds (Furniture Gazette 1875). Raw material was delivered to the cutting shop. Here iron tubes, angle iron and iron rods were cut into the required length for the different components. The rods were studded and the angle iron stamped flat by presses and punched by machines. Rods might be then bent and shaped for scroll work (Shill 2006b, 37-8).

Common to all bedstead works was the casting shop. Such shops were located on the ground floor. At Whitfield's the various metal parts were taken into the casting shop for assembly into the headers and footers for the bed. These cut and shaped pieces of iron were placed on a frame fitted with casting boxes known as chills. Designs for headers and footers varied and in some patterns more than twenty chills were used. The placing of the chills was at the joint point of the pieces (Shill 2006b, 37–38).

Pig iron was melted in a furnace, and when molten transferred into casting pots and taken to the various parts of the shop where the headers and footers were being assembled. Molten metal was poured into each chill and quickly cooled. Soon after the metal was poured, the chill was opened, leaving the shape of a flower, knob or other ornamental shape, which served the dual purpose of decoration and fixing. Each casting was then ready for the work called chipping, where the shape was cleaned of all superfluous bits. Once the header or footer shape was complete, chills were then fitted to the corners of the posts to form the sockets into which the dovetail ends of angle iron were placed. Various bedstead ornaments, such as the tops of pillars were cast in 'stocks', which like the chills, produced a particular shape, but this might be in either brass or iron (Shill 2006b, 37–38).

Some makers like Samuel Whitfield specialised in iron bedsteads, several other bedstead makers made beds in both iron and brass. The foundry in those cases might also make rough brass castings. These were then worked up to a finished product through filing or turning on a lathe. An essential requirement was power supplied by steam engines that turned shafts and pulleys throughout the factory to allow metals to be shaped or cut and lathes turned (Shill 2006b, 37–38).

The next stage was to protect the iron against rust. This was achieved through a process known as japanning, which required the application of a series of tar-based layers and subsequent baking in stoves located on the ground floor. Japanning and painting was carried on in separate shops. Colour was determined through heating in stoves of different temperatures. Common, or black japanning, was the most frequent mode of covering. There were other choices of colours, and each had a special room for the task to be done. Finally packaging and wrapping were conducted in another department. Packing involved a process called strawing up, were the parts were protected for carriage by road, railway or canal to British furniture showrooms or for shipment abroad (Shill 2006b, 37-38).

Comparisons with sales advertisements for bedstead works reveal a standard design of ground floor engine houses, foundries and heating furnaces, and shopping extending usually to a first, second and sometimes a third floor, for all other ancillary processes. Benjamin Cook's Green Street Factory in 1883 comprised a mixture of two- and three-storey shopping (BDP 1883), whilst Cox and Luckman had a three-storey factory in Stanhope Street (BDP 1898). R W Winfield had a four-storey bedstead factory on their site (visible on image; BCA MS 322/197).

THE BEDSTEAD WORKS – A HISTORY

Ray Shill

The bedstead works (Phase 4)

The bedstead works were an integral part of the Cambridge Street works of Robert Winfield. The exact date of construction cannot be confirmed historically, although it certainly occurred by 1855 when it was first depicted on Piggott Smith's Map. A series of major alterations had occurred within the Cambridge Street Works during the 1840/50s. The wharf on Gibson's basin that was the location of Winfield's coal merchant's business had moved by 1852 and the land was acquired for the bedstead works (Phase 4).

Robert Winfield advertised that he was the patentee of metallic, military, travelling and house bedsteads for home use or for export. Robert regularly stated in advertisements and notices that he was the proprietor of the original patent for the metallic bedstead and patentee of other improved principles associated with their manufacture. The reality was probably somewhat different but he was steadfast in enforcing those rights. In particular, there was the patent for the continuous hollow pillar that was so essential to the success of the Birmingham metallic bedstead trade (eg Patent nos 8891, 12268, 12302, 13576). He also used wrought iron, in place of cast iron and this metal was used in the construction of the 'travelling' bedsteads supplied to the army and navy. Innovation in the bedstead industry continued with new patents taken throughout the preceding decades (eg Patent nos 2724, 4489, 9888) but the industry was in general decline from the late 1880s. The fire at the bedstead works in September 1888 damaged the main four-storey block and severely affected production (BDP 1888a).

Change and the arrival of the aluminium works (Phase 5)

With the winding up of the Winfield's operations, the offices, brass foundry and bedstead works had various occupiers, which have been listed previously (see Chapter 4).

Sale and Fragmentation 1898–1910

Sperryn and Co took over a major part of the bedstead works with the remainder being leased to the Player family, railway lamp makers of Broad Street. R Richardson (mechanical engineer) leased buildings adjacent to Baskerville Place. One brief occupant of the premises on the Baskerville Place side of the bedstead works was the Batty Engineering and Spoke Manufacturing Company Ltd, which were one of the many firms that rushed to capitalise on the cycle boom of 1896–1899. The firm went into liquidation in 1901. Plant including milling machines, planning machines, lathes and a Tangye gas engine was offered for sale (BDP 1901).

Sperryn and Co was established first at Hospital Street and comprised a partnership of E C Kemp, Clifford Kemp, H Herbert Wright and George Sperryn. The Kemps left the partnership in December 1890 (LG 1891). George Sperryn became a principal partner in the firm. He had been born in India, and had spent time in Canada. On his return to England George Sperryn gained experience as manager of a chandelier factory before entering business on his own account. Sperryn and Co expanded their business with the lease of a section of the former bedstead factory in Baskerville/Attwoods Passage. They remained here until 1910, when new premises were taken in Moorsom Street.

John Edwin Player had a business as a ship and railway lamp maker in Broad Street. He moved to Attwoods Passage to take over the eastern part of the bedstead works. John took Sydney Weis Mitchell into the partnership and they then traded as Player and Mitchell. John Player left this partnership in 1907, but the firm continued to be known as Player and Mitchell (LG 1907). This firm became a limited company in April 1911 and subsequently removed to new premises at Doris Road during the year 1915 (BBP 26436). Here they continued the manufacture of lamps at what became known as the Sentinel Works.

The aluminium works 1903–1918

Much of the Cambridge Street Works was reclaimed and amalgamated by the Birmingham Aluminium Casting Company between 1903 and 1918. This firm began as the Hydraulic Joint Syndicate in 1896 with head offices in Nottingham. They started to make hydraulic joints for cycles, motor carriages and other purposes. Offices then transferred to London and by 1901 the title was changed to the Birmingham Aluminium Castings Ltd (PRO BT31/7060/49740). Under a special resolution this company was wound up in April 1903 and replaced by a second company, Birmingham Aluminium Castings Ltd. The space in the former Winfield's chandelier and gas fittings factory was used (at the eastern end of Baskerville/Attwoods Passage). Birmingham Aluminium Castings Ltd then gradually acquired the use of the buildings south of Gibson's Basin. By this time the factory occupied only the ground floor and two or three storeys above. These floors were reached by spiral staircases or outer wooden stairs, and although a lift was later installed in 1909 it remained a maze of small and long rooms not best adapted for the trade the new occupiers were engaged in.

Extra space was needed and building plans show they took over the former Sperryn, Player and Mitchell, and Richardson premises in 1915 and in 1917 to have the whole former bedstead works (the western side described above, including the areas of the site). They also took over the former Winfield's boiler house next to Sperryns in 1917 (BCA MS1422 60/1/7/2). Such plans indicate a considerable change to the buildings. The Richardson premises and boiler house formed part of the new extension for the No 3 die casting shop, which occupied part of the ground floor. The first floor was occupied for despatch and the second floor for warehouse space (BBP 26514A; BBP 28315). This plan indicates a major reconstruction rebuild of this section of the site. The die casting shop (number 1) was on the neighbouring plot, formerly Sperryns. In Winfield's time the buildings in this area rose to four stories and one had the upper storey rebuilt after the fire of 1888 (BBP 6354). A section of the original canal-side four-storey buildings and a central core of buildings up to Attwood's Passage were retained, but a large part of the Player and Mitchell buildings were replaced with modern saw-tooth (northlight) roof buildings to house a casting shop and dressing shop. The canal was also built over to form a solid building link between the main works in Cambridge Street, visible in early 20th-century photos of the works and the ordnance survey maps (BCA MS 322/181–5; Ordnance Survey 1902, 1917).

Birmingham Aluminium came to specialise in the supply of aluminium castings to the motor industry - crank chambers, for example, which had an increasing demand for use in engines for motor cars and cycles. Part of the Gas Fitting Department retained a casting shop. The old showrooms fronting Cambridge Street became a 'Dispatch and Finished' shop. There was a sand mixing shop and the offices formerly used by Winfield's on the first floor became the manager's (Mr Gower's) office, clerks' offices and general offices for the Aluminium Company (BBP 28058).

Aluminium was a metal that was first extracted by chemical action by acids on aluminium salts. Local plants at Solihull Lodge and Oldbury had been the initial providers of this metal. New extraction techniques were adopted during the 1890s that used cheap hydro-electric power, which was available in Scotland and North Wales. Greater amounts of aluminium metal became available for manufacturers. The demand for aluminium castings benefited the Birmingham-based company, which extended its Cambridge Street property, taking over other premises on the Easy Hill Estate. A new casting shop was erected on what had been Adams Brick and Tile Wharf and town gas from the Birmingham Gas Department was used to supply a high-pressure gas-fired melting furnace in 1910 (Times 1910). This was the first of its type to be employed in Britain (Metal Industry 1911). Single-storey warehouses were built on the former Baskerville House Mill site (BBP 21502), which had been used as a timber yard since demolition in 1888.

Another industrial development was the process known as die casting, which had been perfected in America during 1907 and was employed in this country to make small castings when needed in volume. The concept involved the injection of molten aluminium alloy (eg Al 92%, Cu 8%) into a die with the aid of compressed air.

With the onset of war, aluminium castings were in demand for engine parts for military vehicles. There also came a new industry that was the supply of aero engine parts. Birmingham Aluminium Castings supplied many thousands of cylinders and pistons for the aircraft (Metal Industry 1923). During the Great War of 1914 to 1918 many of the factories in Birmingham were pressed into serving the war effort. The Cambridge Street works was no different. The Birmingham Aluminium Castings introduced die casting at their factory and contemporary issues of Metal Industry show that the process was initially confined to zinc-rich alloys. The use of aluminium was certainly important to the fledgling Royal Flying Corp and the RAF that succeeded it.

The end of the aluminium works (Phase 6)

During 1919 Birmingham Aluminium Castings acquired the Midland Motor Cylinder Company Co., Dartmouth Road, Smethwick, and announced to their shareholders their intentions to build a new casting plant in Dartmouth Road. Harper Bean and Co, car makers of Dudley and Tipton, had purchased £100,000 of shares in the Birmingham Aluminium Casting Company and contracted with them to supply aluminium parts for 2,000 cars per week (Times 1919). Walter Mausdlay, chairman of the Birmingham Aluminium Castings, had already been approached by Birmingham Corporation regarding the purchase of their property (BC 1920). Birmingham Aluminium Castings completed the move to Dartmouth Road during 1922, and the buildings became vacant and ready for demolition. The new factory was massive compared to the cramped site at Easy Row (Greenslade 1976, 110–111).

THE BEDSTEAD WORKS - THE EXCAVATIONS

Chris Hewitson and Will Mitchell

A full description of every feature contained within the bedstead works is not necessary. Instead an overview of the area has been undertaken. Each room was assigned letters A–G during the post-excavation process, these separations being based upon the stratigraphic relationships and documentary evidence. The buildings shall each be described briefly in terms of their room separations, construction materials, dimensions and phase.

The phases covered by these buildings were phases 4 and 5 (1840s/50s to early 20th century). There were subtle alterations to the building layout during Phase 4, therefore, this phase has been divided into two:

- Phase 4a Main build
- Phase 4b Internal reconstructions (Subtle wall alterations, some of the original walls demolished and built over).
- Phase 5 Further alteration in concrete (Pillar bases - for roofing over of external walkways, additional machine bases. Re-flooring in concrete. It is likely that this rebuilding occurred during the period that the aluminium casting company were in ownership.

The bedstead works (Phases 4; Fig. 5.2, Plate 5.1)

The foundations of the bedstead works were well preserved. Archaeologically, the layout of the works could be clearly seen, as could different phases of construction and rebuilding. The area was defined by a series of workshops constructed around pathways. Inside these workshops there was evidence of machine bases, bench locations, crucible hearths and storage areas. Some of the activities undertaken within each area could be identified by the deposition of artefacts and the structures contained within. The preserved dimensions of the bedstead works were 42.5m (139 ft 5 in) in length x 20.5 in width (67 ft 3 in). This was approximately only a third of the original length of the whole works. The rest of the works were located outside the excavation area, and the surviving foundations had been truncated by the construction of the surrounding buildings (Centenary Square in 1936, Baskerville House in 1938 and the REP in 1971).

The bedstead works was made up of a series of buildings containing individual workshops. These workshops

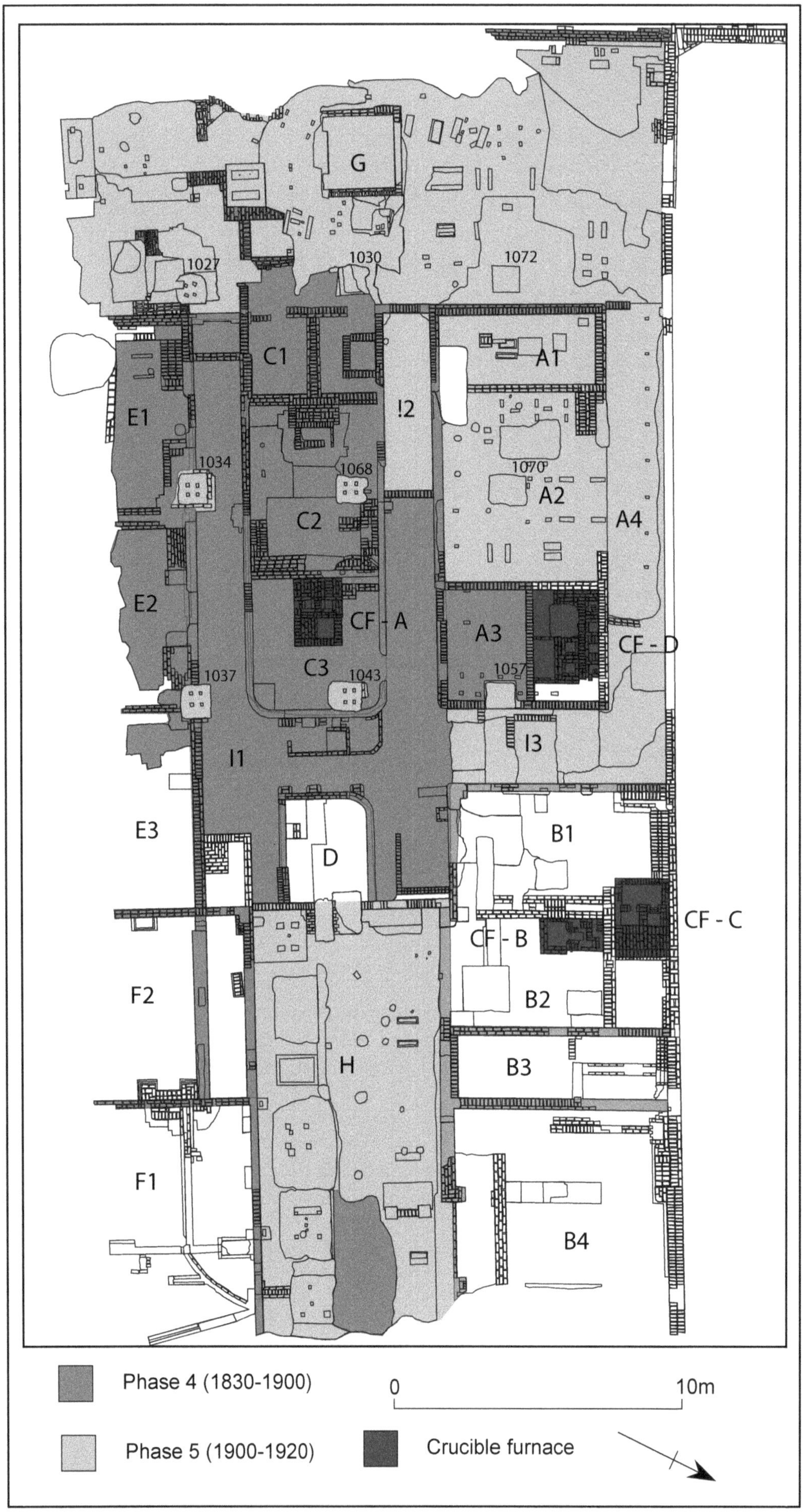

Figure 5.2 The bedstead works, Phase 4 and 5

PLATE 5.1 THE BEDSTEAD WORKS, OVERHEAD SHOT, EAST FACING (IMAGE BY AERIAL-CAM)

housed machines, workbenches, hearths, storage areas etc, the locations of which were identified through machine base and hearth foundations, postholes and artefactual evidence. Between the buildings there was a series of external walkways, some of which had been built over and converted to internal workshops at a later period.

The artefactual evidence contained small items which had been swept into the machine base and bench foundation holes. These items give an idea of the processes being undertaken at these locations. Items included small pieces of stamped brass, off-cuts and glass fragments. These items have been depicted in Fig. 4.14 and Fig. 6.4.

The Buildings

The majority of the works was constructed of standard machine-cut, 9 x 4 ¼ x 3in engineering bricks set in an English bond. There was, however, a close range of variations of brick used. Holes set into the brick and concrete floor were the locations of machine and workbench bases. A full inventory of these locations is not possible here but they were located within every surviving floor surface. The main buildings were constructed during phase 4.

Building A

Dimensions: 13.5m length x 8m width. Main wall 1065, orientated east to west.

Room A1, 2.4m length x 5.5m width; Room A2, 6.4m length x 5.5m width; Room A3, 3.7m length x 5.5m width, there was a red brick floor throughout (1056) and it contained crucible furnace D; Room A4, 13.5m length x 2m width and contained glass residues and off-cut evidence of stained glass working.

Building B

Dimensions: 17m length x 8m width. Main Wall 1125/1105, orientated east to west.

Room B1 3.5m length x 6.8m width; Room B2 3.1m length x 7.4m width, contained crucible furnaces B and C;

PLATE 5.2 BUILDING C, ROOM C3, WEST FACING

Room B3 2m length x 7.4m width; Room B4 6.2m length x 7.4 width.

The largest deposit of crucible bases was recovered from context 1571 contained within room B1 adjacent to crucible furnace C and beneath the floor level. Within the building was a series of drains, 1109 and 1113 running approximately north-south into Gibson's Basin. Crucible fragments were recovered from both of these features (1113, SF35; 1109, SF36g).

Building C

Dimensions: 14m length x 5m width. Main wall 1065 - rounded at E end, orientated east to west (Plate 5.2).

Room C1 2.5m length x 4.4m width, brick floor 1007, split into two rooms; Room C2 5.6m length x 4.4m width, brick floor 1039/1041; Room C3 4.4m length x 4.4m width, brick floor 1044, contained crucible furnace A.

Building D

Dimensions: 3.6m length x 3.4m width. Main wall 1074, orientated east to west.

Building E

Dimensions: 26.5m length x 4m width. Main wall 1066, orientated east to west.

Room E1 6.2m length x 4m width, brick floor 1033; Room E2 6m length x 4m width, brick floor 1035; Room E3 6.5m length x 4m width, brick floor 1036.

From the brick floor 1035 a brass hook was recovered (SFN12b).

Building F

Dimensions: 8m length x 4m width. Main wall 1086, orientated east to west.

Room F1 8m length x 4m; Room F2 6.1m length x 4m width (this room was a western extension to Room F1, Phase 4b).

Building G

Dimensions: 9.5m length x 21m width. Main wall 1501, brick floor 1007/1130, stamped brass and brass off-cuts.

Building H

Dimensions: 14.1m length x 6.5m width (filled in area between Buildings B and F).

Area I

Series of corridors, orientated east to west (17.5m length) and north to south (16m length), width of corridor 1.7m. Area I1 Brick floor 1038/1046/1048.

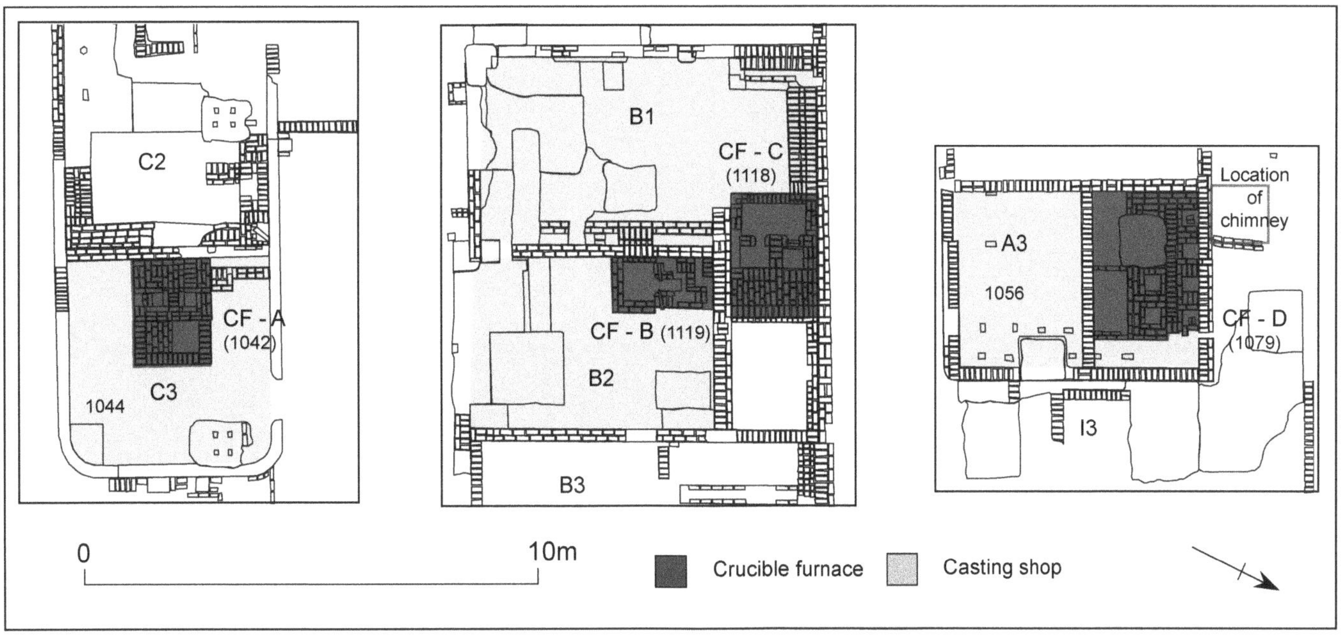

Figure 5.3 The crucible hearths

Although artefacts were associated with the individual buildings, the surface finds almost certainly relate to Phase 5 alteration of the works. It is only in the rooms where the crucible furnaces were located that we may be witnessing Phase 4 artefacts and production residues.

The Crucible Furnaces (Fig. 5.3)

Located within several of the workshops were the foundations of crucible furnaces, the presence of which confirmed the surviving illustrated examples of crucible furnace working on the site. This evidence was supplemented by the presence of broken crucibles and crucible waste (slags). There crucible furnaces survived in different forms, but each would have functioned in the same way. They may however, have performed different functions. The furnaces were essentially fire pits within which there were grills. The fire was set upon these grills, beneath which were flues. These supplied air to the fire. Each furnace would probably have had its own small chimney, constructed above the furnace. The furnaces were probably used for the casting of small items. Off-cuts of brass were placed within the fireclay crucibles and heated until the brass was molten, then the resulting mix was then poured into moulds. Amounts of clinker beneath the fire-pits suggest the fuel used was coal or coke.

Crucible Furnace A (1042)

Dimensions: 1.7m Width x 2.3m Length x 1.2 Depth (Fire-holes 0.25m x 0.25m)

Description: machine-cut 9 x 4 ¼ x 3 in red brick construction. Two fire-holes lined with firebrick (Plate 5.3). The casting pit in front of the furnace was filled by back fill of ash clinker (1317), black coke and ash (1255) and waste casting sand (1254).

Associated artefacts and residues: Fill 1255 contained considerable quantities of material including crucible fragments (SF96), triangular profile iron files, and slag (SF97). One crucible from the hearth had a high lead content whilst a very high value of tin (16% Sn) was present in fragments from another. Four lengths of iron pipe (two have concretions, 1255) may suggest that these are possibly associated with the production of hollow-brass ware after Winfield's patent.

Crucible Furnace B (1119)

Dimensions: 1.2m Width x 2.2m Length x 0.48m Depth (Fire-hole 0.25m x 0.25m)

Description: machine-cut 9 x 4 x 3 in red brick construction. One fire-hole lined with firebrick. Very truncated. (Plate 5.4)

Crucible Furnace C (1118)

Dimensions: 1.9m Width x 2.7m Length x 1.1m Depth (Fire-holes 0.45m x 0.45m)

Description: machine cut 9 x 4 ¼ x 3 in red brick construction. Two fire-holes lined with firebrick. Associated fuel ash pit (1244). Similar in construction to crucible furnace A, surviving metal grill in one of fire-holes (Plate 5.5). The casting pit in front of the furnace was filled by a series of layers of slag, coke and clinker mix (1323, 1324 and 1325). The fuel ash pit (1244) was filled by a hard layer of clay (1256) and ash silt and brick back-fill (1245).

Associated artefacts and residues: Slag was recovered from all of these layers that suggest non-ferrous brass casting. A number of artefacts were recovered including two cast bars that were in the as-cast condition and again

PLATE 5.3 CRUCIBLE FURNACE A, WEST FACING

PLATE 5.4 CRUCIBLE FURNACE B, FULLY EXCAVATED, EAST FACING

Plate 5.5 Crucible Furnace D, under excavation by Erica Macey-Bracken, northeast facing

high in zinc which would allow for zinc loss during further hot working (see McDonnell, Chapter 7). In addition, two further objects, a tube (1245) and a strip of metal (1324), were also in as-cast condition, possibly suggesting that they may have been produced in this state prior to rolling or drawing. In addition it was noted that of a series of cast objects, at least one, a round-headed stopper, had been produced for colour (1245).

Crucible Furnace D (1079)

Dimensions: 3.2m Width x 2.25m Length x 0.8m Depth (Fire-holes 0.35m x 0.35m; Plate 5.4)

Description: cut 9 x 4 x 3 in red brick construction. Four fire-holes lined with firebrick, with surviving metal grills, paired by four additional holes, truncated by later machine base. The flue was backfilled with ash waste (1334) and rubble (1333).

Discussion

Overall the suggestion is that the crucible furnaces, as expected, were used for non-ferrous brass metalworking. It is clear from the evidence that as-cast brass items may have been produced for other areas of the work as as-cast tubes and as-cast sheet have been found associated with Crucible Furnace C. It would appear that different blends of brass were being produced. Some were produced with high zinc concentrations in order to compensate for zinc loss during processing. Others were clearly being produced for variation in colour, purely for an aesthetic effect. Crucible from Furnace A (1255) contained residues of tin and lead which suggested other elements were incorporated into the brass during the casting process to produce varying mixes. The overall impression of the crucible furnaces is that the area was dedicated to cast-brass production in Phase 4, possibly continuing into Phase 5. Large dumps of crucible fragments such as those located in Room B1, beneath the floor level may suggest disuse of the crucible furnaces and discontinuity of this element of the works. However, this is very difficult to prove and it is maybe better to suggest that it is likely that that the area was adopted for casting between 1850 (the start of the bedstead works) with possible continuity into the earliest part of the 20th century (c 1910, prior to the closure of Sperryns).

Painted and stained glass production (Phases 4 and 5; Fig. 5.4)

Part of the business that Winfield's was known to have been involved in was stained glass (see Chapter 4 above). One of the locations of glass working was known to have been the second floor of the building above the muffles (BCA MS 322/32).

Artefacts, including glass off-cuts and waste, were identified on the bedstead works side. Within the postholes located in floor (1063; SF84 and 85) there were accumulations of glass debris (melted glass) suggesting glass preparation was undertaken within this room. A large assemblage of glass, cut and prepared for use in stained glass windows, was found upon the floor in the adjacent room (1051, SF20; 1058; SF21). In addition, an unstratified

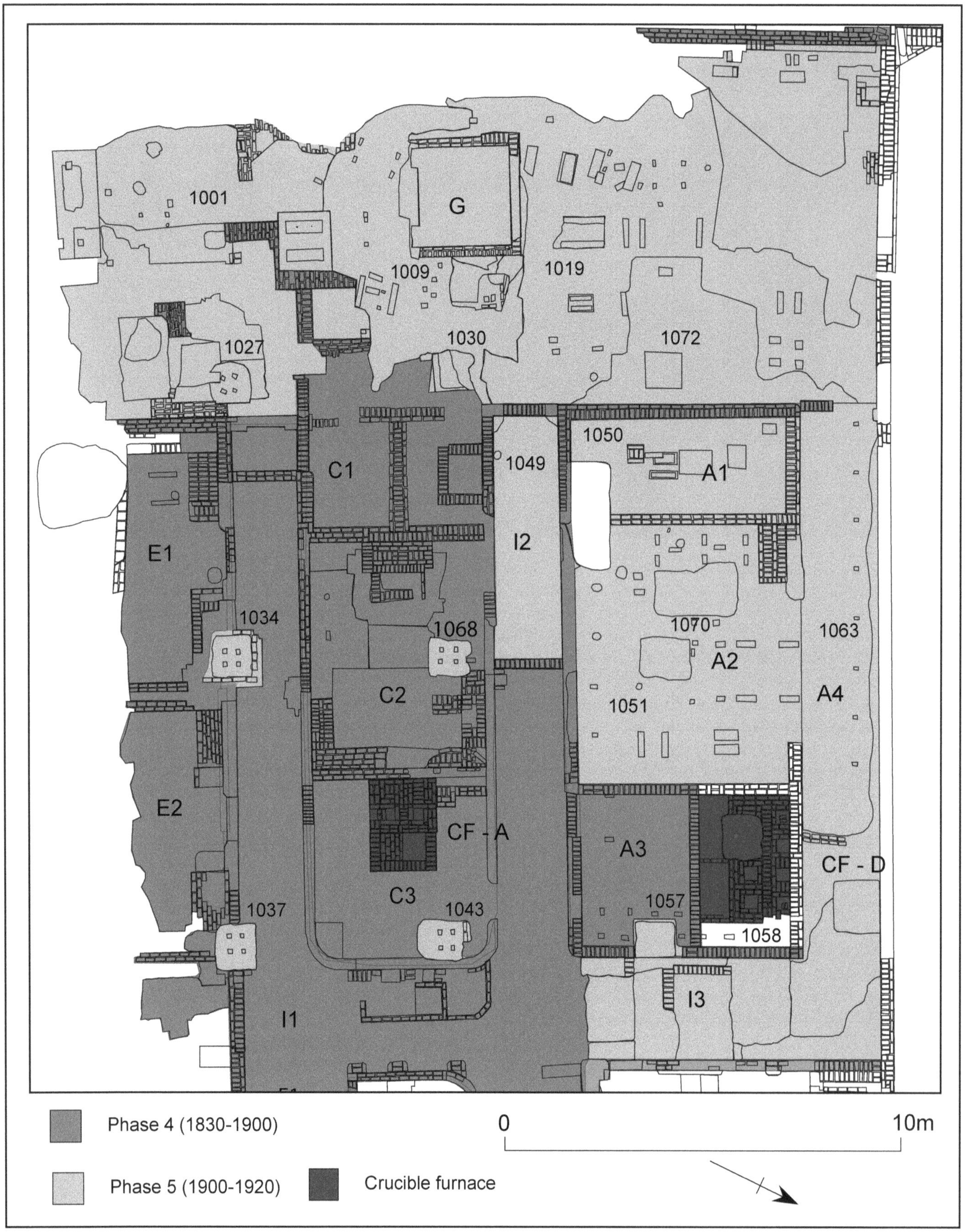

Figure 5.4 The bedstead works, Phase 5 north

leaded composite lead glass window frame (SF 34a, 1137) and a plano-convex colourless glass artefact (1137) were recovered from the backfills of Gibson's Basin.

The locations of these artefacts may suggest that this area of the bedstead works was used for stained glass production at some point. It is not entirely clear if this was the case however. The postholes located through floor 1063, may have derived from Phase 4 or 5. The floor 1063 was a Phase 5 alteration of the bedstead works possibly associated with its conversion to the aluminium works. Likewise floor 1058 of room A2 was a skimmed concrete floor believed to have been part of a Phase 5 alteration.

It is necessary to question whether these derived from a stained glass works located within these rooms or from stain glass production spread from other locations. If the location was the second floor above the muffles some intra-site transfer of material may have occurred, possibly during the late 19th/early 20th century, after the end of Winfield's occupation. This could easily have become incorporated into the postholes. A structure presumed to be a walkway was denoted on the 1890 Ordnance Survey 1:2500 map of the site, in this location over the canal basin (see Fig. 5.6). The postholes could easily be part of a wooden make-up of the walkway that was left in place and subsequently skimmed around with concrete. The cut glass

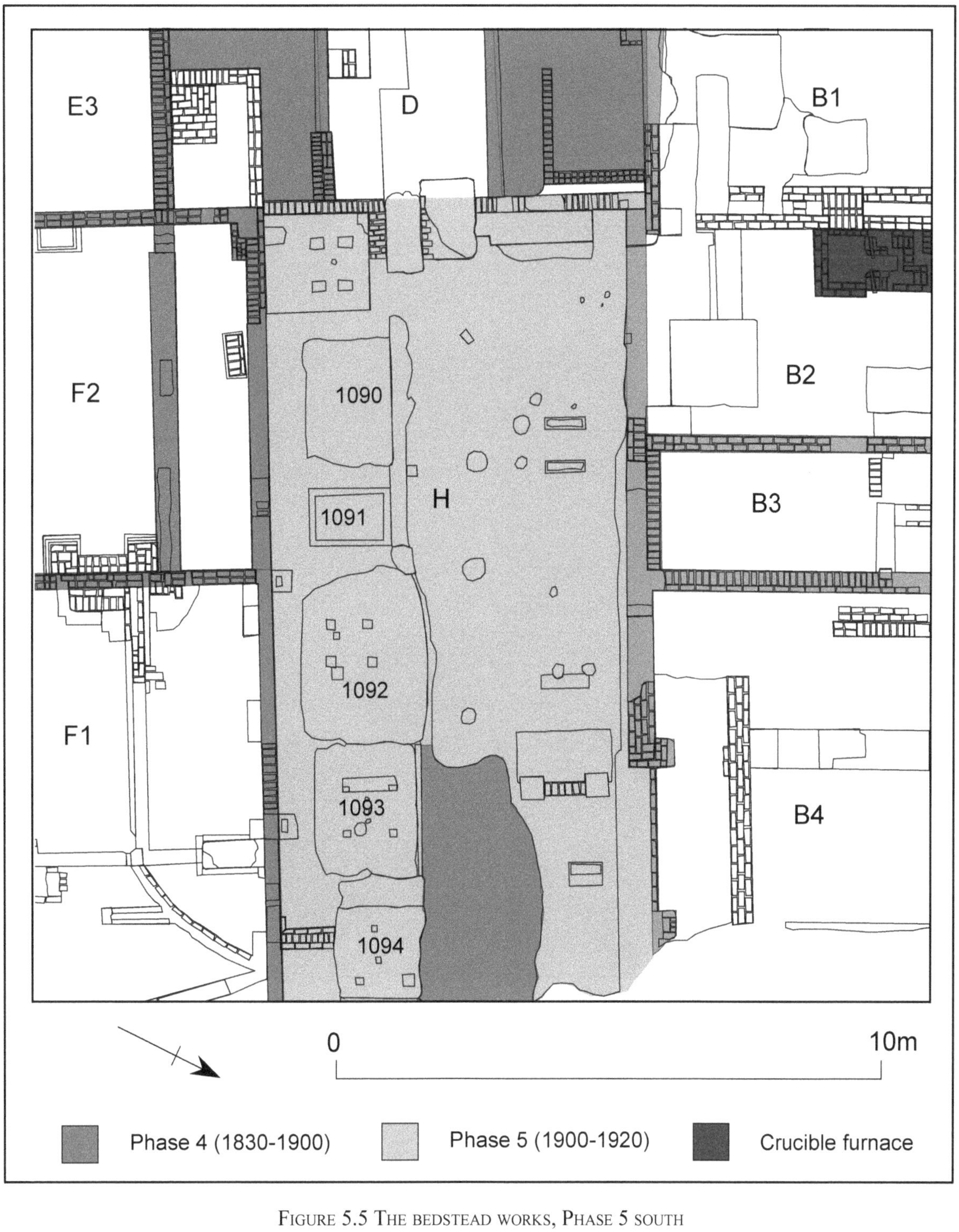

FIGURE 5.5 THE BEDSTEAD WORKS, PHASE 5 SOUTH

is distinctly worked and it seems surprising that it survived intact on the site. This may suggest that it was discarded subsequent to the demolition and is derived from a later period.

The 20th-century works: Sperryn's and the Aluminium Casting Company (Fig. 5.4, Fig. 5.5)

There was significant rebuilding during the later period (Phase 5) A group of pillar bases (1027, 1030, 1034, 1037, 1040, 1043, 1057, 1070 and 1072, Fig. 5.4), which probably supported a rebuilt roof were also from this phase. In addition a new layout was established, identified by the presence of concrete floor surfaces. These were located in only certain areas:

- Building A: Concrete floors 1050, 1051 and 1063.
- Building G: Concrete floor 1001/1009/1019, orientated north to south.
- Building H: Concrete floor 1096; machine bases 1090–1094 (Fig. 5.5).
- Area I: Area I2 Concrete floor 1049; Area I3 Concrete floor 1062; machine bases 1059, 1061.

Alteration to the buildings

The survival of a group of pillar bases (1027, 1030, 1034, 1037, 1040, 1043, 1057, 1070 and 1072) probably supported a rebuilt roof. Three lines of three bases can be identified and these would have corresponded with the steel joists that ran east–west across the site. It is almost certain that the entirety of the western part of the excavated bedstead works, including Buildings A, C, E and G represents the remains of the modern saw-tooth (northlight) roof buildings, which housed a casting shop and dressing shop. The pillars would have supported four pitches of northlight roofs, and corresponding with the location of the gutter. It is probable that the exterior brick walls were retained from the earlier bedstead works and incorporated into the re-built factory. The significant rebuilding during the early 20th century also corresponded with the addition of concrete skimmed floor surfaces (1001, 1009, 1019, 1050, 1051, 1063 and 1096; Plate 5.6). The extent of these is somewhat unclear. They do not appear to cover the entirety of the area - entirely covering Building G, and partially covering Building A.

PLATE 5.6 BUILDING A, ROOM A WITH SKIMMED CONCRETE FLOOR, PHASE 5, SOUTHEAST FACING (IMAGE BY AERIAL-CAM)

Building A

The concrete floor surface, as did room A1, retained a large quantity of parallel slots, some with wood in place, others as four corners of postholes. These appear to relate to a substantial quantity of small machine bases. Changes in the industrial function of the buildings as they adapted to the use of aluminium may relate to smaller machines, such as lathes and drills designed for fabrication of lighter aluminium parts. Within Building A small metal offcuts were recovered from machine bases within the room (1051, SF19; 1063, SF83). These also included fragments of triangular-shaped files (1063, SF83) that may well have been associated with the finishing process of cast objects. In addition there were three small circular pressed blanks that were half-finished washers in Building A4 (SF82) and A3, containing crucible furnace D (1058, SF22; 1056, SF28; 1079, SF30). Room A2 and A4 both had skimmed concrete floors, with a number of slots that may have been associated with small presses and machines in the later stages of the lifespan of the works, and the pressed objects may be remnants of these. Remnants of slags (1051, SF18, SF20) were located on the floor, possibly suggesting metal casting was occurring in the area. However, these were ferrous in nature as opposed to non-ferrous.

Building G

Building G appears to be the most radically altered. It may have extended much further west, but remains have been destroyed by late 20th century re-modelling of Centenary Square. The floors contained a number of machine base slots, 1017 and 1018. Distributed on the southern side of the building were small circular pressed brass blanks (1016, SF1; 1006, SF2; 1025, SF9) and brass off-cuts (1017, SF5; 1018, SF7; 1018, SF8). This may suggest that the machine bases were presses associated with the production of washers and other small brass items. One of the washers (SF1) when tested by XRF contained more than 90% copper, suggesting that a range of washers was being produced. Generally, the washers have an average zinc content lower than the other types (13.5% Zn compared to the overall average of 26% Zn), but this was distorted by the copper washer from 1016 (see McDonnell, Chapter 7). A swivel from context 1017 (SF5) was, in comparison, very high in zinc content (80%).

Some small rectangular cast items were also recovered from the floor surfaces, including a small rectangular cast block (1018, SF3) and small cast brass machine parts (1017, SF5a). The cast block (SF3) was a lead-antimony alloy containing a small amount of tin - a form of pewter. This may suggest some small-scale casting was still being undertaken in the building in Phase 5. Slags were recovered from the area (1018; SF3 and 4), which suggested that brassworking was still occurring. Overall, buildings A and G appear to have been adapted in the 20th century to accommodate presses for predominantly non-ferrous brass and copper small part manufacture, probably by Sperryns.

Building H

The addition of concrete machine bases (1090-1094, Fig. 5.5) at the eastern end of the bedstead works may relate to the addition of new machines as the works converted to the manufacture of aluminium castings. This area was taken over after the demise of the bedstead works by Richardson's Mechanical Engineer. The presence of a continuous wall between the two parts of the works, on the western side Sperryn and Co. and on the eastern side Richardson's, suggests that the structures may have pre-dated the occupation by the aluminium works. The change in use of Passage H, from a through passage for movement of goods, to possibly a covered work area corresponds with the division of the bedstead works during ownership change. It may well be that the machine bases relate to Richardson's occupation as a mechanical engineer and were designed for heavy lathe work or drilling. This could suggest the area was not designed for through passage to the west. It is possible that when the aluminium works took over the area the separation of the two elements of the works was maintained.

Area I

Apart from the addition of concrete machine bases little had changed in the layout of the internal corridors of Area I. Slag, SF100 from the fill (1329) of pit 1073, had copper content below 50% and a very low non-ferrous alloying content. These are very likely to have related to some of the final processes on the works or back-filling.

Discussion

The floor surfaces appear to respect previous sub-divisions. This may suggest that interior walls were largely retained and that their function continued after the re-build. This would particularly suggest that Buildings C and E continued to operate as a series of workshops, with room C3 possibly still operating as a furnace. This may in part indicate that some elements of the works were retained and for example that brass casting continued past the demise of the bedstead works and subsequent occupation by Sperryn and Co.

The development of the ferrous box-girder originated in the early 19th century, with its earliest origins in the Ditherington Flax Mill at Shrewsbury (*c* 1796–1797; Trinder 1992, 189–223). However, universal acceptance of the construction technique did not occur until the late 19th and very early 20th century. Even then it was used in combination with brick in a composite technique where the box-frames would be infilled by stretcher laid brickwork, the strength of the structure, however, being reliant on the frame.

Northlight or saw-tooth factory roofs were a distinctive feature of early 20th century industrial architecture, typical of 'Lowry-esque' scenes of factory complexes. The design of an acute-angled, almost vertical roof pitch on the northern side and a shallow-angled pitch on the southern side spread large quantities of diffuse light to the workspace. The near-vertical northern pitch would be entirely glazed and indirect light, away from the southern sun, would enter the workspace via this face. Developed in the southern hemisphere the reverse principal was true and these systems were referred to as southlight (Fleming *et al* 1998, 405). The date of the redesigned complex of *c* 1915–1917 fits perfectly into the chronology of factory architecture.

How much the transition to the aluminium production altered the works is unclear as there was minimal material evidence of production. Brass production clearly occurred up until the very last years of the works, *c* 1915, when the Aluminium Casting Company took over the site. There is some evidence of a re-build of the works in the early 20th century but whether departments changed function is unclear.

DISCUSSION

Chris Hewitson and Will Mitchell

The buildings south of the canal, which formed the bedstead works, were used to do the lighter elements of work such as casting, fixing and finishing. The main construction phase was during the 1850s (Phase 4) and the footprint of the buildings stayed intact albeit with major alterations in the 20th century. Phase 5 was the next major phase of building, during the 1900–20 period, when many of the brick floors were covered by a thick secondary concrete floor. The roof of the workshop was almost certainly demolished in 1915 and this probably also resulted in the demolition of internal walls to make way for a new Aluminium Casting Shop. The pillar bases are likely to have supported a northlight or saw-tooth roof as mentioned in the historical sources.

Phase 4 - The bedstead works, layout and function

In understanding the bedstead works, we are indebted to the discussion by Edward Peyton of his bedstead works (Peyton 1866, 624–627), the discussion of brass founding by Aitken (1866, 271–274) and the more detailed descriptions of the brass-foundry and tube works of William Tonks and Sons of Moseley Street, Birmingham in '*England's Workshops*' (Strauss *et al* 1864, 51–59). Although the account related to brass foundry and tube making alone, much of the detailed description of the processes applied of all brass trades and as such can form the basis of an understanding of the bedstead works at the peak of the trade, from 1843–1897. The production of rolled brass and wire and the sheathing of tubes would have been carried out on the northern side of the canal. It is probable that a number of processes involved with bedstead production were carried out on the southern side. An inventory of Peynton and Peyton's factory from 1884 (BCA MS 119) gives us an idea of the various departments within a bedstead works;

'Carpenter's shops, paint cellars, lacquering rooms, smithies, rail fitting galleries, brass bed finishing, iron and sand casting shop, core making shop, sand brass casting shop, cot fitting gallery, case fitting shop, brass department with tube cutting machines, angle iron casting shop, piercing angle department, brass strip casting shop, tube department, mattress department, sacking department, turned brass department, rough warehouse.'

Detailed examination of the cartography, and the graphic image from the Great Western Railway Guide (dated c 1861, see Shill 2006, 36; BCA MS 322/197, Fig. 5.6) give a clear indication of the layout of the works. The archaeological excavations have helped confirm the western element of the works. These were a series of single or two-storey buildings with pitched roofs set around a central courtyard (Buildings A, E and G), with a series of single-storey workshops running part-way along the centre (Buildings C and D). The main element of the bedstead works was located to the east and consisted of three- or four-storey blocks, the majority of which lay east of the excavation but two (B and F) formed the westernmost elements of the excavated site.

FIGURE 5.6 IMAGE OF THE CAMBRIDGE STREET WORKS, GREAT WESTERN RAILWAY GUIDE *C* 1861 (COURTESY OF BIRMINGHAM ARCHIVES AND HERITAGE)

Of all these buildings only three (A, B, C), collectively four rooms (A3, B1, B2, C3), can be positively assigned a function. These were associated with the crucible furnaces located in the rooms and appear to be general purpose casting shops. The description of the workshops of William Tonks (Strauss *et al* 1864, 54–55) enables us to understand their function:

All brass used for casting is made on the premises. There are two qualities made;

- *First quality for fine castings, where the moulded surface is preserved in the finished article. This consists of three parts of best selected copper and two of spelter, melted into ingots, with a proportion of best scrapped brass and a little tin.*
- *Second quality, for ordinary work. This consists of two parts of ordinary copper and one of spelter, melted into ingots, with a proportion of scrap brass and iron filings.*

The shops are referred to as 'smelting or meal-mixing shop', where pig brass is made, but also as 'casting shop'. It would appear the differentiation is between fine and ordinary work. In examining the workshops in the bedstead works, it could be assumed either was occurring. The rolling mills however, had their own casting shop located east of the excavated area, only partially excavated. This would have undoubtedly been associated with ordinary work to be utilised for rolling, wire drawing and tube making. It would seem that the workshops on the bedstead side were associated with 'fine casting'. It is probable that these were the same workshops that caught fire and resulted in the closure of the works temporarily in the late 1880s.

The description of the metal mixing room at Tonks' works (Strauss *et al* 1864, 54–55) was not dissimilar to that of the casting rooms and Aitken also describes the process (1866, 272). This forms a basis for our understanding of how they functioned. The works used direct mixing, ie strip casting of brass. The crucible furnaces consisted of clay melting pots of normally Stourbridge or plumbago (London) clay. Given the known attribution of Winfield's with Stourbridge for firebricks (see Chapter 6 below) it would seem probable they used this clay source. Tonks' was probably isolated if not unique in his use of plumbago crucible pots, which he believed to be more efficient, but which Aitken suggests was only adopted by one or two manufacturers (Strauss *et al*, 1864, 53; Aitken 1866, 267). The other equipment required was a sand trough, various qualities of sand (and also loam and meal flour for finer work), cast iron or wood-moulding frames or boxes, moulding boards, clamps to hold the frames together (Aitken 1866, 272).

As discussed above, the evidence clearly points to the production of brass artefacts. The presence of crucible pots found in large numbers and the residues that adhered to the interior and exterior testify to this. The layout of the casting shops was crucial. B1, B2 and C3 appear to have adopted a two-furnace system set side by side. The crucible set on the grill would then be surrounded by coke with the draught of air drawn from beneath. An image of the works produced c 1861 (BCA MS322/197) clearly shows the location of chimneys associated with the furnaces and this would provide the draw for the furnaces in order to increase air flow and oxygen intake for the fuel. Room A3 is curious in that it contains eight furnaces. This suggests that this room had been set aside for much larger castings, which would involve multiple furnaces to produce sufficient brass. It may be that a workshop like this was being adopted for casting larger items - where greater quantities of brass were required, these could include the use of the casting boxes known as chills that used brass as opposed to iron.

A trough was constructed for all the furnaces set directly in front. The example in Room A3 still contained considerable quantities of sand, presumably from its final backfilling. The reality is that the casting box would have been laid in an open space within the shop. Aitken (1866, 272) describes it thus;

The process of moulding consists in filling the first half of the box with sand; when filled the pattern if flat are laid on the surface; if circular, they are driven in to half their diameter; dry-parting sand is dusted all over the surface of the first half of the box; this is in order to separate the two halves of the box more readily; the upper half of the box is then dropped on, and is held there by three dowels. The sand is then filled in and beaten down; a moulding board is placed on the back and then the box separated. The patterns are lifted out and gets or connections are formed by cutting away the sand and connecting these with apertures of the box provided for the introduction of the metal. The mould is then dusted over with bean flour (meal flour); the two parts of the box closed together, and held in that position by clamps; the melted metal is poured in, and a perfect copy of the original pattern of the mould is produced.

The process of preparing the brass was described as follows;

Great care is required in melting the brass that the temperature is not too high, as the zinc oxides with great rapidity, burning with a bright bluish-white lambent flame, and pouring forth volumes of dens white fumes, which sublime in the form of a fluorescent white substance (oxide of zinc), familiarly known as philosopher's wool. ...When the caster judges the melted metal to be of the proper temperature for pouring, he lifts the crucible, or pot, out with a long pair of crucible tongs, and carries it to the skimming place, where the lose dross is skimmed off with an iron rod. He then proceeds to fill the mould, which operation in is attended with a rushing or hissing sound from the flow of metal and escape of the air, and generally also by small but harmless explosions of gases which escape from the seams of the mould.

The layout of each of the casting rooms appears to have had sufficient space for casting to occur away from the furnaces. It is unclear when the furnaces fell out of use but it probably coincided with the end of bedstead production. It is possible that some casting continued after this as a private enterprise or for small-scale production of tools or fittings during the occupancy of the aluminium works.

The other rooms in the works are more mysterious. Building G retained a number of subterranean structures, notably a rectangular pit. The workshop saw most alteration when the building was converted for aluminium production and the floor was subsequently concrete skimmed. The presence of a number of slots in the floor for machine bases and the distribution of small pressed finds in the workshop, particularly brass and copper washers, suggests that in its final form it was used as a workshop for pressing metal from brass sheet.

The description of Tonks' works mentions a number of rooms that must have been closely associated with the casting shops and it is possible that it had a different function at an earlier date. It could be that Building G acted as the casting pattern room or the rough warehouse. The casting pattern room contained a huge number of patterns '*of every description from the plainest to the most ornamented*' (Strauss *et al* 1864, 52). The quantity of patterns was huge at Tonks' works and it seems likely they were stored close to the casting shops in storage rooms. The second area described as being close to the casting shops was the '*rough warehouse, an immense room ninety feet long*' that was used for finishing items made by the casters (*ibid*).

Buildings A, C or E could also have served a number of functions including those of storing patterns, hot brass soldering, or other hot workshop use, something particularly possible for the three rooms of Building E which were crucially truncated by 20th-century activity where the furnaces could have been located at the rear of the block. Conjecture of function will always be difficult with so little evidence but the characteristic of these buildings on the western side of the bedstead works, visible on images from c 1861 (BCA MS 322/197), was one- or two-storey buildings served by chimneys. It would suggest that the area had been deliberately designated for hot work. A good parallel is Newman Brother's coffin-furniture factory, Fleet Street, dated c 1892. The layout has warehouse, shops and offices lined in two ranges around a central courtyard, with the third side of the courtyard formed by low single-storey casting shops each served by its own chimney (Cattell *et al* 2000, 2–3, fig. 109). This pattern is similar to the bedstead works, with hot work and other functions separated from assembly and finishing.

After brass was cast, in common with the wrought brass of rolling, wire and tube drawing, it was annealed. As previously described, the process of annealing involved heating the brass to high temperature before allowing it to cool in order to change the micro-structure of the metal and make it less brittle. There seems little doubt that the casting shops on the bedstead works probably used the rolling mill annealing furnaces. The evidence for this is provided by footbridges that link the casting shops to the annealing shop and muffles north of the canal. The locations of these can be seen on the 1890 Ordnance Survey Map (transcribed on Fig. 5.7). Two walkways crossed the canal basin - one led from room I3 between building A and B, the other led through building B, called room B3. The western led to the western muffles building, and later the gas fired muffles, the second led to the eastern muffles, but also to the rolling mill casting shop and the ingot store, suggesting it acted for transporting goods in both directions. These footbridges were presumably for the use of pedestrians and trolleys. There must also have been a means by which the rolled brass was transferred from the mill to the bedstead works side. It is likely that the road network running around the outside of the Cambridge Street works was used for this purpose. Cranes, located in the loading areas adjacent to the canal, may also have been used to transfer materials. The buildings were arranged generally east to west along the length of Gibson's Basin and were separated into ranges of buildings, within which were workshops.

The final process appears to have been pickling and dipping. There remains the possibility that this was done on the rolling mill side of the works. Notably the pickling shop at the Newman Brothers site is housed centrally in a small unassuming block (Cattell *et al* 2000, 92–93, fig. 109). This may suggest that a similarly unassuming block D and the adjacent in-filled pit may have served the same process.

The actual assembly of the various parts of the bedstead does not appear to have been the principal function of the excavated block. The three- and four-storey blocks located east of the excavated area may have served this function. The nature of this work appears to have been less skilled than that of the general brass founder. Peynton suggests '*there is nothing in the trade except the ornamental japanning that requires an apprenticeship. Any man who can use a hammer and a file can be a bedstead maker'* (1866, 627). Peyton's factory at Bordesley was equipped to supply both the domestic and foreign markets. Their premises included (NA BT 31/32228/145098):-

(a) Ornamenting and fitting-up shop
(b) Brass fitting shop
(c) Lacquering shop, stoves, enamelling shops
(d) Paint rooms
(e) Smiths' and blacksmiths' shops
(f) Polishing and plating shop
(g) Packing warehouse

It is likely the three- and four-storey blocks depicted on the c 1861 image (BCA MS 322/197) related to shopping in which ornamenting, fitting-up (assembly), lacquering (the function was also undertaken above the muffle house on the northern side of the canal), painting, polishing, and packing were undertaken. Neither buildings B nor F, contained floor surfaces (except casting rooms B1 and B2)

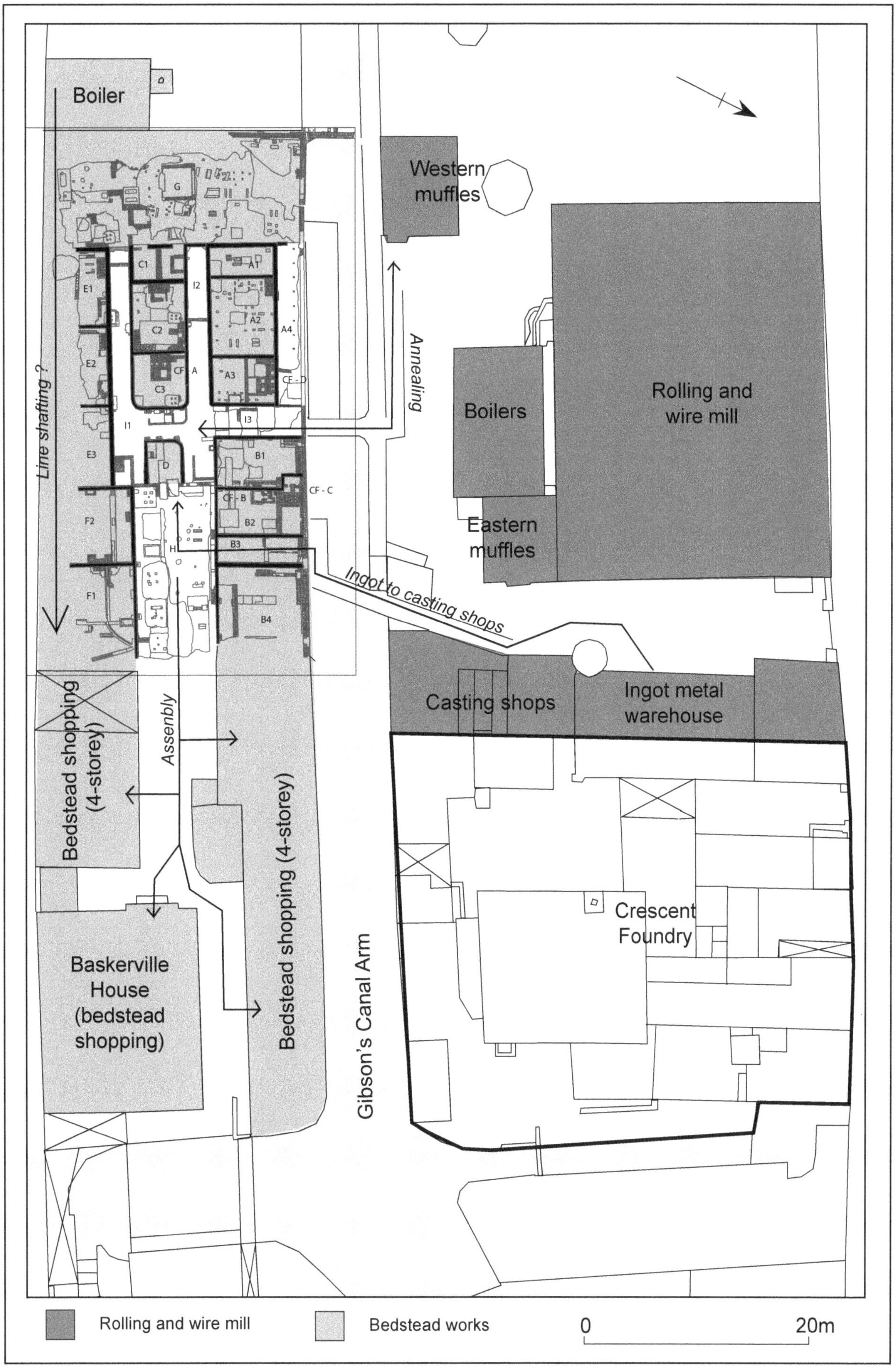

Figure 5.7 The bedstead works, flow of goods

and this may suggest that they were raised higher than the remainder of the levels around. It could also suggest that later activity resulted in them being stripped out.

Areas for other functions, such as black-smithing cannot be discerned. Although the crucibles excavated did contain residues of iron, it is unclear if the area of the site excavated was also adopted for black-smithing using iron. It seems unlikely as the majority of the crucibles contained high levels of zinc and copper (see McDonnell below). A location on the bottom floor of the shopping may be possible as this would link easily with the Crescent Foundry, located north of the canal and east of the excavated area, where the majority of iron forging would have taken place.

It is also not clear where the patent iron-brass composite elements of the bedstead were produced. These consisted of iron rods, wrapped in brass tubes. The possibility remains that they were produced directly at the tube works, using rolled brass from the adjacent works and wrought iron from the Crescent Foundry. A natural transfer of materials may have existed using Cambridge Street as a thoroughfare before the final assembly piece was delivered to the bedsteads works via Attwood's passage. The use of public space to transfer items may seem alien but was still occurring in the manufacturing districts of Birmingham even in the last ten years, in Deritend, Hockley and Selly Oak.

A final area of confusion exists. Excavation uncovered no recognizable power source. Two further engines are known to have operated, in addition to the one at the rolling mill. It was suggested that '*motive power for the gas-fittings and bedstead departments is supplied by two engines of 150 and 50 horsepower*' (IMechE 1897).

Peyton's Factory included three steam engines of 12, 16 and 20 horsepower to supply power for the lathes and smith shops and a boiler house to supply the steam. William Tonks employed a 12 horsepower steam engine that was subsequently replaced by one of 20 horsepower. These powered most machinery:

In this factory steam [power] is in fact made to do all the work of which it can, shafting being carried into every workshop in the building; great care has been taken in this to erect the shafting everywhere out of the reach of the workmen, and to fence all dangerous bands, so as to render accidents all but impossible, except from gross or wilful carelessness (Strauss *et al*, 1864, 56).

Altogether the suggestion is that a smaller 50 horsepower steam engine was used at the bedstead works. It is not clear where it was housed. The adjacent Baskerville Mill lay on the opposite side of Attwood's Passage and was never incorporated into the bedstead works (although the house associated with the mill was workshops associated with the bedstead works by the later 19th century). The only other possibility was the use of the power from Avin's Grist Mill, to the west of the bedstead works. It is possible that power was initially shared with the mill. By 1873 the mill had been incorporated into the Cambridge Street Works and used to install a new boiler (presumably for a steam engine) incorporated into the bedstead works. This opens the possibility that the boiler was placed here to serve a pre-existing engine, possibly located within Building G. It would be a much smaller engine than that seen on the northern side of the canal basin. It was also more likely to be a horizontal engine by 1873, with lines of overhead shafting that could have extended along the shopping on both the southern and northern sides of the bedstead works (see Fig. 5.7).

Phase 5 – Sperynn's or the Aluminium Casting Company?

Phase 5 was visible as preserved concrete flooring and pillar bases. The extent of the change of the site in the final stage is debatable, but it is more likely that the role of the Aluminium Casting Company in the shaping of the bedstead area of the site was relatively small. The period it functioned, from 1915 to 1921, was relatively short and it seems likely that the overall scale of change was less than expected. Although the above-ground remains were almost entirely removed, below-ground remains and the footprint of the earlier buildings appear to have survived. The change was possibly more cosmetic, the original sub-division of the rooms survived, as interior walls were retained but the structure was replaced with a saw-tooth or northlight roof. It is possible that the area became a single large open space but the excavated evidence suggests this this is less likely - some rooms have concrete skimmed floors, others do not, and this suggests that the wall divisions were left in place.

The function of some of the rooms in Phase 5 may be surmised by studying the deposition of artefacts and types of structures present in the foundations. Building G, in common with some others, had small fragments of discarded or dropped brass found in several of the postholes set into the concrete floors. Postholes set into the concrete flooring, surrounding and cutting these rooms, may indicate the positions of workbenches. Holes located in more central positions within the rooms may have been for fly presses, drop stamps and other types of machines. The artefacts may have been swept into or may have fallen into the holes, suggesting the likely location of their production. Lumps of stained glass were found also within postholes, although from a different location. Different rooms were used for different purposes but the chronology of this change is unclear. It could be suggested that the structures on this part of the site survived as brass manufacturing or for a much longer period.

The significance of the bedstead works

The manufacturing role of the Cambridge Street Works cannot be seen as an innovative technological development in the same way as that of the Union Rolling Mill. It adopted a model of work begun at the Soho Manufactory and Foundry that was copied throughout the city.

This great increase of the quantity of work, which, in consequence of the division of labour, the same number of people are capable of performing, is owing to three different circumstances: first, to the increase of dexterity in every particular workmen; secondly, to the saving of the time which is commonly lost in passing from one species of work to another; and lastly, to the invention of a great number of machines which facilitate and abridge labour, and enable every man to do the work of many (Smith 1776).

The development of systems of work that decreased the necessity for skilled artisans in the workplace was one of the principal advantages of Birmingham. These included the development of division of labour that had been noted in the work of gun-making in the late 17th century (Wise 1949; Rowlands 1975) which had been extended to different manufactures.

The principle of the factory was first undertaken at the silk mills of Derby, but was applied to the metal industries by Matthew Boulton at his Soho Manufactory, which adopted the concept of manufacturing in a single location. His concentration of a number of branches of the hardware trade in a single location together with the necessary commercial organization, enabled the division of labour to be applied on a greater scale. The Soho Manufactory had water power, machinery and newly-erected workshops, along with the necessary artisans to produce a range of manufactured goods including jewellery, buttons, buckles and other small articles collectively known in the Birmingham trade as 'toys' (Gale 1946, 3). The skills and lessons learnt in the development of the Soho Manufactory could then be applied to a variety of new trades, of which the brass trade was one. The Cambridge Street Works was a natural progression in the development of the factory, but it also shared a common link with the more traditional small master system of work. Indeed, its growth followed the early development of many industries in the city:

Beginning as a small master, often working in his own house, with his wife and children to help him, the Birmingham workman has become a master, his trade has extended, his buildings have increased (Aitken 1866, 223).

Robert Winfield quickly passed from a small manufacturer to much larger scale working. In many respects the Cambridge Street works bridges the gap between the domestic development and the industrial factory. Domestic workshop-cum-industrial unit, a piecemeal development with a domestic house at the front and shopping to the rear, that can be seen in the Jewellery Quarter (eg 27–32 Mary Street; 93–94 Vyse Street; JW Evans 54–57 Albion Street; Cattell *et al* 2000, 54–62, fig. 59, figs 63–64, fig. 66) and adopted former houses and their yards to create small businesses. Robert Winfield's works mimic small workshop development in the Jewellery Quarter in many respects such as the domestic scope of development, acquisition of neighbouring businesses and expansion in a single location. However, the scale and development of the Cambridge Street works shares more in common with the factory.

The Cambridge Street works has more in common with the larger works in Hockley. It adopted a variation of the courtyard plan visible at the Victoria Works, Graham Street of 1839–1840 (Cattell *et al* 2000, 27, fig. 74) which had three-storey shopping surrounding a central courtyard. The plan of the Cambridge Street works had shopping and offices located around the perimeter of the site, with the courtyard within facing onto the canal basins. Separate departments were kept in separate areas of the site. The bedstead works itself had its own department along Attwood's Passage, separate from the rolling mill, whilst other departments were in two courtyards at the east of the site facing Easy Hill. Excavations revealed an entirely separate area for casting shops located within the bedstead works. These single-storey structures located at the western end of the site were located away from the main shopping in order to reduce the risk of fire. The outbreaks of fire recorded at the bedstead works in the late 1880s proved that this was less than successful. Excavation revealed that at least some of the casting rooms were located in the tall shopping blocks in contrast to the general habit of the mid 19th century which housed them centrally in the courtyard or to one side of the ranges (eg Thomas Aston and Sons, 12–15 Regent Place, 1855–1861; Newman Brothers, Fleet Street, *c* 1892; *ibid*, 92–96).

The adoption of simple machines by Birmingham manufacturers, including the battery stamp, the fly-press and the circular lathe, meant products could be produced cheaply on a large scale using only semi-skilled labour. Stamped products were being produced in the town from the 1760s when Lord Shelburne visited (Court 1938, 244). They were not solely used by any one trade, but became ubiquitous to the manufacture of a wide range of articles that included elements of the bedstead works. All relied on semi-processed metal materials in order to produce the goods but the most important one was sheet metal. As Court states; '*The rolling-mill, more than any other mechanical invention, became the machine upon which the Birmingham trades depended*' (*ibid*, 241). Here in a nutshell was why the Cambridge Street works was such a successful model for a mid 19th century business: it not only produced the semi-raw processed metal for an expanding market, it created a product, in the bedstead, which required that raw material. It controlled both elements of the market.

Yet by the end of the century changes in fashion meant that the bedstead had become less popular as an item. In part this may be seen as one of the weaknesses of the economic model of the Cambridge Street works. Because all elements of the works were in a single place, as part of a single business, it was difficult to remove one element of the business. The rolling, wire drawing and tube drawing

PLATE 5.7 COLLECTED FINDS FROM THE DEMOLITION DEBRIS

mill continued to be relatively successful until the 1920s and 1930s, because they had alternative markets. In comparison, the bedstead works stopped manufacturing at the site, when the Cambridge Street works was sold off as lots, a possible sign that it was no longer a successful part of the business.

The fact that the bedstead works eventually became the location of an aluminium manufacturer also reflects the development of new non-ferrous metals. In the mid 19th century, when the Cambridge Street Works was at its height, aluminium had been a valuable commodity, comparable to silver in price. By the early 20th century it was competing directly with copper in price, with new hydro-electric production works at Falls of Foyers and Kinlochleven in Scotland. Yet, although aluminium was the newest non-ferrous metal, it was not until after the Second World War that it replaced copper, tin, lead and zinc as a principal metal. In the inter-war years it was useful as a specialist material (Page 1976, 170–175). The demise of the Cambridge Street works should be seen within the broader economic changes of industry, that will be discussed in greater detail in Chapters 8 and 9.

CHAPTER 6: THE ARTEFACTUAL EVIDENCE

Emma Collins, David Dungworth, Chris Hewitson, Samantha Hepburn and Erica Macey-Bracken

INTRODUCTION

The excavations from Cambridge Street did not produce a diverse artefactual assemblage in the conventional archaeological sense. Small quantities of artefacts were recovered from the site, including pottery, clay tobacco pipe, glass, worked bone and shell. In addition a series of brick samples were taken from the site for each separate structure where possible. A number of larger and contaminated items were recovered but they were not removed from the site either due to their size or for health and safety reasons. They were recorded on site by means of written description and photography and from part of the archive samples of metallurgical residues, fragments of scrap and waste have also been recovered and analysed.

The recovered finds were washed and marked and then quantified by count and weight for each context. The finds were then examined macroscopically for the purposes of this report. The assemblage is very small and largely non-diagnostic. The early modern date of the site has meant that finds have not provided significant dating evidence. The greatest value of the artefact assemblage is the facility it offers to elucidate details of the development of the works, this being due to the nature of the artefact (eg discarded product, machine part or tool or metalworking residue).

The general absence of artefacts may be due in part to the site clearance undertaken at the end of the life of the works and throughout its lifetime. Periods of changed ownership, reorganisation and rebuilding would have been incentives to clearance. There would also have been the general and practical need to keep work areas clean and organised and this was probably regularly carried out. It is known from sales catalogues that items were regularly sold off; upon the closure of the plant, large items of machinery would have been offered for sale to other owners, or as scrap. Many of the machines and tools would have been moved wholesale to the new premises at Icknield Port Loop. Industrial sites often lack artefactual evidence due to these reasons.

POTTERY

Emma Collins

The pottery assemblage consisted of 62 sherds weighing 1054 grams and was recovered from a minimum of six contexts. The assemblage dates to the late post-medieval period to the modern period. The pottery was examined macroscopically and was counted, weighed and the fabric/type and form identified.

The majority of the pottery recovered came from layers and contexts of 19th-century origin (see Table 6.1 below; distributions are shown in Fig. 6.4). The small quantity of 17th- and 18th-century Blackwares or Coarsewares were either residual or came from a fragmentary cobbled

TABLE 6.1 CATALOGUE OF THE POTTERY ASSEMBLAGE

Finds No	Context	Type	Form	Qty	Wgt	Rim	Base	Spot date**
SF10b	1027	Blackware		1	5g			17–18th
SF22b	1058	Creamware	Bowl	1	12g	1		M18–E19th
SF26b	u/s	Industrial slipware	Bowl	2	24g			19th
SF36c	1109	Modern glazed ware		1	1g			19th–E20th
SF38	1109	Blue transfer print	Bowl	1	6g	1		19th–E20th
SF38	1109	Blackware		3	26g			17th–18th
SF40c	1101	Coarseware	Jar	1	56g		1	17th–18th
SF40c	1101	Creamware	Scalloped edge plate	1	3g	1		M18th–E19th
SF45	?	Modern glazed ware	Tea cup	3	19g	1		19th–E20th
SF54	?	Brown salt glaze		37	848g			19th–E20th
SF54	?	Creamware	Plate	7			1	M18th–E19th
SF74	?	Modern glazed ware		2	52g	1	1	19th–20th
SF88	1384	Industrial slipware		1	2g			19th
-	1456	Blackware		1	<1g			17th–18th

*Weight in grams.
**Dates in centuries; E – early, M –mid.

surface on the northern side of site (1456). Fragments of a Creamware bowl, plate and scalloped edge plate date to the 18th to early 19th century; likewise they may represent fragmentary residual remains of the 18th-century occupation of the site. In comparison the 19th- and 20th-century assemblage was more utilitarian, containing modern glazed ware, brown salt glazed wares and slipwares mostly from unstratified contexts, reflecting the demolition or final use of the site. More decorative wares are only represented by a fragment of a blue transfer-printed ware bowl (1109) and a modern glazed ware teacup (unstratified). There was also a small amount of non-vessel ceramics in the form of two fragments of toilet (SF60) and one fragment of sink (SF48).

The nature of the deposits and the size of the assemblage meant that they cannot be used to derive dating evidence or social information about the site. The large-scale absence of pottery reflects the industrial nature of the site.

Bricks

Emma Collins and Chris Hewitson

The brick assemblage consisted of 37 fragments (predominantly single whole brick samples); weight 112,070g; recovered from a minimum of 22 contexts, comprising material from the early modern period (dated c 1820–1936). The majority of the assemblage was collected on-site as a representative sample of each brick-built structure on-site. Some samples were collected from rubble deposits and specifically selected due to the presence of brick stamps. The assemblage was studies macroscopically with the occasional use of a 6x magnification hand lens. Each brick/fragment was measured, weighed, photographed and the fabric examined. No reference collection was available so a fabric series was created. The brick samples came from a series of structures including areas of the bedstead works, the canal wall, the boiler bases, the engine house, the wheel races and the flue structures. There were three types of brick in the assemblage, firebrick, red brick and engineering brick. A fabric series was created and was as follows:

Fabric 1. Red brick. Light orangey- pink slightly soft fabric with occasional large rock fragments.

Fabric 2. Red brick. Light brownish-orange fairly hard fabric with occasional fragments of red iron stone and occasional small sub-rounded lumps of red grog.

Fabric 3. Red brick. Mid brownish-red hard fabric with very occasional sub-rounded light yellowish brown grog and occasional grit.

Fabric 4. Engineering brick. Mixed dark grey and greyish-red hard fabric with dark brown rock fragments.

Fabric 5. Engineering brick. Dark brownish-red core turning dark purplish grey exterior. Inclusions as Fabric 3 but with frequent lumps of light coloured grog.

Fabric 6. Engineering brick. Mid brownish-red very hard fabric with dark purplish blue surfaces. Has occasional 1–2mm limestone fragments, very occasional sub-rounded dark grey grog and very occasional angular voids inclusions.

Fabric 7. Engineering brick. Dark reddish-brown fabric with dark greyish brown surfaces. Inclusions include varying sized lumps of grog and very occasional 2–4mm rock fragments.

Fabric 8. Firebrick. Light brownish-pink hard fabric with a rough feel and varying sized irregular voids and small sub-angular rock inclusions.

Fabric 9. Firebrick. Light yellow fabric with frequent varying sized irregular voids and frequent dull white angular rock fragments.

Fabric 10. Firebrick. Light brownish-yellow hard fabric with frequent 2mm black ironstone inclusions.

The assemblage is summarised in Appendix 2.1.

The red brick

The majority of the brick on the site was red brick used as a standard construction material in industrial buildings. The period of the works spanned the 1820s to the 1936 but the majority of the brick sampled probably comes from a period in the earlier half of that time frame 1820–1860. This would suggest that the majority of the brick was hand-made as opposed to wire-cut brickwork that did not come into use until after 1841 (the date of its invention) and was not ubiquitous until the 1870s (Cooksey 2003, 75). Later development such as the Bradley and Craven steam-operated brick presses appear to have become common by the 1880s (Stuart 1995, 139).

Red brick was used in combination with the more resilient engineering and refractory bricks on both phases of the boiler bases and the water tanks. The inner lining would be of specialised brickwork whilst the outer lining would be of normal red brick. In addition structural elements such as the engine house base and the large wheel pit were also constructed in red brick. The majority of the red brick was of a reasonably standard size being 8¾–9 x 4–4¼ x 2¾–3in, the only exception being a brick stamped '1839' that was slightly larger but was presumably designed to form a decorative element of the works and was found unstratified (Fig. 6.1c).

The origin of the red brickwork is hard to pinpoint. The majority of the specialised brickwork (see below) had a Stourbridge region origin and it would be reasonable to suggest that red brick was also imported from here. But local brickworks were also available at California (Weoley Castle) and at Saltley. Both could have produced red brick (Hodder 2004, 143–145). The fabrics suggest a fairly limited source as only three are present. As always brick fabrics are very difficult to source and it could be that grog or temper within the bricks are highly localised. There does not seem to be any variance between the Phase

Figure 6.1 The ceramic building material

3 structures (1820–40, water tanks, Cornish boiler bases, steam engine base, flues) and Phase 4 structures (1840–50, Lancaster boiler bases, wheel pits) later re-builds of the canal wall (later 19th century). Re-use of earlier brickwork in construction may have occurred but the presence of the dated brick to 1839 suggests a single episode of building possibly associated with an expansion of the works. Historical events from this period may involve any number of expansions to the rolling mill, or the tube works; the bedstead works was not built, however, until the 1840s.

The engineering brick

Engineering, or blue, brick was highly fired and subsequently more durable than red brick. It was usually designed for structures that would be placed under greater stress, be it from weathering or general wear, but was also used as a decorative devise in conjunction with red brick, combining strength with polychromatic designs. These are particularly seen in canal and railway architecture of the 19th century.

Examples of the use of engineering brick are seen in the canal walls. The wall is multi-phase having been repeatedly re-built throughout the 19th century. Examples vary in size, often between 2 ½ – 3 ½in thick. They were used in the construction of the Phase 3, Cornish boiler bases, in place of firebricks. This may suggest a technological change in boiler base design from non-refractory bricks for Cornish boilers being replaced by refractory bricks for Lancashire boilers in order to retain and recycle heat. The engineering bricks were also used for the workshop walls of the bedstead works, possibly reflecting a reduction in the cost of brick by the time the buildings were expanded. One of the unstratified examples was a single bull-nose brick, commonly used for rounding factory wall corners, particularly where traffic flow through a space could have damaged the wall.

The origin of the bricks is unclear. A single example (SF64; unstratified) is stamped with the name 'TIMMIS and CO. LTD.' from Hay Green, Lye (Cooksey 2003, 117; Fig. 6.1d) and may suggest that much of the brickwork was being imported in common with the refractory bricks (see below) from the Stourbridge area. Four fabrics have been identified with small geological variations that are most likely present within a single clay source. Fabric 5 was the most common, and was associated with the 'TIMMIS' mark above.

The chronology of the bricks suggests many of the engineering bricks came from the earlier Phase 3 period of the site between 1820 and 1840. As discussed above they were used in the original canal wall (Phase 2) and later re-builds of the wall. This possibly corresponded with the construction of the bedstead works in the 1840s.

The refractory brick

Six examples of refractory (fire) bricks were recovered from the assemblage were recovered. The use of refractory bricks was essential in the majority of 19th-century industries where high temperatures were required within the processes.

The brick samples were recovered from five separate structures. These included the base of the Lancashire boilers (BN3; 1165), the associated chimney and flues (BN18; 1183), the Phase 3 water tanks (BN16; 1381), the gas-powered later annealing hearths (BN5; 1186) and the floor of the brass crucible hearth from the bedstead works (BN14; 1079). The role of these probably varied slightly. Those around the Lancashire boilers were designed to retain heat within the boilers, although the presence of the earlier muffles set against the side of the boilers suggest that they were also designed to recycle heat. Those in the annealing hearth and brass crucible hearths were designed to resist heat as well as retain heat.

Two bricks bearing stamps revealing the origin of purchase were recovered. 'BEST STOURBRIDGE' (BN 14; 1079; floor of Bedstead Works; Fig. 6.1a) probably dates to the 1840s, but does not correspond to a known works of the era and would appear to be a mark of origin and quality. An example of similar date from one of the chimney flues is marked 'KING BROTHERS, STOURBRIDGE' (BN 18; 1183; Fig. 6.1b) and refers to a known manufacturer King Brothers, Park Lane, Cradley (Cooksey 2003, 117). Three separate fabrics were identified, all of fired light yellow-grey. The inclusions varied but were all small, as is to be expected from the well ground and sieved firebrick clay. They probably represent localised geological variation within the region of Stourbridge.

The chronology of the bricks is not extensive, the earliest being from the water tanks (BN16) dating possibly to the 1820s, whilst the majority were associated with the 1840s re-build of the Cambridge Street works (BN3; BN14; BN18) and the latest associated with the gas-fired muffles c 1900 (BN5). The sizes, like most of those from the 19th century, varied little being 8¾–9 x 4–4½ x 2½in.

Other ceramic items

Two fragments of a rectangular ceramic item were recovered from unstratified backfill layers. They were of light yellow fire-clay construction. The form consisted of a flat base on one side, thick walled, with external tongue and groove arrangement to allow two of each item to fit together. Internally they consisted of two parts each with a circular holder 75mm in diameter. It is unclear what the items were designed for. They may have served a decorative function or been holders for crucibles.

Discussion

The assemblage represents a wide variety of bricks, reflecting the multi-phases of the site, and also bricks for specific purposes, for example, the refractory brick lining of the boiler bases. The majority of the structures on the site were constructed in a period from the 1820s to the 1860s.

The majority of the bricks appear to be hand-made with very small variation in size typical of the 19th century. It is only in the later re-building of the 19th century and 20th century that wire-cut bricks are visible.

Brick as a material was first adopted in the later medieval period to 17th century in the West Midlands, but was predominantly used in higher status houses, (eg Blakesley Hall, Cheldon Hall, West Bromwich Manor House and Sandwell Priory Hodder 2004, 156–157) or in chimneystacks (eg Saracen's Head, Kings Norton; Hislop *et al* 2011). Its use in a domestic setting is first visible in the service ranges of Aston Hall dating 1618–1635 (Driver *et al* 2004, 124) but it was not until the late 17th century that the timber-framed local vernacular building stock was replaced by brick (McKenna 2005a, 25–30). The process of brick production in the evidence for brick and tile kilns in Birmingham is relatively poor and the only documented brick fields and kilns are shown by street and field names from the 18th century such as Old Brick Kiln Close (SMR 20507) on the corner of Broad Street and Paradise Circus at the top of New Street, to the south of Cambridge Street. Brick-making was a seasonal agrarian occupation in the rural periphery at Yardley to the southeast of Birmingham (Nash 1799, 479; Skipp 1970).

By the 19th-century the brick industry in Birmingham grew. The Mercia Mudstone band that runs along the eastern side of Birmingham led to the development of the brick manufacture on a very large scale in the 19th century. A series of settlements traditionally associated with the industry included Saltley, Yardley and Acocks Green (Hodder 2004, 143). Two sites have been investigated for the 19th century in recent years at Blakesley Hall, Stechford Road, Yardley (Cook 2008) and Coachman Road, Saltley, (Duncan and Jones 2010). The examples at Blakesley Hall reflect the medium sized production for the 19th-century speculative housing markets (Cook 2008, 1–6). A large concentration of brickworks was located at Adderley Park extending to Saltley southwards, utilising clay deposits revealed during the construction of the railway and canals in the 1830s–1840s (Shill 2006b, 134). These yards were established from the 1840s and continued through the mid to late 19th century and included the Globe Brickworks and the City Brickworks and Tileries and the College Brick works that may have included the Coachman Road site (Duncan and Jones 2010). Elsewhere the area along the Bournbrook from Harborne to Bartley Green, in particular the California area, to the southwest of the city centre was also adopted for brick manufacture in the 19th century, with a possible earlier origin suggested by field names (Hodder 2004, 144).

The second source of bricks for the construction of the works is the Stourbridge region in an area extending from Cradley Heath, Brierley Hill and Lye in the east, Amblecote, Pensnett and Gornal to the north and Halesowen further south. There were over 30 refractory manufacturers in the Stourbridge area, but not all specialised in bricks, producing a wide range of industrial ceramics including gas retorts, glass pots, and blast furnace linings as well as a range of other linings that required heat-resistant material that could retain heat. The exceptional clay was well known from at least the 16th century when it was used by Stourbridge glass makers but was adopted for the use of refractory materials with the development of Abraham Darby's blast furnace (Cooksey 2003, 11–13).

The presence of more than one brick mark on site suggests that the bricks used in the construction of the site were not all from one source. It is almost certainly the case that the refractory bricks came from Stourbridge, and the presence of engineering bricks with brick marks from Lye may suggest that much of the brick was actually obtained from the Black Country as opposed to Birmingham. There is a considerable variation in the fabrics yet none display the sub-angular, pink argillaceous inclusions described by Cook (2008, 4).

Overall the majority of the brick from the site comes from phases that date from the 1820s to 1850s. Although Birmingham's expanding settlement was served by brick factories in a broad zone from Saltley to Acocks Green, it is clear that the nature of the clay from these deposits was best suited for the production of red and engineering brick. The Stourbridge zone of brickworks was more suited to producing the high quality refractory bricks but the seams of clay also contained substantial quantities of lower quality clay that went to making red and engineering bricks. Without being definitive, there remains the possibility that much of the brick was derived from the Stourbridge area.

1. Firebrick, sooted on stamped side. Stamp: 'BEST STOURBRIDGE', floor brass furnaces - bedstead works, Fabric Type 5, dimensions 9 x 4½ x 2½in, weight 2850g. BN14, 1079. Fig. 6.1a.
2. Firebrick, stamped: 'KING BROTHERS STOURBRIDGE', chimney culvert/flue, dimensions, fabric type 9, Dimensions 8¾ x 4 x 2in, Weight 2400g, BN18, 1183, Fig. 6.1b.
3. Red brick, brick stamped with '1839', fabric type 2, dimensions 9½ x 4½ x 3in, weight 3775g, BN27, U/S, Fig. 6.1c.
4. Engineering brick, stamped with 'TIMMIS and Co LTD', dimensions ? x 6 x 2¼in, fabric type 5, weight 4850g, BN64, U/S, Fig. 6.1d.
5. Ceramic fire clay item, dimensions length 370mm, 160mm, height 90mm, two internal circular holders, diameter 70mm, weight 5200g, U/S, Fig. 6.1e.

CLAY TOBACCO PIPES

Erica Macey-Bracken

Three clay tobacco pipe stems were recovered from 19th- and 20th-century contexts. The stems were insufficient to demonstrate their age and provide context. The details are listed in Table 6.2, distributions are shown in Fig. 6.4.

TABLE 6.2 CATALOGUE OF THE CLAY PIPE ASSEMBLAGE

Finds No	Context	Description	Material	Measurements
SF9c	1025	1 x pipe stem	Clay Pipe	19mm
SF36	1109	3 x pipe stems	Clay Pipe	38mm, 39mm, 22mm
-	1005	1 x pipe stem	Clay Pipe	25mm

WINDOW, BOTTLE AND VESSEL GLASS

Erica Macey-Bracken

The assemblage of vessel and window glass recovered from the site was small. The assemblage was uncovered from all excavated contexts but also included a selective sample from the machine excavated upper horizons that resulted in only the best-preserved examples of complete vessels being removed. The majority of the assemblage was therefore fragmentary although a few individual vessels did survive and were included. The majority of the assemblage can be seen to derive from the later 19th- and 20th-century contexts associated with the demolition and demise of the site, distributions are shown in Fig. 6.4. The assemblage was insufficiently large to make statistical analysis based on count, type or composition.

The window glass

A small quantity of window glass was uncovered from the site. This included an assemblage of stained glass (SF20; 1051) and a leaded window glass frame (SF34; 1137) discussed in greater detail below (see Dungworth, this chapter). The remainder of the assemblage was composed of thick opaque window glass or frosted glass and is detailed in Table 6.3.

The bottle and vessel glass

The small assemblage of bottle and vessel glass contains only a few identifiable bottles. Details of the assemblage are given in the Table 6.4.

The three identifiable vessels included a 'DADDIES' sauce bottle. The Daddies brand name was first launched in 1904, but the ketchup was launched in 1930, and was one of the trade names of HP Sauce, a former Birmingham company located in Aston (Thring 2010). Another bottle was from the Smethwick based 'Mitchell and Butlers' brewing company first formed in the 1860s and still trading today. Another Birmingham based brewing company was represented by the Ansell's Brewery, originally located in Aston Cross (McKenna 2005b). Both breweries are now based in Burton-upon-Trent.

The collection, although small, would appear to reflect the final demolition of the works. It comes as no great surprise to see three of Birmingham's major food and drink manufacturers represented in the vessel glass assemblage, but what is surprising is that there are no other examples of defunct manufacturers. This is perhaps a reflection of

TABLE 6.3 CATALOGUE OF THE WINDOW GLASS ASSEMBLAGE

Finds No	Context	Description	Measurements
SF3	1018	7 x fragments modern window glass, ridged on one side so that it is not see-through 1 x scrap of modern clear window glass	Various sizes
SF20	1051	1 x frosted window glass 1 x clear window glass	Various sizes
SF75		1 x piece of thick opaque window glass	109mm x 97mm x 5mm

TABLE 6.4 CATALOGUE OF THE BOTTLE AND VESSEL GLASS ASSEMBLAGE

Finds No	Context	Description	Measurements
SF20	1051	3 x red vessel glass 4 x amber vessel glass 8 x clear vessel glass 2 x light green bottle glass 1 x dark green bottle glass	Various sizes
SF22	1058	1 x neck and part of body from a clear green Daddies sauce bottle. The name DADDIES is embossed on one side and remains of the red, yellow and white printed label are visible on another side.	Bottle: 137mm x 37mm Shaped glass: 26mm x 25mm
SF39	1105	1 x complete clear glass drink bottle, embossed with the name Mitchells & Butlers Ltd. Early 20th century.	200mm high
SF75		1 x neck of a clear glass bottle	45mm
	1245	Glass bottle rim	
-	1406	1 x fragment of clear green bottle glass, embossed with the letters DI	30mm x 31mm x 3mm
-	U/S	1 x complete dark green glass bottle, embossed at the shoulder with the name ANSELLS BREWERY LTD. The rubber stopper is present, and is also embossed with ANSELL'S BREWERY LTD BIRMINGHAM and the AAA trademark. Mid 20th century.	278mm high

the period at which the vessels may have been deposited (towards the end of the depression in 1936) when firms that had survived to the mid 1930s became large national firms.

Other glass

There was also a small assemblage of glass fuse fragments from the exterior of tungsten fuses (SF20, 1051; SF55).

STAINED GLASS AND GLASS RESIDUES

David Dungworth

A small assemblage of glass and related materials was submitted for examination. This was believed to be associated with the manufacture of bedsteads, including the production of coloured-glass decoration for the bedsteads.

All items have been visually examined and selected samples have been analysed using EDXRF. Used in this way, EDXRF is non-destructive but yields only qualitative results which indicate the chemical composition of the surface. Therefore, these analyses will simultaneously indicate both the original composition of the analysed material as well as surface corrosion or contamination. Unfortunately, used in this way, the technique cannot distinguish between compositions due to original manufacture or later alteration. The glass residues are listed in Table 6.5, distributions are shown in Fig. 6.4.

Table 6.5 Description of glass residues

Finds No	Context	Description	Count	Weight (g)
SF21	1058	Flat glass	22	109
SF84	1063	Glass fragments	4	5
SF85	1063	Glass fragments	10	74
NA	1137	Glass 'lens'	1	71
SF34a	1137	Glass in lead calme	1	63

Glass Fragment - SF21; 1058

The glass from this context comprises 22 fragments of colourless and coloured flat glass (the colours include blue and green). The glass mostly comprises small circles (or fragments of circles) or pieces of glass from which circles have been cut (Fig. 6.2). The size of the circles is comparable with those from context 1137, although all of the glass from context 1058 is flat. All of the glass from this context is a soda-lime-silica glass. The colourless glass contains low levels of iron and arsenic; the green contains chromium, iron and copper, the blue contains manganese, iron, cobalt, nickel copper and arsenic.

Figure 6.2 Coloured and cut glass from context 1058

Glass Fragments - SF84 and 85, 1063

This context yielded small quantities of colourless and red glass (small find 84), comprising only colourless glass, and small find 85, comprising both colourless and red glass. The five small fragments of colourless glass recorded as small find 84 include one piece of flat glass, one very small piece of vessel glass with a moulded surface comprising small roundels (1mm diameter), a possible moil fragment and two small shards of glass. The colourless glass is a soda-lime-silica glass with a wide range of other elements detected (Fe, Mn, As, Sb, Pb). The moil fragment is significant as it implies that glass blowing took place. The colourless glass recorded as small find 85 comprises small chunks of glass with fracture surfaces. The pieces are thicker than most finished window or vessel glass but are comparable in thickness to the plano-convex glass artefacts from context 1137. These fragments may represent large chunks of working waste or fragments of large objects. This colourless glass has the same composition as that recorded as small find 84. The red glass recorded as small find 84 comprises small chunks of glass, which like the colourless glass, are too small/ fragmentary for us to be certain whether they represent broken artefacts or working waste. The red glass is a soda-lime-silica glass but with zinc, cadmium and selenium (the last two elements responsible for the red colour).

Plano-Convex Glass Artefact (1137)

The first consists of a plano-convex colourless glass artefact which was initially identified as a possible lens (see Fig. 6.2). The overall diameter is 63mm, the height is 19mm and the plane surface has a small lip. The convex surface has striations, but these do not appear to be the remains of any grinding or polishing action. There are no indications from the surface texture that the glass has ever been polished. The presence of the striations and lip on the plane surface suggest that the glass was formed by press moulding. The glass appears to be a soda-lime-silica glass. It contains small amounts of iron (but much greater than would be expected for an optical glass) as well as some manganese and arsenic which were probably added to the glass to counteract the colouring effects of the iron.

Composite Lead Glass Item (1137)

This comprises one piece of flat glass and two plano-convex glass components (SF 34a; Fig. 6.3), held together by I- or H-sectioned lead (calme). The use of lead calme for holding fragments of different coloured glass is well known for the production of decorative glass for windows (stained windows). The Cambridge Street example is unusual in several different ways. The glass components are generally rather small and the lead calme rather thick. In addition the two circular glass components are plano-convex rather than flat. The plano-convex shape of these glass components has a similar shape to the larger plano-convex glass artefacts from the same site (see above). The glass in this artefact (Fig. 6.2) appears to have the same

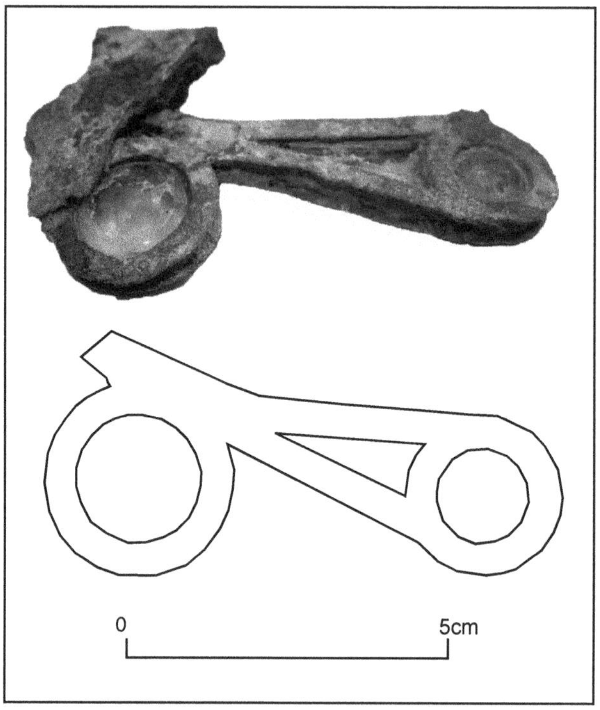

FIGURE 6.3 THE LEAD-GLASS DECORATION FROM CONTEXT 1137 (IMAGE BY DAVID DUNGWORTH)

composition as the larger plano-convex artefact. Lead was detected and it is likely that the arrangement of plano-convex and flat glass into decorative schemes and held together with lead calme was part of the production of bedsteads. The rather thick lead calme would be sensible in this context as beds would be subject to more movement and stress than windows.

Discussion

The small assemblage of glass from Cambridge Street suggests the working of both colourless and coloured glass. Much of the glass comprises finished glass artefacts. However, the presence of a moil fragment indicates that some blowing of glass occurred. Glass blowing suggests that there must have been a furnace for melting glass on site. The glass lumps are difficult to interpret: they may represent fragments from relatively large finished artefacts or broken fragments of working waste. These lumps are too small and fragmentary to allow a positive identification of the process(es) which led to their creation. The size, shape and colours of most of the remaining glass fragments are consistent with the historically documented production of glass decoration for stained glass.

The coloured glass (blue, green and red) was coloured using carefully selected metal oxides. In all three cases at least one metal oxide was used to achieve a particular colour quality. The blue glass was coloured with cobalt, however, manganese, iron, nickel and copper are present. The green glass was coloured with chromium, although this was modified by the presence of copper. The red glass used both cadmium and selenium to achieve the red colour.

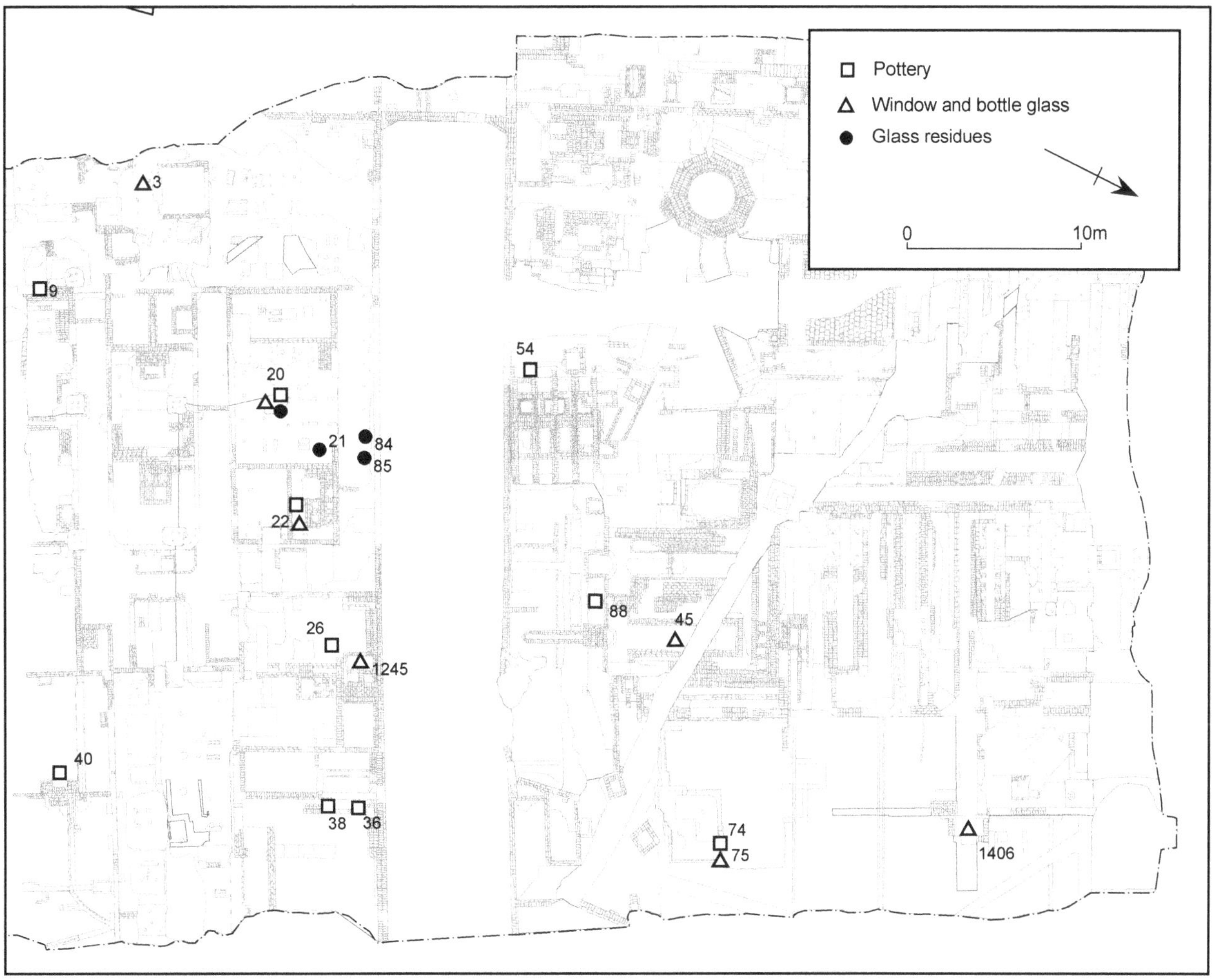

FIGURE 6.4 DISTRIBUTION OF POTTERY AND GLASS ITEMS

The use of chromium as a colourant in a glass suggests a late 19th-century date at the earliest. The use of cadmium and selenium for the manufacture of a red glass suggests a very late 19th-century (or early 20th-century) date at the earliest. Glass manufacturing technology in Britain in the late 19th century was generally uniform throughout the country and there is nothing about this glass assemblage which could demonstrate its manufacture in Birmingham.

SYNTHETIC ITEMS

Erica Macey-Bracken and Chris Hewitson

A series of 19th- or 20th-century Bakelite or plastic items were recovered from unstratified layers. None was sufficiently significant to warrant detailed reporting or offer insight into the function of the works. The items uncovered included a doorknob, two vessel base fragments, a vessel rim, a circular threaded fitting and a shallow cup possibly made to protect carpet from furniture castors. All these were made in Bakelite or plastic.

Although invented in 1909, Bakelite was available from the late 1920s, with the first British Company Bakelite Limited formed in 1926 from the amalgamation of three suppliers of phenol formaldehyde materials: the Damard Lacquer Company Limited of Birmingham; Mouldensite Limited of Darley Dale and Redmanol Chemical Products Company of London which opened a factory in 1928 in Tyseley, Birmingham, (subsequently demolished in 1998; Dring 1940, 201–202; American Institute of Chemical Engineers 1977). It suggests that the majority of these items can be closely associated with the final demolition of the works in the 1930s.

LARGE FERROUS/NON-FERROUS METAL AND COMPOSITE ARTEFACTS

Sam Hepburn and Chris Hewitson

The large ferrous/non-ferrous metal and composite artefacts included a series of structural elements of various machinery parts. The location of their recovery has been recorded and marked on plan. The size, weight, intrinsic value and petrochemical material contamination of the assemblage meant that it was not removed from site. Some of the large ferrous items were initially set aside on site, with a view to retaining them. Unfortunately during the process of moving from Area 1 to Area 2 the majority of the finds were lost. Each object was assigned an individual finds number, a detailed

TABLE 6.6 NON-DIAGNOSTIC AND OTHER METAL ITEMS

Finds No	Description	Material	Measurements (in m)
LF1	Part of a metal sign painted red with white lettering, the letters 'TRY' are visible.	Fe/Tin	0.20 x 0.20
LF9	Heavily corroded section of bent Fe pipe	Fe	0.17 x 0.10 x 0.10
LF10	Small Fe plate with 4 bolt attachments	Fe	0.05 x 0.05 x 0.01
LF11	Possible Fe machine component	Fe	0.25 x 0.13 x 0.05
LF13	Square Fe vent with small circular holes.	Fe	0.25 x 0.25 x 0.025
LF28	1 wooden stake, and 2 large metal bolts, possibly used to secure large machinery	Fe/ Wood	0.05 x 0.05 x 0.25; 0.05 x 0.05 x 0.30; 0.10 x 0.10 x 0.50

description was undertaken and a scaled photographic record was taken of each item. Due to their size the items were not weighed. A number of miscellaneous items were recovered from the site; they were either of no clear function or do not appear to have had specific functions associated with the machinery in the works. Details of all large finds are described in the Table 6.6.

The small cast-iron pulley wheel assemblage (LF27; Fig. 6.5a) had a track for the rope or chain and may likewise suggest a device, attached to a pre-existing beam or fixed crane, for the movement of larger goods or raising of items. Pulleys may have been used to remove vats from furnaces.

In the light of the identification of the hinged machine arm (see below) as a simple jib crane, the purpose of the large circular stone weight with a metal fixing (LF34; Fig. 6.5b) may suggest it was a counter-balance weight. The large hinged machine arm comprised a tapering cast iron beam connected by a pin to a hooked end (LF23; Fig. 6.6a). The function of the hook and arm is unclear but it may have acted as a simple crane system with the cast iron beam acting as the jib and the hook being used to carry and distribute loads around the factory. No clear attachment for chains is visible. Uses could well have been to lift wire or metal plate during the rolling process.

The two machine components (LF22 and LF24; Figs. 6.6b and 6.6c) are roughly the same size and shape, consisting of two ferrous metal pins connected together by a small cast iron beam. They may be long linear bolts designed to hold two elements together, the short arm acting as a plate, a further one screwed on the opposite end (now missing).

The cast ferrous metal machine arm (LF31; Fig. 6.6d) would appear to be part of a working machine component. It may be one of a pair of connecting rods from a horizontal steam engine. The connecting rod was the element of the steam engine that allowed the vertical motion of the beam to be transferred to the rotational motion of the flywheel. Examples suggest they may have been sited singly or in pairs on either side of the beam.

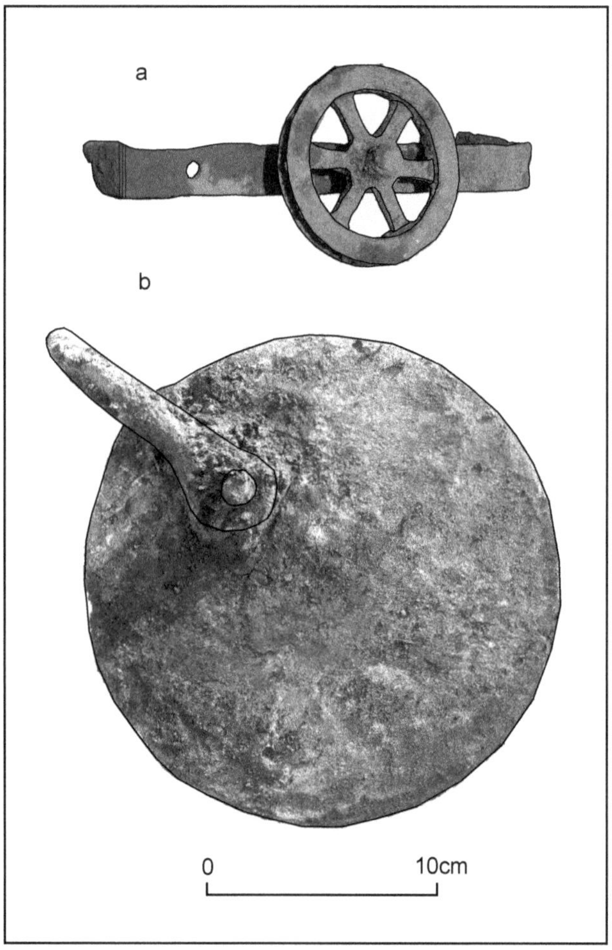

FIGURE 6.5 LARGE FERROUS METAL AND COMPOSITE ITEMS 1

Catalogue of Illustrated Items

1. Large ferrous metal hinged machine arm, comprising a tapering cast ferrous metal beam connected by a pin to a ferrous metal hooked end, LF23, dimensions 3 x 0.20 x 0.06m, Fig. 6.5a.
2. Ferrous metal pulley wheel with a track for the rope or chain, it is attached to a ferrous metal bracket with its bolts still attached, LF27, dimensions 0.15m diameter, 1.25 x 0.15 x 0.15m, Fig. 6.5b.
3. Large cast ferrous metal machine arm with decorative ribbed designs along the length, at each of the rounded ends is a 0.05m diameter bolt hole, one square the other hexagonal, LF31, dimensions 1.45 x 0.20 x 0.08m, Fig. 6.6a.
4. Large circular yellow sandstone (geological identification unknown) weight with a metal fixing

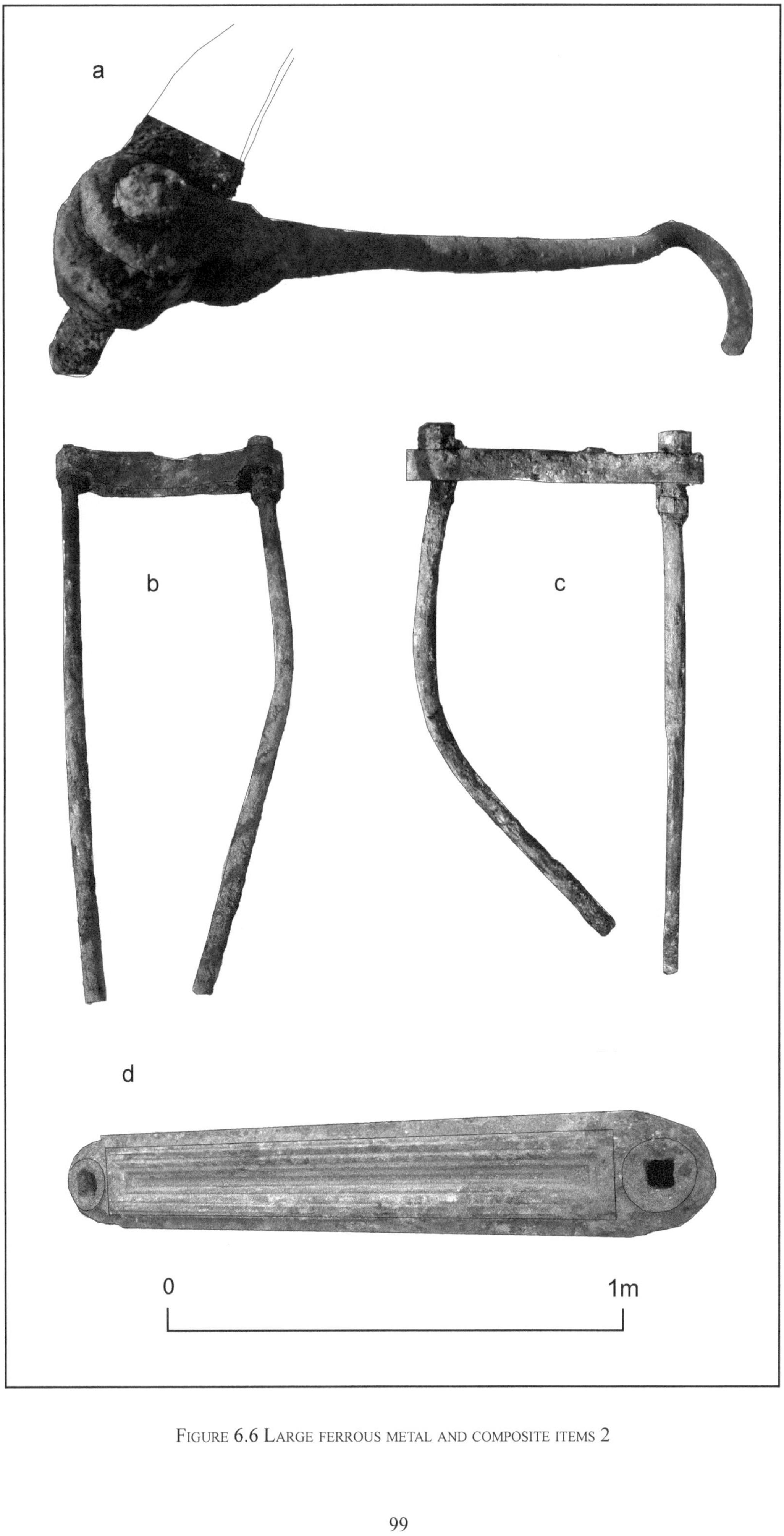

Figure 6.6 Large ferrous metal and composite items 2

used to hang it LF34, dimensions 0.44m diameter, 0.15m, Fig. 6.6b.

5. Machine component consisting of two ferrous metal pins connected together by a small cast iron beam LF22, dimensions 1.10 x 0.50 x 0.03m, Fig. 6.6c.
6. Machine components consisting of two ferrous metal pins connected together by a small cast iron beam, bent out of shape LF24, dimensions 1.10 x 0.50 x 0.03m, Fig. 6.6d.

SCRAP METAL AND CRUCIBLE FRAGMENTS

Sam Hepburn and Chris Hewitson

The assemblage consisted of 209 fragments of ferrous and non-ferrous scrap metal fragments from 69 separate stratified and unstratified contexts dated from the period 1820–1936. The assemblage had a combined weight of 29.61kg. The collection policy was not to take a 100% sample, but to take a representative example of material for further analysis. Unsampled material was discarded on site or remained uncollected. The material was examined visually, weighed, quantified and catalogued. Descriptions were based on the authors' experience and cross-reference to the material contained within the JW Evans archive (located at English Heritage stores in Atcham). Material typologies were based on visual analysis only at this stage of assessment.

The assemblage contained a mixture of items from waste products (slag and crucibles) to unfinished objects. All of the metal finds recovered and assessed here fell in to three types: tools, formed items either cast or pressed but not finished and metal waste/off-cuts. The formed metal items are produced by using one of several processes, the sheet metal items have been pressed using drop stamps or fly presses to cut out the blanks and to shape them, the brass studs and many of the fe items have been cast, and the wire present has been drawn from the brass/cu alloy strip found on the site. Context 1404 has several examples of this strip with signs of it being drawn in to wire.

The assemblage was also collected in order to establish the processes undertaken within specific areas of the site. The potential of the assemblage for aiding understanding of the composition of the different material used within the site, and thus the different processes undertaken, is detailed in the archaeo-metallurgical assessment (McDonnell, Chapter 7). The scrap wire and rolled brass also allows us to understand the nature of the final produced material from the factory. The distribution of the artefacts is shown in Fig. 4.14.

Cast iron artefacts

A small number of cast iron (fe) artefacts were recovered from the site. These related to tools associated with the finishing process. They included a number of cast iron files associated with removing casting burrs and fragments. There were also the partial remains of a pair of pliers.

Selected Catalogue

1. Seven fragments of triangular-shaped files 1042, SF86a, unstratified
2. Two pieces of triangular-shaped file, 1255
3. Cast iron pincers (one half), unstratified
4. Cast iron poker/pry bar, weight 800g, unstratified, SF73

Cast iron items

A number of cast iron items were also recovered. These were predominantly scrap metal. There were three lengths of cast iron strips, possibly from machinery, and a number of iron pipes, possibly derived from hollow casting of bedsteads, but equally likely to have been associated with plumbing. In addition, there were a number of pieces of hollow iron wire (not catalogued).

Selected catalogue

1. Three cast iron strips (from a machine?), weight 902g, unstratified, SF73
2. Three lengths of iron pipes with holes, covered in concretions, weight 1800g, SF86a, 1042
3. Four lengths of iron pipe (two have concretions), weight 1970g, 1255

Cu alloy wrought items

The cu alloy wrought items predominantly consisted of brass metal strip and off-cuts and wrought brass wire.

The brass sheet consisted of three types: Strips of rolled brass - this may have been part processed ready for production into wire, or it may have been part rolled ready to be further rolled and subsequently used for pressed brass items. A further category was off-cuts and, finally, waste from brass pressing. A number of pieces of the sheet brass were analysed to consider their metallurgical composition (see McDonnell, Chapter 7). The sheet brass is listed in Table 6.7.

The brass wire was of a range around 2mm gauge and derived from different stages of the wire drawing process. Of particular interest is a shaped piece of brass with flattened ends (SF78) where it was placed in clamps that must represent the start of the drawing process. Two strips of brass (1406) also appear to represent an early stage of the drawing process. Like the sheet brass, the wire has been subject to metallurgical analysis by Gerry McDonnell in Chapter 7. Elsewhere the wire is clearly at a complete stage. Details of the wrought-brass wire are provided in Table 6.8.

Cu alloy pressed items

There was a large range of cu alloy pressed items. Their composition was of copper/zinc alloy mixes that would categorise them as brass. The majority of the items were

TABLE 6.7 WROUGHT BRASS SHEET AND OFF-CUTS

Finds No	Context	Description	Weight
SF5a	1017	7 small sheet metal off-cuts (brass?)	40g
SF7	1018	Small rectangular brass sheet metal off-cut	1g
SF8	1018	5 brass sheet metal off-cuts	5g
SF12c	1035	3 corroded brass sheet metal off-cuts	50g
SF19	1051	Small brass sheet metal off-cut	1g
SF25	-	Large brass thin sheet metal off-cuts	569g
SF49	-	4 brass strip off-cuts	60g
SF56	-	Piece of brass sheet metal off-cut	90g
SF58	-	Brass sheet metal off-cut	300g
SF63	-	4 brass strips, off-cuts	69g
SF67	-	2 pieces of brass sheet metal off-cuts	42g
SF71	-	4 brass strip off-cuts	78g
SF83	1063	Small piece of brass sheet metal off-cut	25g
SF94	1245	Sheet brass fragment	30g
SF99	1327	2 brass sheet metal off-cuts	12g
SF101	1333	Brass sheet metal off-cut	10g
SF104	1406	Large pile of brass strip off-cuts, a small pile of brass sheet metal off-cuts, 2 strips of brass that have started to be drawn into wire	780g
	1324	Sheet metal from sample no 3 assoc. with Crucible Hearth 1118	20g

TABLE 6.8 WROUGHT BRASS WIRE

Finds No	Context	Description	Weight	Metal Sample No.
SF49	-	3 pieces of drawn wire	40g	11
SF52	-	7 pieces of cu alloy drawn wire off-cuts		
SF56	-	6 pieces of drawn wire of different gauges	300g	
SF58	-	2 pieces of drawn wire	70g	
SF67	-	2 pieces of brass drawn wire and 2 pieces of brass sheet metal off-cuts	77g	
SF71	-	3 pieces of brass drawn wire, a coil of copper electrical wire	72g	
SF78	-	A shaped piece of drawn wire with flattened ends (a small handle)	12g	
SF99	1333	2 pieces of thick gauge brass drawn wire	18g	
SF102	1347	A length of thick gauge drawn brass wire, twisted to form a loop	288g	
SF104	1406	Large pile of drawn brass wire lengths/off-cuts, 2 strips of brass that have started to be drawn into wire	350g	16

circular brass blanks, either for washers or buttons. A number of buttons were also found. It would appear clear that button blanks were either being pressed on site, or waste was imported for re-melting. The alternative is that brass washers were being produced on site. This was associated with the gas and water fitting trade.

In addition a number of pressed items may have been machine parts. These were small in number and suggest they were associated with specific machines.

Selected catalogue

1. Small circular pressed brass blank with hole punched in the centre, weight 2g, 1016, SF1
2. Eleven circular pressed brass blanks of various sizes, weight 30g, 1006, SF2
3. Five circular brass pressed buttons, weight 10g, 1058, SF28
4. Small circular brass pressed blank and washer, weight 1g, 1056, SF28
5. Broken circular pressed cu alloy/brass, button, weight 1g 1079, SF30
6. Piece of brass sheet metal off-cut (waste from button making), 4g, 1109, SF36b
7. Six pressed brass machine parts, weight 20g, 1333, SF101
8. Four pressed cu alloy vessels/machine parts, weight 240g, 1406, SF104

Cu alloy cast items

The number of cast brass items was relatively small, reflecting the re-use of brass on the site during the casting process. A small cast brass block may have represented a small waste ingot (1018, SF3). Other items included a cast circular stud (1035, SF5a), an unusual item by the later 19th century as the majority of studs would have been produced by pressing of the metal. The most significant collection of material was a series of cast metal fittings associated with the water/gas fitting trade. These included remnants of an eroded brass tap, stop cocks, brass washers

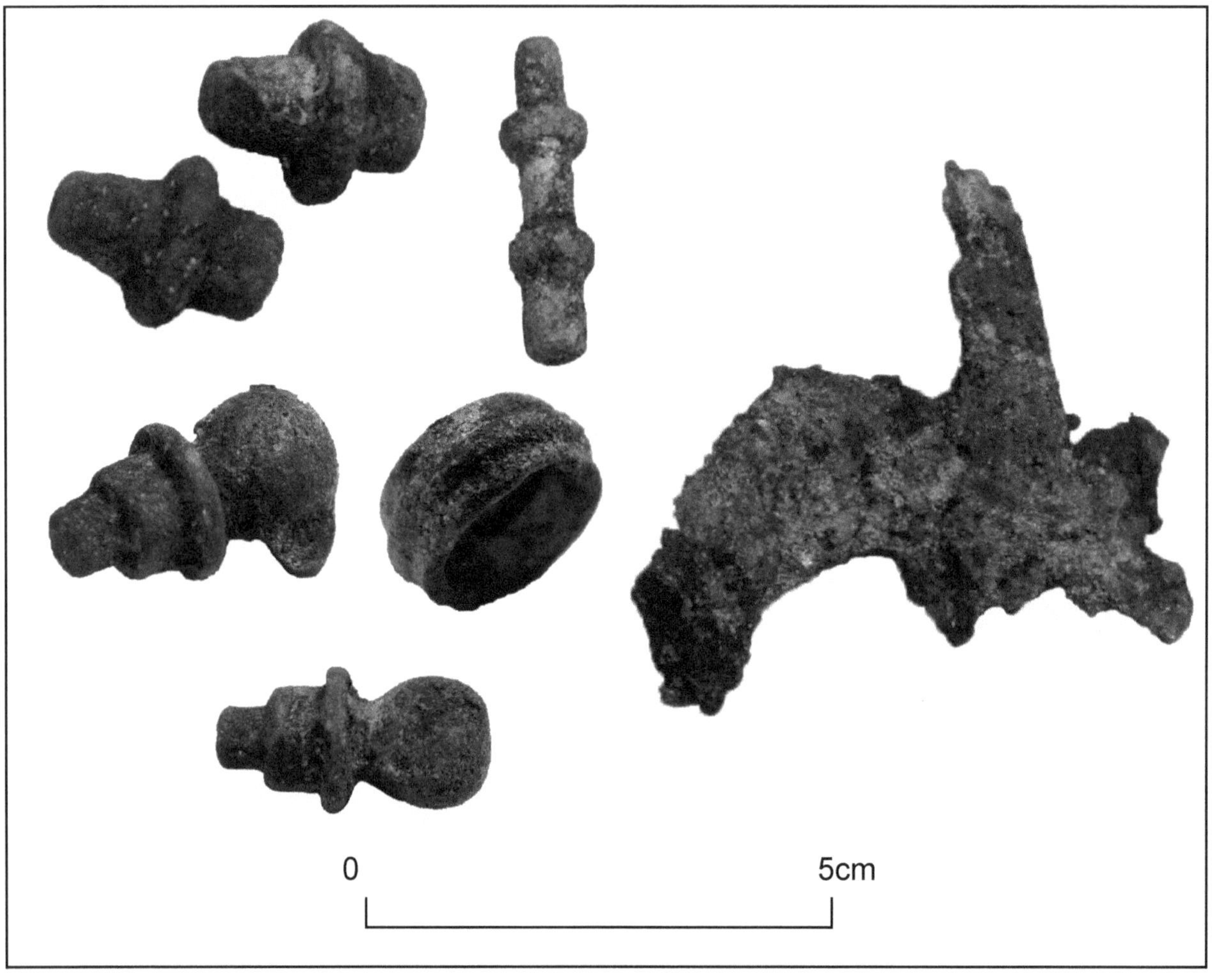

FIGURE 6.7 CU ALLOY METAL ITEMS

(1245; Fig. 6.7). This provides proof that items from other departments (ie the gas fitting department) were recycled in the crucible furnaces of the brass casting foundry of the bedstead works. The cast items have been subject to metallurgical analysis in Chapter 7.

Selected catalogue

1. Small rectangular brass cast block, weight 16g, 1018, SF3.
2. Small cast machine part, weight 20g, 1017, SF5a.
3. Brass decorative cast handle, weight 11g, SF12b, 1035
4. Decorative cast circular brass stud, 15g, 1035, SF12c.
5. Nine brass cast machine parts, ribbed rods, pipe fittings, rivets, knob, weight 180g, 1245, SF94, Fig. 6.7.
6. Two cast brass items, weight 120g, 1406, SF104.
7. Three pieces of concreted cast brass, weight *c* 4000g, U/S, SF58.

Slag deposits

A number of slag deposits were collected throughout the site. These varied in type and variety. Detailed analysis of the slags is discussed by Gerry McDonnell in Chapter 7. Table 6.9 lists and details the slags recovered from the site.

Crucible fragments

The crucible bases were recovered from a number of areas, but the majority were recovered from areas associated with the four crucible furnaces at the bedstead works. The largest deposit of crucible bases was recovered from context 1571 of which only a small sample was retained. Details of the crucibles sampled are listed in Table 6.10.

The majority of the crucible fragments were thick-walled and had metal concretions adhering to both the interior and exterior of the crucible bases. Typical width of the wall was 20–25mm. The estimated height of the crucibles was around 250 to 300mm or 10 to 12in in height. Aitken (1866, 267) suggests that the majority of Birmingham crucibles were made of Stourbridge clay although there was a movement to introduce plumbago clay from London at William Tonks' works. An example of the

Table 6.9 Slag deposits

Finds No	Context	Description	Count	Weight	Metal Sample No.
SF3	1018	Small piece of slag	2	6g	
SF4	1018	Small piece of slag	3	10g	2
SF17	1051	6 small pieces of slag	6	134g	5
SF18	1051	2 small pieces of hard and light slag	2	44g	6
SF20b	1051	4 pieces of light and rough slag	4	388g	7
SF26	u/s	2 large pieces of hard, rough, globular looking slag	2	801g	9
SF29	u/s	16 pieces of slag	16	-	10
SF36a	1109	Small piece of slag	1	55g	
SF40	1101	Small bag of pale fragmentary slag		35g	
SF58	u/s	3 pieces of concreted slag	3	5,500g	12
SF72	-	Small piece of slag	1	2g	
SF80	1151	Large pieces of slag	1	385g	
SF89	1396	2 small pieces of slag	2	15g	
SF90	1388	2 small pieces of pale coloured slag	2	32g	
SF91	1411	Small pieces of slag	1	23g	
SF94	1245	4 small pieces of slag	4		
SF98	1279	Medium sized bag filled with small pieces of slag	-	-	15
-	1317	Slag from sample bag 1, crucible hearth 1042	-	-	
-	1323	Slag from sample bag 2, crucible hearth 1118	-	-	
-	1324	Slag from sample bag 3, crucible hearth 1118	-	-	
-	1325	Slag from sample bag 4, crucible hearth 1118	-	-	
-	1329	Slag from sample bag 5, crucible hearth 1118	-	-	
-	1334	Slag from sample bag 6, crucible hearth 1118	-	-	
-	1406	Slag from sample bag 9, machine base	-	-	

Table 6.10 Crucible fragments

Finds No	Context	Description	Count	Weight
SF36g	1109	3 crucible fragments	3	1142g
SF35	1113	2 pieces of crucible	2	496g
SF86e	1042	Small circular crucible base	1	1000g
SF96	1255	4 pieces of crucible	4	3400g
SF105	1547	Part of a crucible base	1	856g
	1571	Deposit of crucibles	20+	*c.* 20,000g
-	u/s Area C	Small circular crucible base	1	1396g

crucible fragments (from 1255, Crucible Furnace 1042) is illustrated in Fig. 6.8.

Discussion

The assemblage also has the potential to aid understanding the distribution and typology of small machines operating within specific areas of the site, and the distribution of the work within different areas of the site.

The distribution of wire production has an unsurprising bias towards the northern side of Gibson's basin and the wire drawing mill. Likewise, rolled metal strip was also found more on this side of the basin, although there is also a distribution on the southern side. This may in part suggest that recycled brass was being incorporated during the casting process. Alternatively, the prevalence of pressed items on the northern side of the canal may explain why strip brass and off-cuts were present. Particularly in rooms A, C and G large numbers of pressed circular items, possibly washers or buttons were found, which may suggest that the area included the production of small pressed components like washers, or also had rooms in which pressed buttons were produced. This suggests that they relate to the occupancy of Sperryn and Co. from 1900–1915 and not the period from 1915–1922 when the Aluminium Company took over the works.

The distribution of slags and crucible fragments suggest that they can be associated with the casting process. The slags in particular are either located in the bedstead casting shops or in an area distributed around the rolling mill casting shop. These items are unlikely to have travelled far as they would not have been re-cycled like waste metal. The presence of a large number of cast brass items that can be associated with plumber's brass foundry (ie taps, washers, stop-cocks) from context 1245 supports the interpretation that manufacturing was varied in the works. There is no

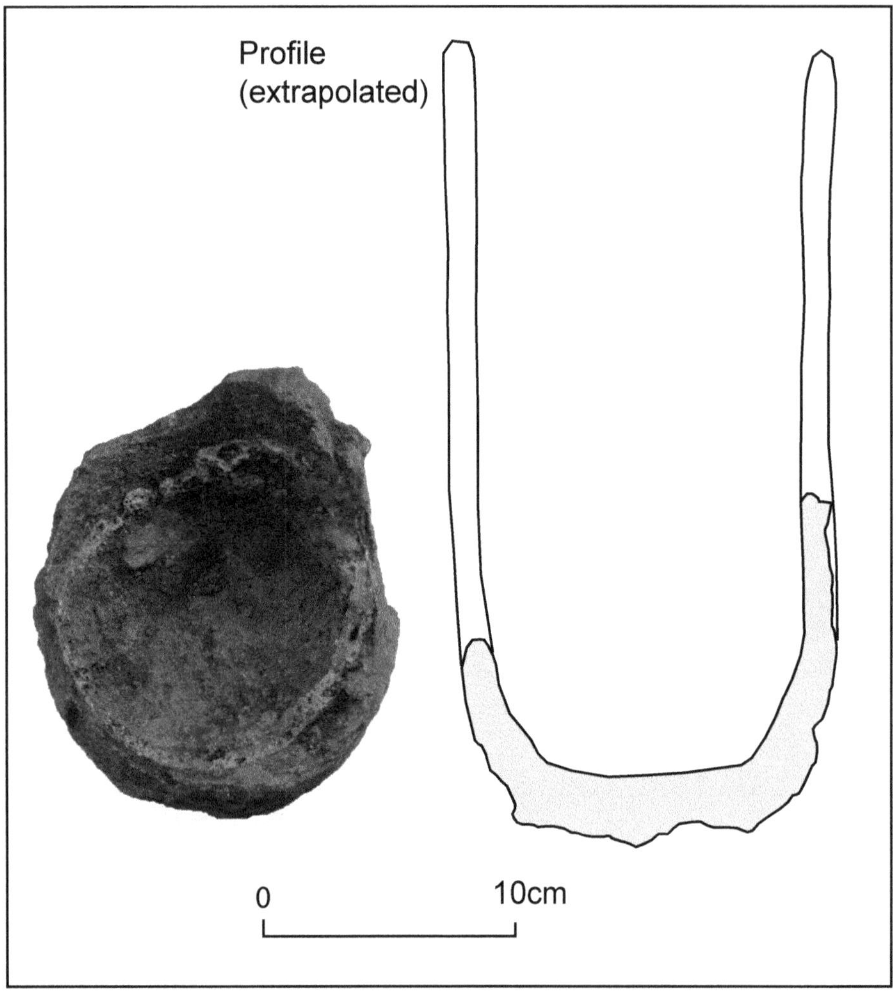

Figure 6.8 Brass crucible fragment

reason to suppose these were necessarily produced on site but they may again relate to scrap material being recycled.

Overall the assemblage supports the known historical use of the works but there are some suggestions that the bedstead works was adopted for light manufacturing, for example, pressed metal items, as well as the casting already suggested by the presence of crucible hearths in three rooms.

WORKED SHELL AND BONE

Erica Macey-Bracken and Chris Hewitson

The cut-outs from button-making are common finds on Birmingham sites and represent part of an increasing body of evidence for the shell button industry. The shell button blanks were uncovered from demolition layers and are representative of the early 20th-century industry.

The shell button industry flourished from the 18th to the early 20th century. The industry employed some 2,000 people by 1866, mostly on a piecemeal basis (Turner 1866, 441). It declined in part due to the loss of American markets during the Civil War and subsequent competition from the continent, but its eventual demise was due to the replacement of bone by plastic in the 20th century (White 1977, 72). The process involved the cutting of blanks with a tubular saw (producing the distinctive circular patterns), through-drilling of the blanks and the decorating or engraving of the buttons (*ibid*, 71).

The main areas of mechanised production appear to have been bounded by Colmore Row, Snow Hill, Mount Pleasant and Congreve Street (White 1977, 75–76) to the east of the site. This does not discount production in the area because much was done by small premises under the guise of 'garret-masters' (*ibid*, 72). Shell was used to produce buttons and more exotic types, such as mother-of-pearl buttons, were a speciality of Birmingham that involved the importation of shell from the East Indies, the Philippines, the Red Sea and the Persian Gulf (*ibid*, 71). An identified example of Indian Ocean turbo mamoratus shell was uncovered from a back-filled well at Floodgate Street, and dated to the 18th century (Sommerville forthcoming). Further examples of waste from workshops in the Birmingham city centre were uncovered at Edgbaston Street (Bevan *et al* 2009, 179–180) and the Coach Station, Digbeth (Levers 2010), but have also been uncovered within the suburbs at Saltley (Duncan and Jones 2010), demonstrating the ubiquitous nature of the industry in Birmingham.

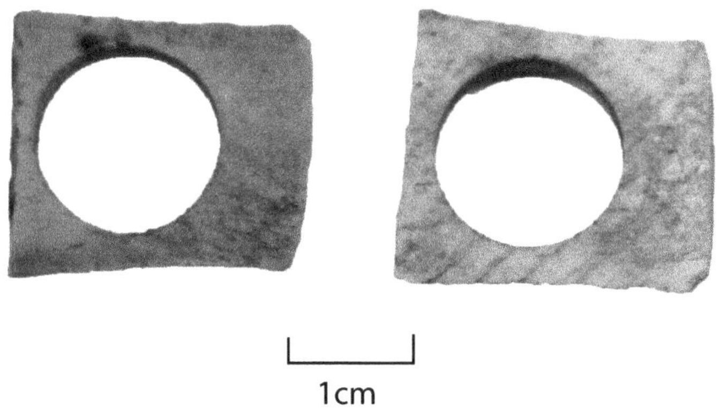

Figure 6.9 Worked bone button blank

A single fragment of sawn bone from working waste was recovered (1025, SF9b). It has been demonstrated that worked bone was used for buttons, knife handles, cutlery and other small tools in Birmingham (Bevan *et al* 2009, 180–181).

Catalogue

1. One x cutout from button-making, dimensions 25 x 22 x 4mm, weight 5g, 1001, Fig. 6.9.
2. Two x pieces of bone-working waste - cutouts from button making, dimensions 23mm x 15mm x 3mm; 21mm x 11mm x 3mm, weight 10g, SF36d, 1109.

LEATHER

Chris Hewitson

The three sets of leather shoes and fragments of a shoe sole were recovered from the site (see Table 6.11). The three complete sets of shoes were heavily contaminated with petro-chemicals and were recorded on site by photograph and subsequently discarded. Three fragments of shoe sole were also recovered from the site from the backfill of the Gibson's Basin (SF 34b; 1137). All the items were found in unstratified layers that dated to the demolition phase. The items have been photographed and are included as part of the site archive.

WOOD

Sam Hepburn and Chris Hewitson

The wooden items were either too large or heavily contaminated with petro-chemical residues to be retained. Their location has been recorded and marked on plan. Each object was assigned an individual finds number, a detailed description was undertaken and a scaled photographic record was taken of each item. Due to their size the items were not weighed. No species identification was undertaken on the wooden items. Details of all large finds are described in Table 6.12.

The large wooden beam (LF30) is likely to have been a cross-beam from the roof structure of the rolling house or an adjacent building. The rolling house would have been almost entirely open to allow ease of access around the rolling machines. Examples from the Hafod Works,

Table 6.11 Leather items including complete shoes

SF No	Context	Description	Measurements (approx m)
LF2		Small pair of leather boots	0.15 x 0.05
LF20		Large pair of leather work shoes	0.18 x 0.07 x 0.06
LF33		Leather shoe	0.26 x 0.11 x 0.09
SF34b	1137	3 x fragments of shoe sole. Stitching holes visible on all pieces	Various sizes

Table 6.12 Details of the wood assemblage

SF No	Description	Measurements (approx m)
LF21	Wooden shelf with rounded corners, with two Fe metal triangular fixing brackets	1.30 x 0.25 x 0.30
LF26	Wooden beam with a small (0.10m x 0.025m) rectangular hole cut through it	0.09 x 0.09 x 1m
LF29	Wooden beam/plank	1.50 x 0.40 x 0.20
LF30	Large wooden beam rotten in places, with part of a wooden plank bolted to it on one side	2.54 x 0.25 x 0.25
LF32	2 wooden chocks, the ends of each chock taper to a point	0.45 x 0.10 x 0.06 0.30 x 0.075 x 0.06

Swansea (Hughes 2000, 37–40, fig. 50 and 54) have an iron-frame roof structure, but these works were rebuilt in 1910. It is possible the original works had a timber roof structure, but the hazards of fire would suggest a cast-iron structure. Cast-iron superstructures were developed from the very early years of the 19th century, following the development of the Ditherington Flax Mill, Shrewsbury (Rimmer 1960, 49) and it is open to question whether a cast-iron roof structure would have been in place when the Union Rolling Mills were built in the 1820s. The earliest example known in Birmingham was the 1822 retort house, Gas Street (Linnane 1998; Hodder 2004, 146) but the span was considerably smaller than the rolling mill. The gas retort house dated 1849–1850 at Walsall, Wolverhampton Road (Hewitson 2005) was of a more sophisticated iron-frame design and spanned a larger distance than the rolling mill, and suggests that the technology was in place by the 1940s rebuild of the works. The presence of the beam does, however, suggest that the trusses for the earliest elements of the work were wood-built as opposed to cast-iron or iron-frame.

Catalogue

1. Large wooden beam rotten in places, with part of a wooden plank bolted to it on one side, LF30, dimensions 2.54m x 0.25m x 0.25m, not illustrated.

NEWSPAPER FRAGMENTS

Chris Hewitson

Twenty-four fragments of newspaper were uncovered from the site from context number 1406. Although fragmentary, and under normal circumstances discarded, these items were retained as they had small quantities of readable print. No attempt has been made to conserve the items as they are too fragmentary and only provide a minimal glimpse of the times at the final demise of the site. The majority of fragments are unreadable. However, the fragmentary remnants that did survive give an impression of the era. Some displayed the economic period of the 1930s when the Winfield's Rolling Mill moved to Icknield Port Loop:

UNEMPLOYMENT
[Word lost]... FIGURES UP
NEARLY 180,000.77

Others displayed the changing nature of the metals trade in particular the motor car trade exhibitions:

...be said that our manufacturers hold themselves well against these competitors. Indeed...[word lost]...display shows that despite the general industrial depression the motor trade takes a very optimistic view of its...[words lost]...

The newspaper fragments also included fragments of comic strips. Social events were also announced:

BIRMINGHAM ENTERTAINMENT
Children in Uniform [perform before the Prince] of Wales.

A number of adverts were preserved. These included such products as 'Monkey Polish' and general soap. One of the more interesting was this partially preserved advert that shows a very different attitude to today's regarding the use of alcohol, in this case Bass beer:

The purest drink –
[lost words]... nature's way –
[lost word] artificial is added – it is [lost word] concentrated nourishment.
[Which?] is why people feel [lost word] after a glass of BASS.
[lost words] has such health- [lost words] health-maintain-
[lost words] BASS. If you... [lost words]... start today.
[lost words] preventative... [lost words]...further ailments.

Overall the newspaper fragments give an indication of the social state of Birmingham in the 1930s when the works transferred. This was an era in the inter-war years when poverty and depression were widespread, yet optimism was still to be found and new trades flourished.

CHAPTER 7: ARCHAEO-METALLURGY

Gerry McDonnell

A sampling strategy for the metallurgical residues was agreed at the outset of the project. This was due to the recognition that large amounts of residues are present on industrial sites of this type and that a targeted sampling of these residues would be important to keep quantities manageable. Samples were taken to recover information on their origin, type, composition and process by which they were created. Analysis undertaken on these residues included techniques such as X-radiation fluorescence (XRF), metallographic examination and Scanning Electron Microscopy (SEM) analysis.

X-RAY FLUORESCENCE ANALYSIS OF THE COPPER ALLOY METAL

Introduction

The Assessment Report (McDonnell 2009) recommended a programme of non-destructive X-Ray Fluorescence Analysis (XRF) of a selection of copper alloy artefacts to address the following research questions:

- To assess the range of copper alloys present on the site.
- To assess whether different alloy compositions were selected for different artefact types, eg cast objects, wrought sheet, wrought wire.
- To address the research objectives outlined in the Research Framework for Archaeo-metallurgy (Bayley 2008, 69), specifically to establish archaeo-metallurgical data for the brass industry.

It was also noted in the Assessment Report (McDonnell 2009, 5) that brass may suffer de-zincification during the natural corrosion process in a terrestrial environment. XRF is a surface technique (depth of penetration ~<10µm) hence the spectrum gained is dominated by the corrosion layer composition eg presence of iron (Fe) trapped in the corrosion layer. Therefore the data produced by XRF may underestimate the zinc (Zn) content of the alloy, and this needs to be investigated.

The total number of copper alloy artefacts analysed was 67. To assess any variation in composition the artefacts were subdivided into three groups: as-cast objects - artefacts assumed to have been cast and not subjected to further working; sheet metal - copper alloy cast and rolled into sheet; wire - copper alloy probably cast as rod, and drawn down to wire, alternatively worked to sheet, cut into strip and then drawn to wire.

XRF analyses of the cast artefacts

Sixteen artefacts were considered to be cast objects. They included small ingots, specific artefact types, plus some failed castings. The artefacts were analysed to generate data that included the iron content of the corrosion layer (Table A3.3.1). The iron content was removed and the data normalised to generate a new data set (Table A3.3.2), exotic analyses, ie compositions lying outside the normal brass range (10–90% Zn), were removed and presented in decreasing zinc content (Table A3.3.3). The most distinct composition is one of the ingots (1009 SFN 3) which was a lead-antimony alloy containing a small amount of tin, a form of pewter. Another cast 'ingot' was c90% copper (1245 Ingot 2, Table A3.3.2). The remaining cast artefacts were leaded brasses (Table A3.3.3). The zinc content groups around 30% (Figs 7.1 and 7.2), with one, a swivel (1017 SFN 5a sample 3) containing 80% Zn. The lowest zinc content of a brass was a hook (1035 SFN12b) containing 13% Zn.

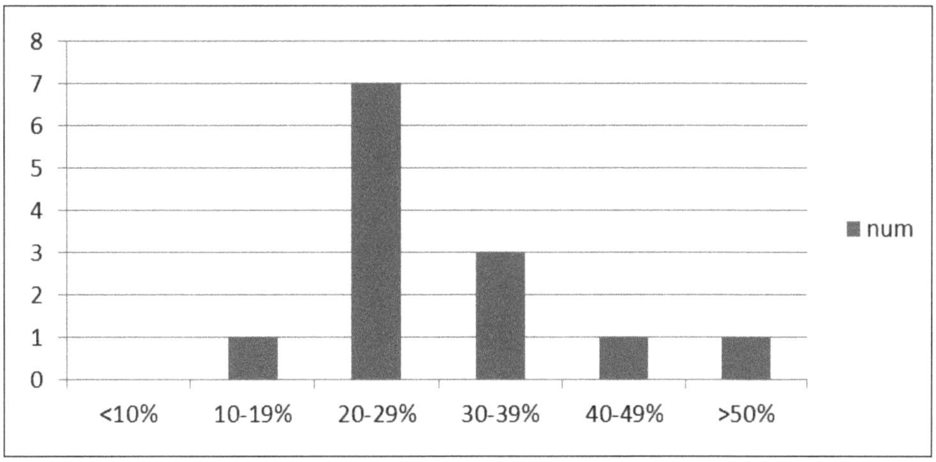

FIGURE 7.1 HISTOGRAM OF ZINC CONTENTS OF CAST ARTEFACTS

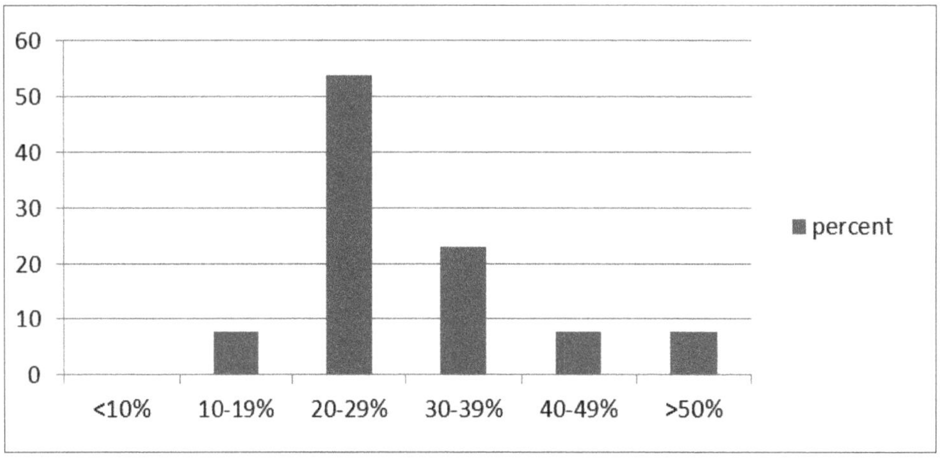

FIGURE 7.2 PERCENTAGE OF CAST ARTEFACTS GROUPED BY ZINC CONTENT

XRF analysis of the sheet metal

Thirty-eight artefacts were identified as sheet. They ranged from identifiable artefacts, eg washers (n=2, 5%), and discs (n=12, 32%), but the majority were scrap off-cuts (n=22, 58%). The compositions ranged widely from high zinc brasses (Zn>50%) to very low alloyed copper (Cu>90%). The main alloying elements are zinc and lead, the highest tin content is 1.3% and cannot be considered as a deliberate alloying process. The compositions can be discussed in broad groups (Tables A3.3.4–5).

The first group are the high-copper, low-alloyed, artefacts. Four artefacts contained more than 90% copper, a washer (Cont 1016, SFN 1) was almost pure copper (98.5%) and would have been very soft. The second group are zinc brasses with greater than 35% Zn. This group would be characterised by specific series of complex microstructures.

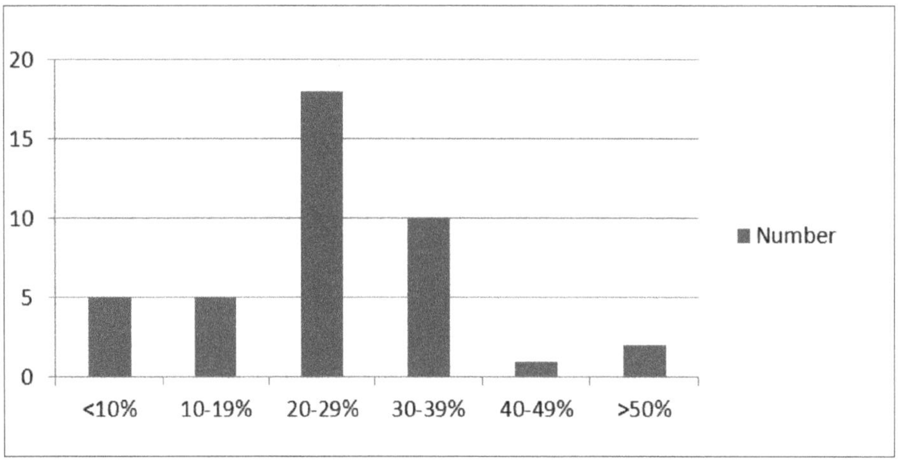

FIGURE 7.3 HISTOGRAM OF THE ZINC CONTENTS OF THE SHEET METAL

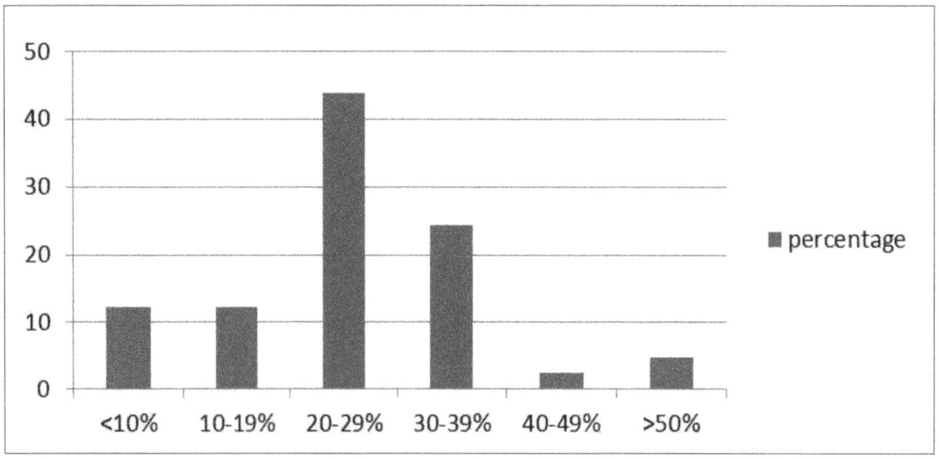

FIGURE 7.4 PERCENTAGE OF ARTEFACTS GROUPED BY ZINC CONTENT

Four artefacts belonged to this group; three had a zinc content higher or equal to the copper. All were alloyed with lead, which ranged in concentration from 0.2–5.8%. The third group (n=29 artefacts) covers the zinc brasses that contained between 10–35% Zn. One artefact was low in zinc (4.4%) but high in lead (8.0%).

The histogram of the zinc contents of the sheet (Figs. 7.3 and 7.4) show a concentration at the 20–30% level, 19 artefacts also had a composition between 25–35% Zn.

XRF analysis of the wire samples

Thirteen wire samples were analysed (Tables A3.3.6 and A3.3.7); all the wire was approximately the same gauge, *c* 2mm in diameter. One sample (context 1406, lump with wire) was pure copper. The remainder were leaded brasses ranging in zinc content from 11–59% (Table A3.3.7) normalised data with iron content removed). No other significant alloying elements were present. The alloy compositions cluster around the high 30s per cent zinc, showing a similar pattern to the sheet (Figs 7.5 and 7.6). This is to be expected based on sample which was a strip that had been partially drawn to wire. However there is a clear discrepancy between the two ends (Table 7.1), the wire being significantly lower in zinc, in the normalised data the difference is 23.5% (equivalent to 60% of the flat end value). This variation exceeds the variations observed in the repeat analyses experiments (Section 3.10, SD=2.4% for Zn). There are two explanations, firstly that

TABLE 7.1 XRF DATA OF STRIP DRAWN TO WIRE

Context	Description	Class	Fe	Cu	Zn	As	Sn	Sb	Pb	Total
1406	strip flat end	s	7.2	38.5	53.6	0.0	0.0	0.0	0.1	99.4
1406	strip wire end	w	7.2	59.4	32.0	0.0	0.0	0.0	1.1	99.6
Context	Description	class		Cu	Zn	As	Sn	Sb	Pb	Total
1406	strip flat end	s		41.8	58.1	0.0	0.0	0.0	0.2	100.0
1406	strip wire end	w		64.2	34.6	0.0	0.0	0.0	1.1	100.0

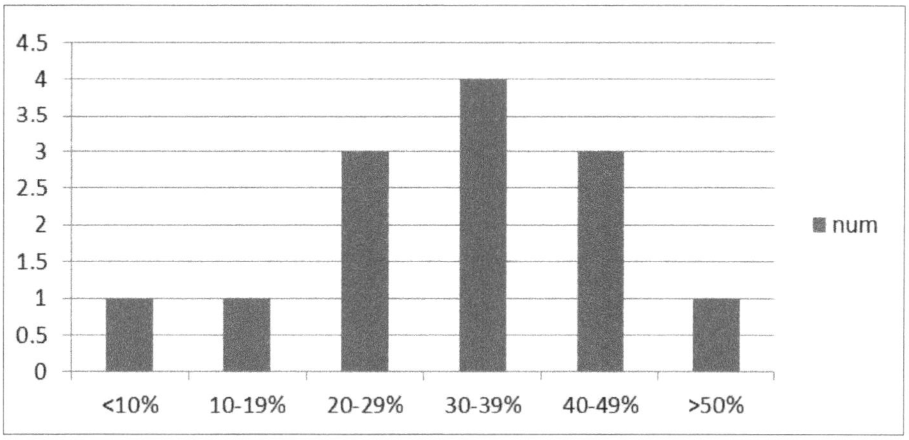

FIGURE 7.5 NUMBER OF WIRE SAMPLES BY ZINC CONTENT

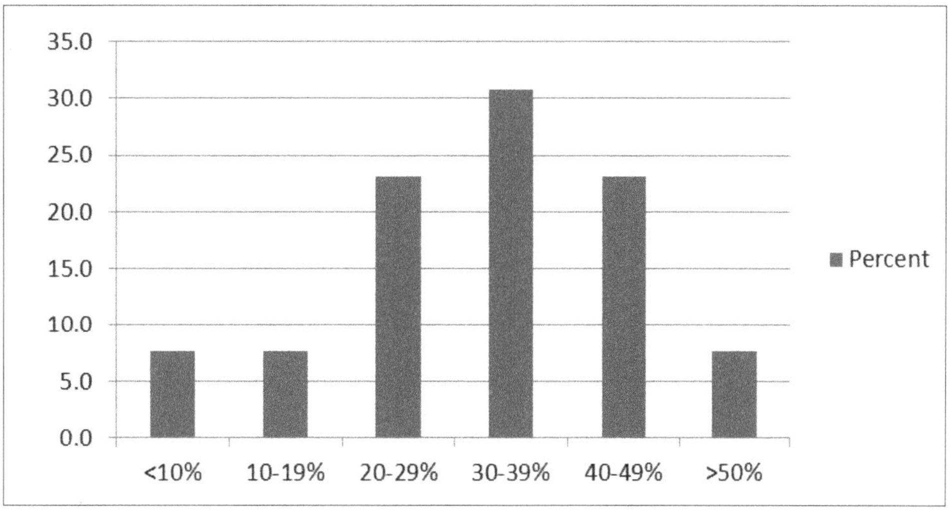

FIGURE 7.6 PERCENTAGE OF WIRE SAMPLES GROUPED BY ZINC CONTENT

the wires end had suffered more severe corrosion resulting in greater de-zincification. However, this is not supported by the iron content which is consistent at both ends, more severe corrosion would have a raised iron content. The second explanation is that the wire drawing was done hot resulting in loss of zinc. However the analysed wires show no consistent lower zinc values compared with the sheet. This discrepancy in the data will be pursued in the metallographic analysis programme.

Discussion

The repeated analysis of the same sample (Section 3.1) demonstrated that the technique successfully analyses corroded brass artefacts. The comparison between the as-received corroded surface and a cleaned surface confirmed that de-zincification of the corrosion products occurs (section 3.2). However the degree of de-zincification is not consistent. If there was a correlation between degree of corrosion and de-zincification then it may be expected that as the degree of corrosion increased, the iron content of the surface would increase, and the zinc content would decrease. Fig. 7.7 plots the zinc against the iron content and shows no such correlation.

The overall mean values for the cast, sheet and wire artefacts are shown in Table 7.2. This confirms that the alloy used was a low leaded brass. The zinc content varies considerably; the average, standard deviation, maximum and minimum for the different artefact types is shown in Table 7.3. This confirms the wide variation of zinc contents. A plot of zinc versus copper (Fig. 7.8) confirms the wide spread of zinc values. The clear linear correlation between the two elements is to be expected given the low level of other alloying present in the artefacts. Figs. 7.9 and 7.10 show the artefacts grouped by type and zinc content. Overall they demonstrate that the composition clusters around 30% zinc. However the plot showing the groups

Table 7.2 Average compositions of the cast, sheet and wire artefacts from the normalised data with Fe removed

	Cu	Zn	As	Sn	Sb	Pb
Cast	66.2	30.4	0.0	0.2	0.0	3.1
Sheet	72.5	25.7	0.1	0.2	0.0	1.6
Wire	66.1	35.3	0.0	0.0	0.0	1.2
Overall	70.1	27.9	0.1	0.1	0.0	1.8

Table 7.3 Average etc inc values for the different artefact types

	Cast	Sheet	Wire	Overall
Average	30.4	25.7	35.3	27.9
SD	16.6	12.3	12.6	13.9
Max	80.3	58.1	58.9	80.3
Min	13.4	0.0	12.1	0.0

as a percentage of each artefact type (Fig. 7.10), shows that the wire artefact have a higher zinc content, having a higher proportion of the artefacts with a zinc content between 40–49%. This data is contrary to expectation. It would be expected that the cast metal should have a (relatively) high zinc content. Subsequent hot working of the metal would be expected to reduce the zinc content through volatilisation of zinc. Hence the most heavily worked artefact type, the wire, would be expected to show an overall decrease in zinc content and the data shows the reverse.

The composition of an alloy determines the microstructure of the alloy; Fig. 7.11 presents a histogram of the zinc values with the major brass microstructures added. Fifty-one of the 67 artefacts (76%) were below 35%, ie alpha brasses; nine artefacts (13%) were duplex brasses having zinc contents between 35-45%. Two artefacts (2.5%) contained between 45–50% zinc and are termed Beta Brass.

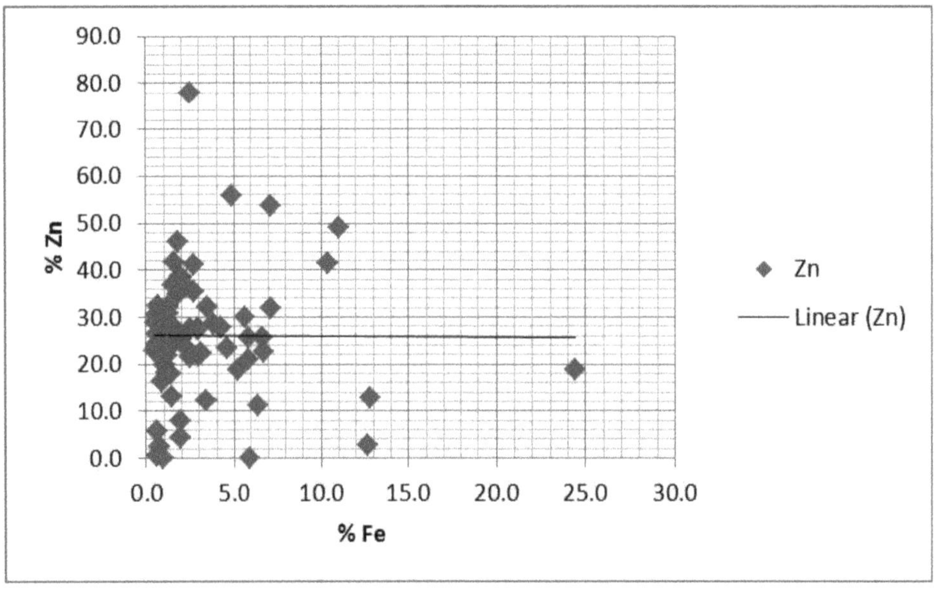

Figure 7.7 Plot of Zn% against Fe% for all analyses

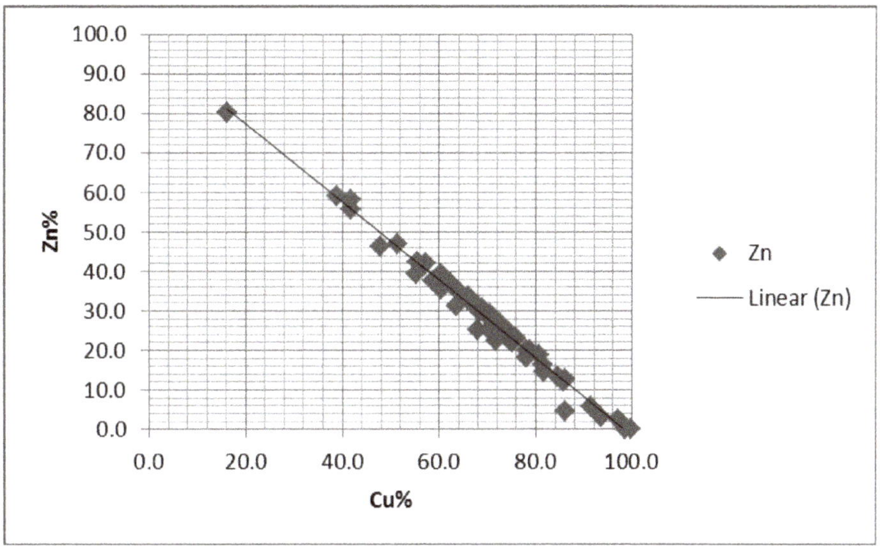

FIGURE 7.8 PLOT OF ZINC AND COPPER VALUES

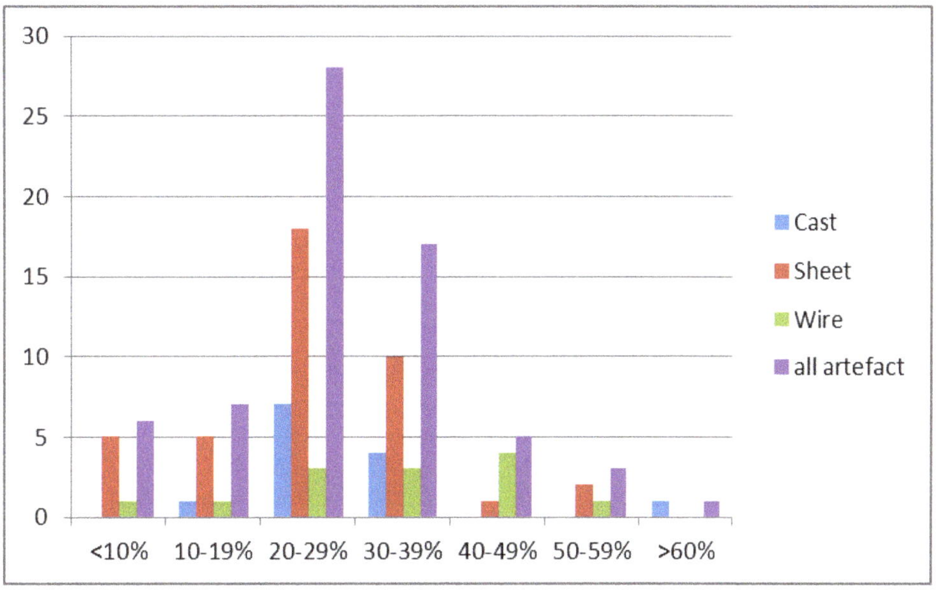

FIGURE 7.9A COMPARISON OF NUMBER OF ARTEFACTS GROUPED BY ZINC CONTENT

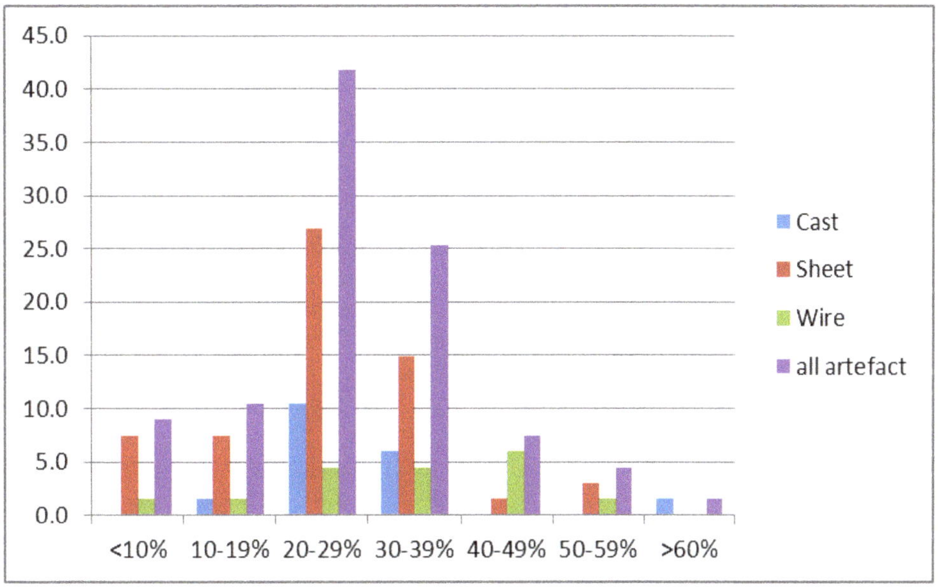

FIGURE 7.9 ARTEFACTS GROUPED BY ZINC CONTENT CALCULATED AS A PERCENTAGE OF TOTAL NUMBER OF ARTEFACTS (N=67)

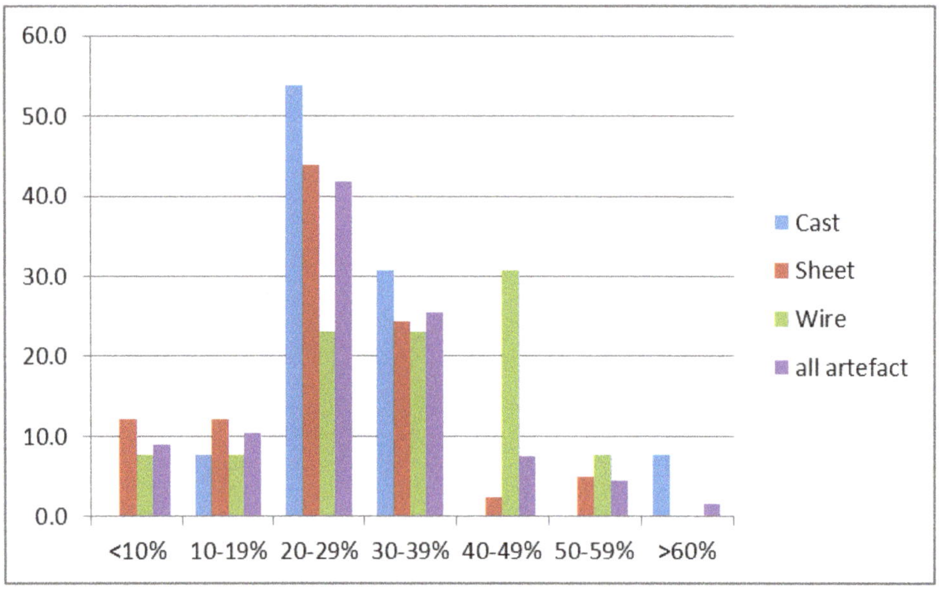

Figure 7.10 Zinc content groups, by type calculated as a percentage of type total (ie total number of cast artefacts = 13=100%)

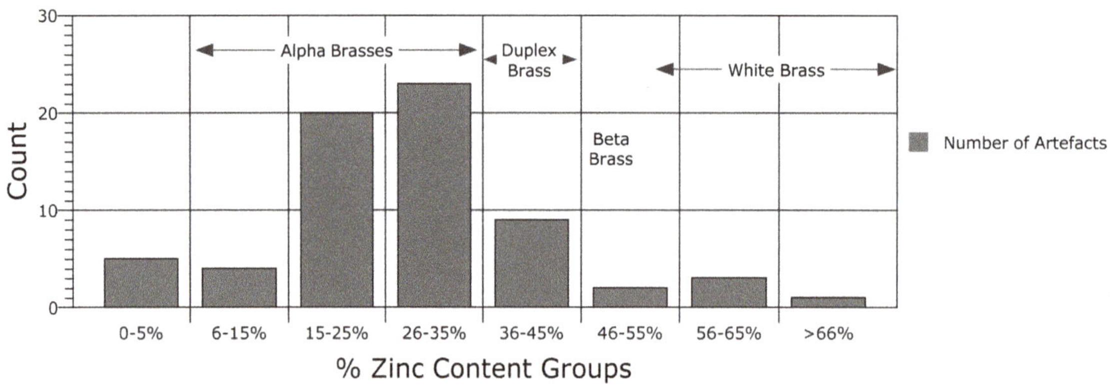

Figure 7.11 Histogram of zinc contents with details of microstructure

The remaining five artefacts (7.5%) were white brass, ie containing more than 50% zinc which in the present day is considered too brittle for use. Compositions containing up to 35% Zn and the microstructure comprises alpha grains. This means that the alloy is malleable and can be cold worked. The addition of low levels of lead will cause the formation of lead particles at the grain boundaries. The higher the lead content the greater its detrimental effect on worked sheet or wire. It may be significant that the highest average lead content occurs in the cast artefacts, ie for cast objects such as the fleur-de-lys shaped artefact, (excluding the ingots), which will not be subjected to further working: lead is then beneficial in the casting process and in any machining required to remove flash etc from the cast. The lead content of the identifiable objects ranges from 1.4–5.3%, with a mean value of 3.2% Pb. In contrast, the bars and lumps ranged from 1.3–2.9% Pb with a mean of 2.0% Pb, with one outlier of 4.2% Pb, the mean then rises to 2.6% if the higher value is included.

Brasses containing 25% Zn appear very golden in colour and could be used as a decorative substitute for gold. It is noted that two of the cast objects (context 1245, the round headed stopper, and perhaps most interestingly 1406 'fleur-de-lys') have close to that composition and may have been produced for the colour. The higher zinc content brasses (35–45%Zn) are known as duplex brasses or beta brasses and the final microstructure is grains of alpha with an additional hard beta/phase. This means that they would have to be hot-worked. Thus any sheet or wire between 35–45% Zn were almost certainly hot-worked. There were four sheet samples having a zinc content of the order of 33–46% Zn, and seven wire samples with a range of 25–47% Zn.

The sheet group of artefacts includes specific objects, eg bowls, washers as well as off-cuts and scrap. Table 7.4 presents the average elemental values for the major types of artefact encompassed within the sheet group. The washers have an average zinc content lower than the other types (13.5% Zn compare to the overall average of 26% Zn), but this is distorted by a copper washer (Context 1016, 99%Cu). With this removed the average zinc content increase to 20% Zn. This would be more ductile than a higher zinc content.

TABLE 7.4 AVERAGE ELEMENTAL VALUES FOR THE DIFFERENT ARTEFACTS IN THE SHEET GROUP (N= NUMBER IN EACH GROUP)

	n	Cu	Zn	As	Sn	Sb	Pb
Strip	14	69.3	28	0.1	0.3	0	2.2
Disc	12	72.2	26.6	0.1	0.1	0	1.1
Offcut	9	74.6	24.2	0	0	0	1.1
Asher	3	84.3	13.5	0.4	0	0	1.8
All 'sheet'	41	72.5	25.7	0.1	0.2	0.0	1.6

Modern brass wire ranges in Zn content from 9–40%, with most wires being manufactured from 35–40% Zn. Those compositions at the lower end of this spectrum 35–37% Zn are more ductile and suitable for severe cold working. There are some wire compositions alloyed with low contents of lead (up to 3.5%Pb), but the presence of lead can lead to tearing as it forms soft particles at the grain boundaries. If the wire with the lowest Zn content (1406, wire 2, 12% Zn) is removed from the data, the average zinc content of the wire rises from 35% to 37% Zn.

X-RAY FLUORESCENCE ANALYSIS OF THE CRUCIBLE FRAGMENTS

Introduction

To produce a wrought iron product, all copper alloy must be melted in a crucible poured into a mould and then either cleaned up as a cast product or hot- and/or cold-worked to produce a wrought product. Hence the crucible is the starting point of all non-ferrous manufacturing processes. All used crucible fragments will generate a signal of the metals used in the alloy. However zinc, the second alloying element of brass presents particular problems because the volatilisation temperature of zinc is 907°C, and the melting point of a 30% Zn brass is *c* 950°C. Hence, all crucible bodies will be heavily saturated with zinc. It is therefore unlikely that an exact answer can be obtained determining the composition of alloys cast in a crucible. However detection of additional alloying elements eg Sn, Pb and Sb would indicate whether a range of alloys was produced or whether they were consistently producing one alloy range in terms of minor alloying elements.

The Assessment Report (McDonnell 2009) recommended a programme of non-destructive X-Ray Fluorescence Analysis (XRF) of a selection of crucible fragments to address the following research questions:

- Was there evidence of elements other than zinc used in the alloying processes?
- Was the crucible XRF data comparable to the metal analyses?

Results

Thirteen crucible fragments comprising ten wall fragments and three base fragments were analysed. The analyses of the interior surfaces (Table A3.3.8) are dominated by the presence of iron from the ceramic fabric, copper and zinc from the melting of the alloy. In addition lead is present on all crucibles, with one crucible (Context 1255, Wall 3) having a high lead content (17% Pb). Tin is present in six fragments with one (Context 1255 Wall 2) having a very high value (16% Sn). Antimony is present at minor/trace levels in three crucible fragments. Table A3.3.9 presents the interior surface analysis data ordered by copper content: this shows an inverse relationship between copper and zinc content. This is to be expected because the copper and zinc content account for an average of 83% of the total, but some exceptions would be expected.

TABLE 7.5 COMPARISON OF THE AVERAGE ELEMENTAL VALUES FOR THE BASE AND WALL COMPONENTS OF CRUCIBLES

	Fe	Cu	Zn	Sn	Sb	Pb
Base all	25.9	37.8	27.8	1.3	1.1	3.4
Wall all	19.3	31.1	41.8	1.6	0.8	3.2
Base External	42.9	10.8	35.6	1.6	2.1	2.9
Base Internal	8.9	64.7	19.9	1.0	0.1	4.0
Wall External	28.6	31.1	41.8	1.6	0.8	3.2
wall Internal	9.9	38.3	43.7	2.0	0.2	3.9

TABLE 7.6 AVERAGE COMPOSITIONS OF METALWORK FROM CAMBRIDGE STREET

	Cu	Zn	As	Sn	Sb	Pb
Cast	66.2	30.4	0	0.2	0	3.1
Sheet	72.5	25.7	0.1	0.2	0	1.6
Wire	66.1	35.3	0	0	0	1.2
Overall	70.1	27.9	0.1	0.1	0	1.8

The exterior surface analyses (Table A3.3.10) are surprising for the extent of metal penetration with the maximum copper content of 61% and zinc of 77%. Lead is present as in the interior surface but both tin and antimony are more common than in the interior surface analyses. This is against expectations. There is a slight inverse trend of copper concentrations on the inner and outer surfaces, ie as the copper content on the internal face increased, the copper content on the external face decreased. This would suggest that the copper on the external face derived from contact with metal outside the crucibles, eg spills etc, rather than the melted alloy inside the crucible. There is no clear relationship between the zinc content on the external and internal faces. This is due to the volatility of zinc; in some cases there was higher zinc content on the external surface compared to the internal surface of the same sherd. The differences in copper and zinc content between the surfaces are presented in Table 7.6. Table A3.3.12 compares the average values obtained from the wall sherds and the base fragments. The internal surface of the base fragment is dominated by copper (65%), which is to be expected, as the molten alloy rich in copper rested on the crucible base.

Discussion

The XRF analyses of the crucibles confirm that they were used for melting brass. Examination of crucibles from earlier periods, suggests that the exterior surface is lower in metallic elements than that of the interior surfaces of the crucibles that were in direct contact with the molten alloy. This is not the case with some of the crucible sherds from Cambridge Street, where examples of high levels of metals on exterior surfaces has been recorded. This is due to two possible causes. Firstly that the exterior surface had been exposed to spillage and molten metal had flowed down the outer surface of the vessel. The second explanation is that the large scale of working caused the working areas to be saturated in the metals; in particular, the volatile zinc absorbed the metals used in the works.

The average compositions of the metalwork from Cambridge Street is provided in Table 7.6, showing that all metalwork is leaded brass with a minor level of tin and no significant antimony levels. The crucible data also shows the presence of tin in the alloys melted in the crucibles, and the antimony level in the crucibles is elevated. However, the antimony derived from the exterior surface analysis, in particular the bases, rather than the inner surface. This would indicate that the antimony detected derives from another source rather than from the alloy compositions.

X-RAY FLORESCENCE OF THE SLAGS

Introduction

The Assessment Report (McDonnell 2009) recommended a programme of non-destructive X-Ray Fluorescence Analysis (XRF) of a selection of slag samples to address the following research questions.

- To assess presence of non-ferrous metal signatures in the slags.
- To assess whether there is evidence to indicate whether the slags derived from copper alloy or ironworking or both.
- To determine the nature of the white slags.

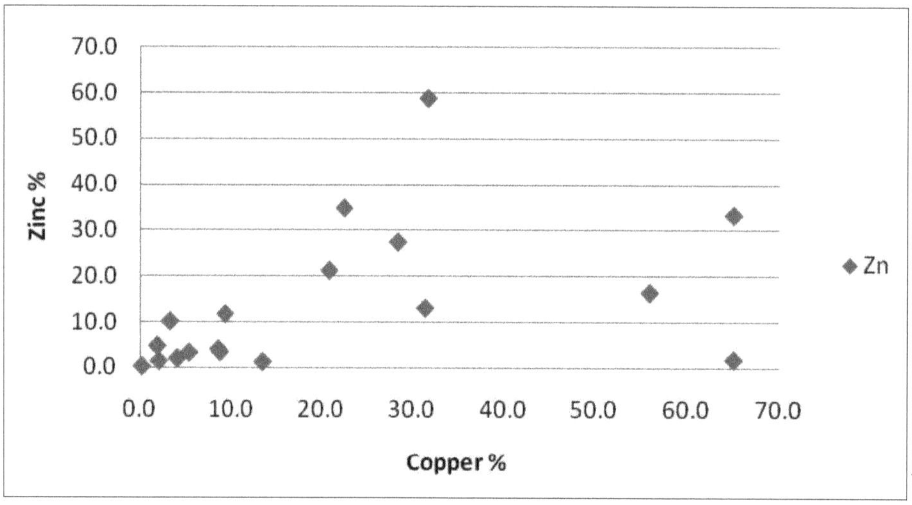

FIGURE 7.12 PLOT OF ZINC CONTENT AGAINST COPPER CONTENT FOR THE SMITHING SLAGS

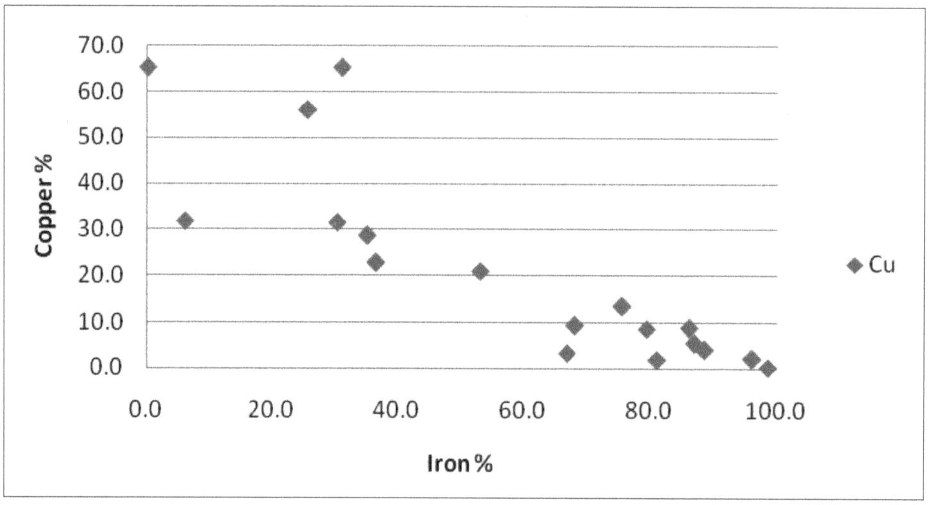

FIGURE 7.13 PLOT OF COPPER PERCENT AGAINST IRON PERCENT

Results

The slags were divided into five broad groups;

Group 1 slags considered typical ironworking smithing slags, including hearth bottoms, displaying no unusual features.

Group 2 typical smithing slags or hearth bottoms but characterised by the presence of copper corrosion products

Group 3 black coloured slags

Group 4 slagged stone or clay structures

Group 5 white coloured slags

Group 1 - smithing slags

The results of the XRF analysis of 18 smithing slag samples are shown in Table A3.3.12 and show that (despite lacking surface evidence of copper alloy) the slags contained significant amounts of copper and zinc with minor levels of lead and tin. Only one sample, identified as a high iron bearing slag (SF26, Sample 9), contained no significant copper alloy content.

Fig. 7.14 is a plot of zinc content against copper content for the smithing slag, and shows no strong trend. Fig. 7.15 shows a negative correlation between iron content

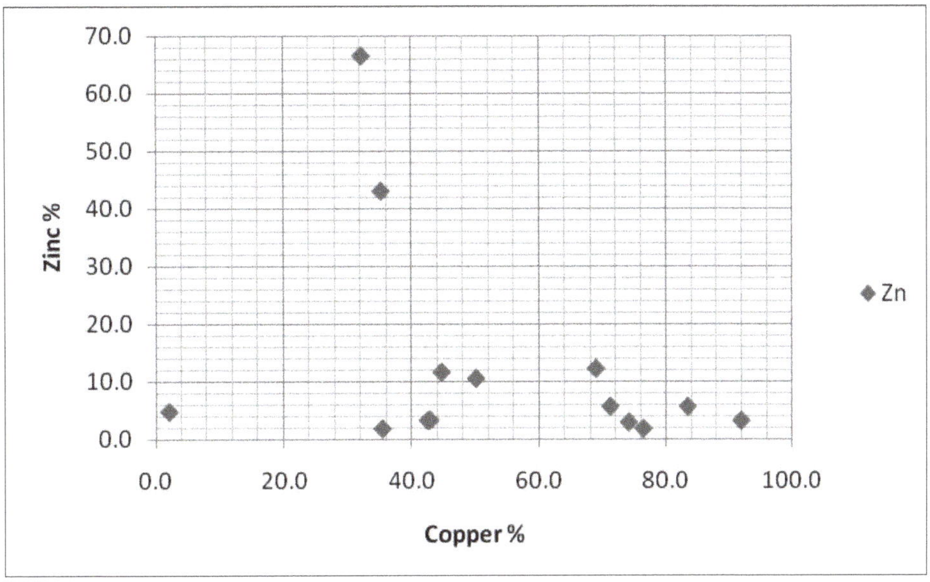

FIGURE 7.14 PLOT OF ZINC CONTENT AGAINST COPPER FOR THE COPPER STAINED SLAGS

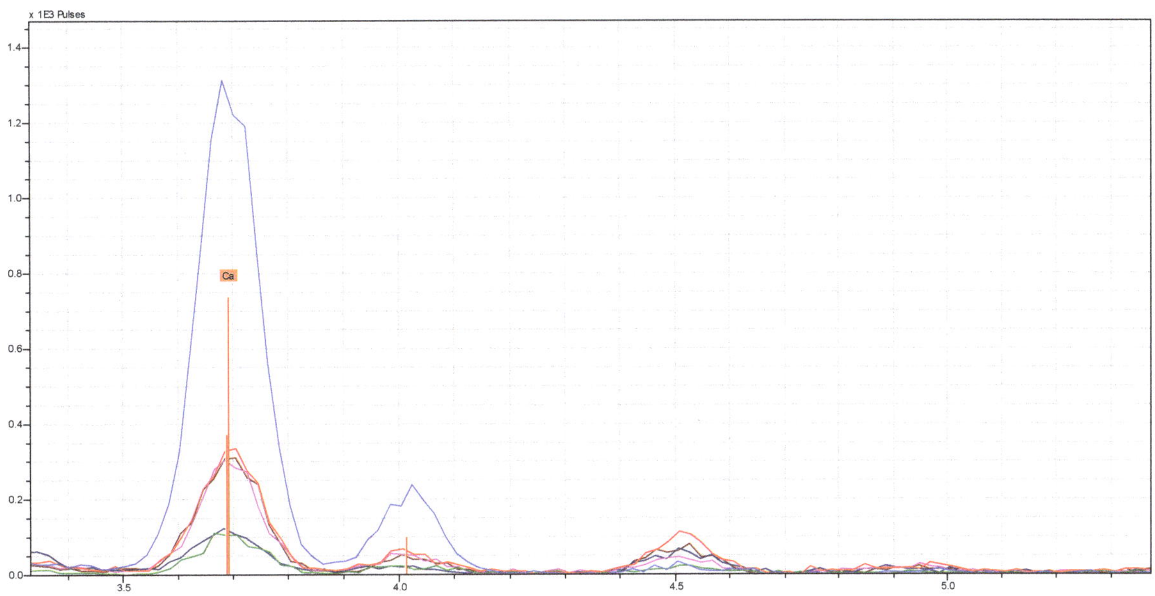

FIGURE 7.15 THE CALCIUM REGION OF THE SPECTRA SHOWING ALL SPECTRA FROM THE HEARTH LINING, THE HIGHEST PEAK (BLUE) DERIVES FROM HEARTH LINING FRAGMENT FROM CONTEXT 1369.

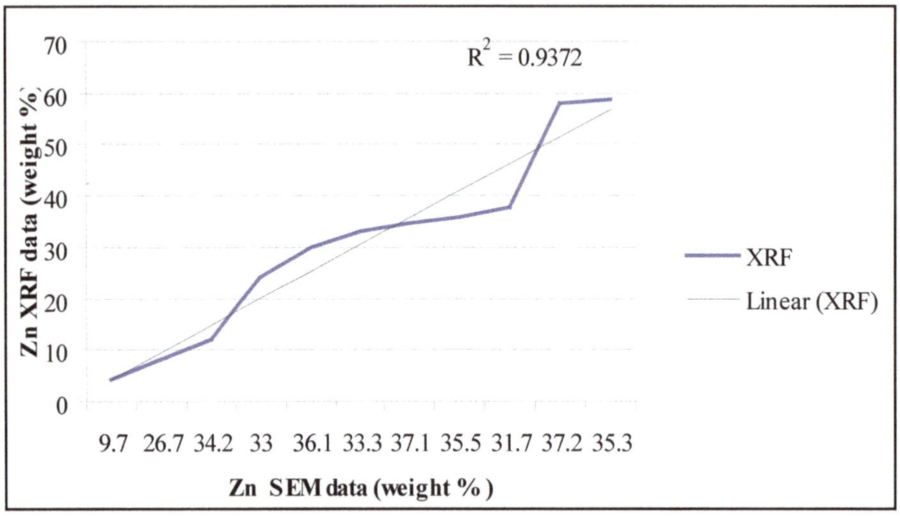

FIGURE 7.16 COMPARISON OF SEM AND XRF DATA FOR COPPER

and copper content, which is to be expected as these two elements dominate the composition. These data confirm that although the slags show no exterior evidence of copper alloy content the slags were derived from working in a copper alloy environment and not exclusively for iron.

Group 2 - smithing slag displaying copper alloy corrosion

Table A3.3.13 presents the XRF analyses of 14 samples of smithing slags and hearth bottoms displaying copper corrosion products. This shows that the average copper content is *c* 54% compared to 21% in the smithing slag group. Only two samples had copper contents below 50%, of which one (Context 1411 SFN 91) had a very low non-ferrous alloying content with copper being *c* 2% and zinc at 5%. Lead is present in nine (64%) samples and tin in eight samples (57%). Despite the higher copper content of these slags the average zinc content is approximately the same as the Group 1 slags, (Table A3.3.12). This is due to the volatilisation of the zinc and the ability of the slag to absorb this element. Hence there is no correlation between the copper and zinc contents in these slags (Fig. 7.16).

Group 3 - black coloured slags

This group of slags were smithing slags with a distinct black colour. The XRF analysis of two samples (Table A3.3.14) show no distinct compositional differences between these slags and those in Groups 1 and 2.

Group 4 - slagged stone or clay structures

One sample of slagged stonework and one of slagged clay structure were analysed (Table A3.3.15), the results demonstrate that these pieces were use as part of copper alloy working hearths.

Group 5 – white coloured slags

These slags form a small but distinct group. The XRF analysis of four samples shows wide variation in composition (Table A3.3.16). The element cadmium (Cd) was included in this data as it occurs at elevated levels in these slags, and due to health and safety reasons these slags should be handled, or disposed of, with even more caution than the others. It was expected that the white colouration would be due to high zinc levels and the formation of zinc

TABLE 7.7 AVERAGE ELEMENTAL COMPOSITIONS OF THE DIFFERENT SLAG TYPES

Slag Group	Description	Fe	Cu	Zn	Sn	Sb	Pb
1	Smithing slag	58.2	21.1	13.8	0.8	1.2	1.4
2	Copper rich slags	29.7	53.7	12.5	0.9	0.0	2.9
3	Black slags	38.4	41.1	15.5	2.2	0.3	1.7
4	Slagged lining	41.0	26.8	26.1	0.6	1.1	2.8
5	White slags	56.1	3.8	25.4	0.0	0.8	0.7
Overall	average	44.7	29.3	18.7	0.9	0.7	1.9
	SD	12.2	19.1	6.6	0.8	0.5	0.9
	Min	29.7	3.8	12.5	0.0	0.0	0.7
	Max	58.2	53.7	26.1	2.2	1.2	2.9

oxide. This white phenomenon is commonly observed on crucibles that have been used for experimental melting of brasses. Although the average zinc content is higher than other slags (see Table 7.7), this due to one sample containing 68% Zn. It is probable that the slags are white due to the surface enrichment of zinc. This will be checked by sectioning a white slag sample (see below).

Discussion

Table 7.7 presents the average elemental compositions of the different slag types. These data confirm that all the slags contain non-ferrous elements derived from copper alloy working. Groups 1 and 2 are morphologically typical of iron smithing slags, in particular Group 2 contains two examples of hearth bottoms, plano-convex accumulations of slag formed in the base of the smithing hearth, (Table A3.3.13, Context 1323). These slags are believed to form during blacksmithing of iron. The exact mechanism of formation is not understood but is thought to be due to reactions between fuel ash, oxidised iron and fluxes used in smithing. The same factors would be present in the hot working of copper alloy, ie copper, zinc, lead and tin oxides, fuel ash and possibly fluxes. Hence it is not unreasonable to expect the formation of copper alloy smithing slags. However they all contain iron, which must play some part in slag formation. It is therefore probable that the hearths were used for both the hot working of iron and copper alloy leading to the formation of smithing slags saturated with non-ferrous metal oxides. Metallographic analysis will clarify the slag composition.

Both the black coloured and white coloured slags were expected to display significant differences in composition to account for their distinct colouration. However the data does not show any such differences, suggesting that the colour variation is due to minor variations in the few microns of the upper surface of the slags, less than the depth penetration of the x-rays, (probably 10–15 microns). The slags attached to hearth structural material are also derived from copper alloy working.

Conclusions

The XRF analyses of the slags demonstrate that all slags were associated with non-ferrous metalworking. If the slags derived solely from ironworking, and they were contaminated by the zinc vapour prevalent in the works due to the casting of brass, then only zinc would have been detected. These data indicate that the slags derive from the hot-working of brass and iron. Archaeo-metallurgical analysis of specific samples will enable some resolution of whether the slags are predominantly iron working slags with some copper alloy present, or are non-ferrous in origin.

X-RAY FLOURESCENCE ANALYSIS OF SLAGGED LINING

Introduction

The Assessment Report (McDonnell 2009) recommended a programme of non-destructive X-Ray Fluorescence Analysis (XRF) of a selection of hearth lining samples to address the following research questions.

- To assess presence of non-ferrous metal signatures in the hearth lining
- To assess whether there is evidence to indicate whether the hearth lining derived from copper alloy or ironworking or both.

Results

The results are given in Table A3.3.17 and show that Context 1018 SFN 4 Sample 2 stands out as having high levels of the non-ferrous metals, with the exception of tin. The samples analysed from Context 1406 area broadly consistent in non-ferrous levels, with a high zinc signature, (higher in one case than the levels detected in the sample from Context 1018). The levels of copper and lead are lower than those detected in Context 1018.

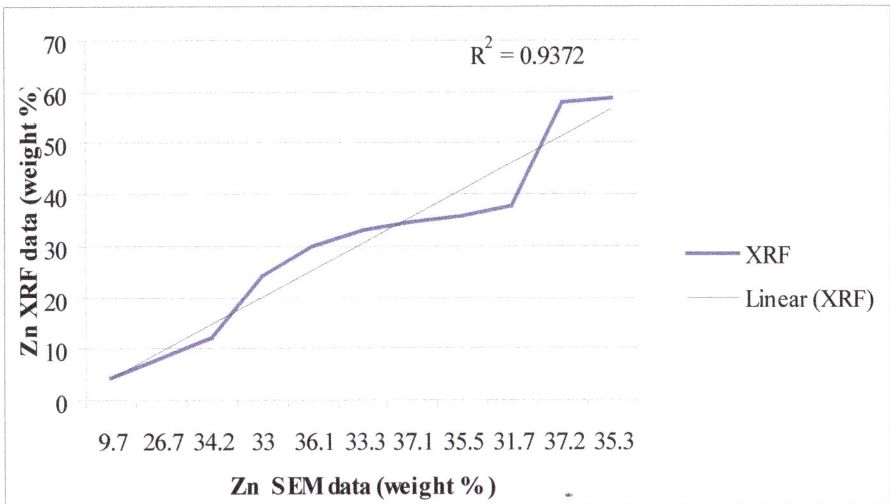

FIGURE 7.17 COMPARISON OF SEM AND XRF DATA FOR ZINC

The samples from Context 1369, SF91 are also consistent with high iron levels and are consequently low in the non-ferrous levels, lead being absent in all three samples. Zinc was present, but at very low levels compared to the other hearth lining samples. Examination of the raw XRF spectra which include peaks of elements not quantified shows that hearth lining from Context 1369 was higher in calcium than the others (Fig. 7.17).

Discussion

The samples from the three contexts are internally consistent in their non-ferrous signature (with the exception of Context 1018 for which there was only one sample), but differ between the contexts.

The overall compositions of the three main types of metal products (cast, sheet, wire) are shown in Table A3.3.18 along with ratios of Cu/Zn, Cu/Pb and Pb/Zn. Similar calculations are shown for the slagged lining in Table A3.3.19. The only figure that is similar in both data sets is the Zn/Pb ratio for the Zn/Pb for the cast artefacts and for the lining from Context 1018. The lining from Context 1369 is distinct and derives from the earliest phase of the works (1820–1840). The XRF data would suggest that these slagged lining fragments were not associated with brassworking.

METALLOGRAPHIC ANALYSES OF COPPER ALLOY ARTEFACTS

Introduction

The report on the XRF of the copper alloy identified a number of artefacts that would benefit from metallographic examination and SEM analysis. The aims of the analysis were:

- To reveal the method of manufacture, eg as-cast, annealed, hot and cold worked.
- To obtain bulk area and phase quantitative data
- To examine the distribution of minor elements.
- Compare SEM data with XRF data
- Assess the suitability of the alloy for its function

Summary Results

A total of ten items were subject to further analysis. The metallographic analyses are ordered into as-cast artefacts, sheet artefacts, and wire artefacts. The results of the analysis are presented in detail in Appendix 3.5 with relevant data tables. The following represents a summary of the analysed items.

SFN 58 Bar Ingot

The metallographic analyses demonstrated that the bar had been subject to some hot working and was not in the as-cast condition. It is probable that it had undergone some hot rolling, possibly with the intention of reducing it strip or wire. The XRF analyses and the SEM analyses are very similar (Appendix 3.5; Table A3.5.1–3) and place the alloy at the high end of alpha brasses. The SEM analyses showed the presence of silica. Analysis of other samples showed that these are sand particles presumably derived from the sand moulds used in casting. This ingot is cast from a high zinc alloy which would allow for subsequent zinc loss as it was worked to a final product eg sheet or wire.

SFN 1245 Bar 1

The bar had clear inter-dendritic porosity concentrated in the centre of the bar, with some non-metallic inclusions present throughout the microstructure. These were distinct dendritic microstructures, possibly due to the lead, alpha dendrites in solid solution with lead droplets, and porosity. Under normal casting conditions the microstructure should have been alpha grains, however rapid cooling may have caused the formation of the dendrites. The XRF data and the SEM data, when compared (Appendix 3.5; Table A3.5.4–6), suggest the SEM data give a higher zinc value than the XRF, which can be explained by de-zincification during corrosion. The bar was in the as-cast condition and again is high in zinc, which would allow for zinc loss during further hot working. The as-cast dendritic microstructure was retained due to the lead, porosity and sand particles and may indicate rapid cooling (quenching) from high temperature.

SFN 1245 Bar 2

Etching revealed an as-cast microstructure of alpha dendrites with the fine lead droplets dispersed throughout the microstructure. The structure shows coarse grains with dendritic microstructure within these grains, possibly due to rapid cooling. The XRF severely underestimated the zinc content, presumably due to severe de-zincification during the burial and corrosion of the artefact. The as-cast microstructure was preserved by the presence of the lead droplets, porosity and sand particles.

SFN 1245 Tube

In the unetched condition the alloy contained numerous non-metallic inclusions and some fine porosity. Etching showed that the tube was in the as-cast condition with dendrites growing in from the inner and outer surfaces. The XRF normalised data and the SEM data (Appendix 3.5; Table A3.5.12–4) show that in contrast to the analyses of the previous artefacts the XRF over-estimated the zinc content, and at *c* 32% Zn the microstructure should comprise alpha grains, with finely dispersed lead particles. The retention of a dendritic microstructure is due to the presence of the lead droplets, the sand particles and porosity.

SFN 1324 Strip 1

Two cross-sections and a longitudinal section were cut from the artefact; all three sections displayed the same microstructures. In the unetched condition the sections showed evidence of inter-dendritic corrosion. When etched the sections displayed fine alpha dendrites with solid solution, plus lead droplets at the inter-dendritic boundaries. The XRF data and the SEM data (Appendix 3.5; Table A3.5.16–18) demonstrate that the alloy is low in zinc with a low tin content. The values of zinc and lead obtained by the different techniques do differ, but this reflects corrosion mechanisms. Analysis of a dendrite and interdendritic phase (Appendix 3.5; Table A3.5.19) indicates that the residual dendritic microstructure is formed by the lead droplets, the porosity and the sand inclusions. The strip is in the as-cast condition, which is unexpected, as it would have been cold-worked and annealed or hot-worked, which should have removed the as-cast microstructures and evened out any compositional differences though diffusion.

To investigate the effect of corrosion phenomena on archaeological and historical brasses an area of corrosion noted in the unetched micrograph was investigated by SEM. The un-corroded metal shows the low zinc content (9%), in contrast the corroded metal is in oxide form and the zinc has been lost. The area analysis of the corroded area is characterised by the high sulphur content and low zinc values. The sulphur is associated with the lead rich phase, where it is occurring as lead sulphide. It is also clear how the lead is retained in the corroded zone and forms a larger volume of the corrosion, hence it is enhanced in the XRF analyses.

SFN 1406 Disc 9

When etched the microstructure showed equi-axed alpha grains with twins, and beta prime at the grain boundaries. Within the micro-structure there were some lighter coloured phases, indicative of either light corrosion or non-alloyed copper and SEM analysis confirms this as copper. The XRF data and the SEM data (Table 3.5.17) show close correspondence. The alloy contains *c* 35% zinc which should result in a final microstructure of alpha grains. However the SEM study confirms that the grain boundary is beta prime. The microstructure in equilibrium conditions should just comprise alpha grains. The manufacturing process induces the formation of the beta prime phase, perhaps by heating to above 250°C and quenching. The presence of twins is expected as they result from cold working and annealing or hot working.

SFN 1406 Sheet offcut 3

Three sections, one flat, one longitudinal, and a cross-section were cut from the sample. In the etched condition the longitudinal section displayed elongated alpha grains with finely dispersed lead particles. The XRF data and the SEM data (Table 3.5.20) show that the XRF analysis reflected the corrosion, with the lower zinc value and enhanced lead. The 'cleaned surface' still clearly retains some corrosion as indicated by the lead content. This is confirmed by the iron values obtained from the XRF analysis, which shows a high value present in the corroded surface, but still present on the 'cleaned surface'. Significantly the SEM analysis did not detect silica, which has been present in the other specimens especially those displaying the remnant dendritic microstructures. This sample is rich in zinc indicating little had been lost during the working process, and displayed the expected microstructure of distorted alpha grains with finely dispersed lead droplets.

SFN 1406 strip and wire

This artefact was a strip that had been drawn to wire, so that one end was classed as strip and the other as wire. The metallography should reflect these differences. The XRF analyses show marked differences between the two ends (Table 3.5.21), which is unlikely but which may represent differences in corrosion.

In the unetched condition the strip and wire ends were very clean with very few non-metallic inclusions present. When etched the strip microstructure comprises grains of alpha with grain boundary beta prime. There is no evidence for distortion or twins. When etched the wire microstructure comprises grains of alpha with grain boundary beta prime, with finely distributed lead particles. The grain boundary phase is concentrated in the centre of the wire, but is not distorted.

The XRF data (Table 3.5.21) shows a difference between the two ends, with a very high zinc value for the strip end (58% Zn) but the SEM data show consistent values between the two ends, and comparison suggests that the wire end XRF data was more representative. An alloy with a zinc content of *c* 37% should result in a microstructure of alpha grains; hence the presence of grain boundary beta prime indicates deviation from the phase diagram, despite extensive working, especially of the wire end.

SFN 1406 Wire 5

In the etched condition the alpha grains are delineated by the inter-granular corrosion, which will have removed the lead droplets. The grains display heavily distorted features due to drawing process. The XRF data and the SEM data (Table 3.5.25) show that the XRF data radically under-estimated the zinc content. This is due to the thinness of the wire and the corrosion effect combining to produce this result.

SFN 58 Wire

When etched the microstructure showed alpha grains with some annealing twins present. The XRF data and the SEM data (Table 3.5.27) show that the XRF over-estimated the zinc content which is in contrast to SFN 1406 wire, which under-estimated the zinc content. The wire was cast annealed and hot worked.

Discussion

The metallography is summarised in Table 7.8. The most striking and unexpected result is the presence of remnant

TABLE 7.8 SUMMARY OF MICRO STRUCTURES

Metal Type	Context/SF	Description	Alpha Dendrites	Alpha Grains	Gb Beta Prime	Twins	Pb incls	Si incls?	Si%	Pb%	Si + Pb	other
Cast	SF58 sample 12	Ingot		y		y	Y	y	1.7	0.1	1.8	
Cast	1245	Ingot 1	y				y	y	0.9	0.9	1.8	
Cast	1245	Ingot 2	y				y	y	0.6	0.8	1.4	
Cast	1245	Tube	y				y	y	1.0	1.0	2.0	
Sheet	1324	Cu alloy strip 1	y				y	y	0.5	0.8	1.3	
Sheet	1406	Disc 9		y	y	y	y		0.3	0.2	0.5	
Sheet	1406	Offcut 3		y			y		0.2	0.0	0.2	fine grained
Sheet	1406	Strip flat end		y	y				0.1	0.2	0.3	
Wire	1406	Strip wire end		y	y		y	y	2.1	0.2	2.3	
Wire	1406	Wire 5		y				y	1.9	0.2	2.1	distortion
Wire	SF58	Wire		y				y	1.2	0.2	1.4	distortion

TABLE 7.9 COMPARISON OF XRF AND SEM DATA FOR THE THREE ARTEFACTS CLEANED BACK

	SF58 Bar			SFN 1406 Disc 9			SFN 1406 Offcut 3		
	XRF (corroded Surface)	XRF (Cleaned Surface)	SEM	XRF (Corroded Surface)	XRF (Cleaned Surface)	SEM	XRF (Corroded Surface)	XRF (Cleaned Surface)	SEM
Fe	1.2	0.1	0.1	2.72	0.2	0.3	3.8	0.1	0.1
Cu	66.2	69.5	66.4	65.7	64.0	63.7	68.4	64.5	63.2
Zn	32.9	29.9	33.3	33.9	35.7	35.5	30.0	34.3	36.1
Sn	0.0	0	0.0	0.0	0.0	0.1	0.0	0.0	0.2
Sb	0.0	0	0.0	0.0	0.0	0.1	0.0	0.0	0.1
Pb	1.2	0.6	0.1	0.4	0.3	0.3	1.6	1.1	0.1

as-cast dendritic microstructures. Brasses are noted for the absence of dendritic microstructures due to the high solubility of zinc in copper and the short freezing range between liquidus and solidus. This suggests that the alloys deviated significantly from equilibrium conditions, such as rapid quenching, on being cast. The presence of dendrites in the presumed as-cast objects, the ingots and the tube, can be explained but their presence in one of the sheet artefacts (SFN1324 Cu alloy Strip 1) is totally unexpected and difficult to explain. The production of sheet would involve a lot of hot-working/cold-working and annealing which in normal circumstances remove any dendritic microstructures.

The artefact SFN 58 Sample 12 was identified as an ingot/bar but displays a worked microstructure due to the presence of twins. This indicates that it has undergone some initial working, presumably to reduce the ingot to a smaller size. Two samples (SFN 1245 tube and SFN 1406 Disc 9) deviate from the equilibrium phase diagrams with dendritic/grain boundary beta prime being present. It is probable that the remnant dendritic microstructure is due in part to the presence of the sand and lead inclusions. Table 7.8 gives the silica content detected by the SEM analysis; it is given as elemental silica but would be present as SiO2. This shows that although those samples displaying the remnant dendritic microstructure do contain high Si+Pb values, it is not exclusive, eg artefact SFN 58 also contains high Si+Pb values but did not display the remnant structure. This may indicate that this sample had been subjected to heavy working. If the cause of the remnant dendritic microstructure is the presence of the silica inclusions, these must have derived from the casting sand moulds or contamination in the crucible. The porosity and inclusions indicate poor casting technology. This may be a result of the ramping up of production from proto-industrial levels to industrial level production, and a loss of quality control. The presence of the beta prime is evidence of high zinc values which suggest excessive use of zinc, again perhaps indicative of poor quality control.

The results obtained by XRF analyses of corroded surfaces was dependent on the extent of corrosion. Two major effects are influencing the results, firstly the corrosion mechanism (Campanella *et al* 2009) and secondly, the thickness of the corrosion layer. The results obtained from the samples studies by XRF and SEM provides a valuable opportunity to assess the effectiveness of the XRF method. Firstly, the data from the three samples was analysed in the as-received condition, ie corroded, then they were cleaned back to bright metal and re-analysed and subsequently analysed by SEM. All artefacts show higher lead and iron in the corroded surface compared to the cleaned surface XRF data and the SEM data (Table 7.9). This is illustrated by the SEM analysis of the corrosion layer of SFN 1324 Strip 1 discussed above. The data from SFN 58 bar is at odds with the others in that the zinc content of the corroded surface is higher than the cleaned surface and similar to the SEM data. In the case of SFN 1406 offcut 3 there is clear increasing zinc content from cored surface to cleaned surface to SEM data. It is similar in the case of SFN 1406 Disc 9, but the cleaned surface XRF data and the SEM data are similar values. This demonstrated that cleaning to bright metal gives good quantitative data of the bulk metal and can be monitored by the lead and iron levels.

Table 7.10 Comparison of XRF data and SEM data ordered by increasing XRF Cu%

Sample	XRF	SEM	Difference	% diff
SFN 58 wire	38.8	64.2	-25.4	65.5
1406 strip flat end	41.8	62.7	-20.9	49.9
SFN 1245 tube	58.9	67.2	-8.3	14.1
1406 Disc 9	64.0	63.7	0.3	0.4
1406 strip wire end	64.2	62.5	1.7	2.6
SFN 58 ingot	66.2	66.4	-0.2	0.3
1406 Offcut 3	68.5	63.2	5.3	7.7
SFN 1245 Bar 1	71.1	65.4	5.7	8.0
1406 Wire 5	85.9	65.3	20.6	24.0
1324 Cu alloy strip 1	86.1	87.8	-1.7	2.0
SFN 1245 Bar 2	89.8	72.2	17.6	19.6

Table 7.11 Comparison of XRF data and SEM data ordered by increasing XRF Zn%

Sample	XRF	SEM	Difference	% diff
1324 Cu alloy strip 1	4.4	9.7	-5.3	121.5
SFN 1245 Bar 2	8.1	26.7	-18.6	229.6
1406 Wire 5	12.1	34.2	-22.1	182.4
SFN 1245 Bar 1	24.1	33	-8.9	36.9
1406 Offcut 3	29.8	36.1	-6.3	21.3
SFN 58 ingot	32.9	33.3	-0.4	1.2
1406 strip wire end	34.6	37.1	-2.5	7.3
1406 Disc 9	35.7	35.5	0.2	0.6
SFN 1245 tube	37.7	31.7	6.0	15.9
1406 strip flat end	58.1	37.2	20.9	36.0
SFN 58 wire	58.9	35.3	23.6	40.1

Table 7.12 Comparison of XRF data and SEM data ordered by increasing absolute difference in SEM F Zn%

Sample	XRF	SEM	Difference	Absolute difference	% diff
1406 Disc 9	35.7	35.5	0.2	0.2	0.6
SFN 58 ingot	32.9	33.3	-0.4	0.4	1.2
1406 strip wire end	34.6	37.1	-2.5	2.5	7.3
1324 Cu alloy Strip 1	4.4	9.7	-5.3	5.3	121.5
SFN 1245 tube	37.7	31.7	6.0	6.0	15.9
1406 Offcut 3	29.8	36.1	-6.3	6.3	21.3
SFN 1245 Bar 1	24.1	33	-8.9	8.9	36.9
SFN 1245 Bar 2	8.1	26.7	-18.6	18.6	229.6
1406 strip flat end	58.1	37.2	20.9	20.9	36.0
1406 Wire 5	12.1	34.2	-22.1	22.1	182.4
SFN 58 wire	58.9	35.3	23.6	23.6	40.1

The XRF data and the SEM data for copper and zinc for all samples are compared in Tables 7.10 and 7.11. As both data sets are dependant, ie if the zinc content in an alloy increases the copper must decrease as they constitute greater than 95% of the alloy composition, then the zinc data is considered. It would be expected that the zinc content of the corroded surface would be lower than the data obtained from the SEM; hence the difference should be negative. However, seven out of the eleven artefacts show negative values, whereas four show higher zinc values in the corrosion compared to the SEM data. The differences are not consistent, varying from near 0 to over 20% difference. This means that no standard correction can be applied easily to the data to correct XRF surface data to a true bulk metal composition. This is illustrated by Table 7.12, which orders the artefacts by the absolute (ie ignoring whether the value is negative or positive), difference. The data show that there is no correlation with artefacts' type, for example it may expected that the wire artefacts may show greater variation than the ingot/bars, due to ratio of corrosion volume to artefacts volume. Seven of the artefacts show differences of less than 10%, again there is no correlation with artefact type.

Table 7.13 summarises the SEM data for all artefacts (counting SFN 1406 wire/strip only once), this shows that all the artefacts are brasses with an average content of 31% zinc, and minor levels of tin and lead, with the exception of SFN 1324 Strip 1 which is low in zinc (9.7%), and contains 1.4% tin. If this artefact data is excluded the average content is 34% Zn, with a standard deviation of 3%. This suggests that a standard alloy composition was in use. The lowest zinc content of this group was present in artefact SFN 1245 Bar 2, which contained 27% Zn. If this artefact is excluded, the average rises to 35% with a standard deviation of 2%. Table 7.14 ordered the artefacts by zinc content. If artefact SFN 1324 Strip 1 is excluded, only two artefacts (SFN 1245 Bar 2 and SFN 1245 tube),

TABLE 7.13 SUMMARY OF SEM DATA FOR ALL ARTEFACTS

		Ni	Cu	Zn	As	Sn	Sb	Pb
Cast objects	SFN 58 ingot	0.1	66.4	33.3	0.3	0.0	0.0	0.1
	SFN 1245 Bar 1	0.2	65.4	33.0	0.1	0.3	0.1	0.9
	SFN 1245 Bar 2	0.1	72.2	26.7	0.1	0.2	0.0	0.8
	SFN 1245 tube	0.0	67.2	31.7	0.2	0.2	0.0	0.8
Sheet objects	SFN 1324 Strip 1	0.1	87.8	9.7	0.1	1.4	0.0	1.0
	SFN 1406 Disc 9	0.1	63.7	35.5	0.3	0.1	0.1	0.3
	SFN 1406 Sheet offcut 3	0.1	63.2	36.1	0.2	0.2	0.1	0.1
Wire	SFN 1406 wire end	0.0	62.5	37.1	0.2	0.1	0.1	0.1
	SFN 1406 Wire 5	0.1	65.3	34.2	0.2	0.1	0.1	0.1
	SFN 58 wire	0.0	64.2	35.3	0.3	0.1	0.0	0.1
	Average	0.1	67.8	31.3	0.2	0.3	0.1	0.4
	Standard deviation	0.1	7.6	8.1	0.1	0.4	0.0	0.4
	Max	0.2	87.8	37.1	0.3	1.4	0.1	1.0
	Min	0.0	62.5	9.7	0.1	0.0	0.0	0.1

TABLE 7.14 SUMMARY OF SEM DATA FOR ALL ARTEFACTS, ORDERED BY ZINC CONTENT

		Ni	Cu	Zn	As	Sn	Sb	Pb
Sheet objects	SFN 1324 Strip 1	0.1	87.8	9.7	0.1	1.4	0.0	1.0
Cast objects	SFN 1245 Bar 2	0.1	72.2	26.7	0.1	0.2	0.0	0.8
Cast objects	SFN 1245 tube	0.0	67.2	31.7	0.2	0.2	0.0	0.8
Cast objects	SFN 1245 Bar 1	0.2	65.4	33.0	0.1	0.3	0.1	0.9
Cast objects	SFN 58 ingot	0.1	66.4	33.3	0.3	0.0	0.0	0.1
Wire	SFN 1406 Wire 5	0.1	65.3	34.2	0.2	0.1	0.1	0.1
Wire	SFN 58 wire	0.0	64.2	35.3	0.3	0.1	0.0	0.1
Sheet objects	SFN 1406 Disc 9	0.1	63.7	35.5	0.3	0.1	0.1	0.3
Sheet objects	SFN 1406 Sheet offcut 3	0.1	63.2	36.1	0.2	0.2	0.1	0.1
Wire	SFN 1406 wire end	0.0	62.5	37.1	0.2	0.1	0.1	0.1
	Average	0.1	67.8	31.3	0.2	0.3	0.1	0.4
	Standard deviation	0.1	7.6	8.1	0.1	0.4	0.0	0.4
	Max	0.2	87.8	37.1	0.3	1.4	0.1	1.0
	Min	0.0	62.5	9.7	0.1	0.0	0.0	0.1

fall below 32.5% Zn, which on cooling should form alpha grains once the temperature dropped below the liquidus temperature, but in fact due to the presence of the sand and lead particles contain a remnant dendritic microstructure. All the other artefacts would initially freeze as alpha dendrites surrounded by beta phase, increasing the likelihood of a remnant dendritic microstructure surviving, which could cause the formation of beta prime which is more brittle.

Why such high zinc content was in use is unclear. The theory is that by starting with high zinc content it would mean that subsequent zinc loss during hot working would not dramatically alter the properties of the alloy. However examination of Table 7.14 does not show this, as if this was the case, the wire artefacts which must have received the most working should contain the lowest zinc content, for example, SFN 1406 wire contains the highest zinc content. If zinc content had been lost during the production of the wire it would have had to start with a much higher zinc content and this is not supported by the data as no artefact shows higher zinc content and the sheet end of this piece contains the same zinc content.

Table 7.15 summarises both the compositional data and the metallographic data and clearly demonstrates that the lower zinc content artefacts contain the remnant dendritic structure, whereas the higher zinc content artefact are all granular. The explanation for this is not clear and in a sense is completely contrary to expectation, in that the higher zinc values, eg greater than 32.5%Zn, initially freeze as alpha dendrites surrounded by beta phase, whereas the

TABLE 7.15 SUMMARY OF SEM DATA AND METALLOGRAPHIC DATA FOR ALL THE ARTEFACTS

		Cu	Zn	Sn	Pb	Microstructure
Sheet objects	SFN 1324 Strip 1	87.8	9.7	1.4	1.0	Remnant dendrites
Cast objects	SFN 1245 Bar 2	72.2	26.7	0.2	0.8	Remnant dendrites
Cast objects	SFN 1245 tube	67.2	31.7	0.2	0.8	Remnant dendrites
Cast objects	SFN 1245 Bar 1	65.4	33.0	0.3	0.9	Remnant dendrites
Cast objects	SFN 58 ingot	66.4	33.3	0.0	0.1	alpha grains + twins
Wire	SFN 1406 Wire 5	65.3	34.2	0.1	0.1	alpha grains
Wire	SFN 58 wire	64.2	35.3	0.1	0.1	alpha grains
Sheet objects	SFN 1406 Disc 9	63.7	35.5	0.1	0.3	alpha grains + beta prime + twins
Sheet objects	SFN 1406 Sheet offcut 3	63.2	36.1	0.2	0.1	alpha grains
Wire	SFN 1406 wire end	62.5	37.1	0.1	0.1	alpha grains + beta prime
	Average	67.8	31.3	0.3	0.4	
	Standard deviation	7.6	8.1	0.4	0.4	
	Max	87.8	37.1	1.4	1.0	
	Min	62.5	9.7	0.0	0.1	

TABLE 7.16 AVERAGE VALUES DERIVED FROM XRF DATA (WEIGHT %)

	Cu	Zn	As	Sn	Sb	Pb
Cast	66.2	30.4	0	0.2	0	3.1
Sheet	72.5	25.7	0.1	0.2	0	1.6
Wire	66.1	35.3	0	0	0	1.2
Overall	70.1	27.9	0.1	0.1	0	1.8

lower zinc contents freeze as alpha grains. It is therefore possible that the high zinc values used in these alloys was a method devised to overcome complications caused by the formation of remnant dendritic microstructure. It is noticeable that although this structure survives in the exceptionally low zinc content, ternary alloy artefact. SFN 1314 Strip 1, none of the other strip or wire contained these remnant microstructures.

Table 7.16 gives the average compositions of the as-cast artefacts, the sheet and wire artefacts as determined by XRF. This shows that the alloys' average ranges between 25–35%, given that although there are no methods to correct the XRF data to true bulk alloy composition, in most cases the zinc content will rise. This would suggest that the majority of artefacts produced at the Cambridge Street works were manufactured using an alloy of *c* 30% zinc.

METALLOGRAPHIC ANALYSES OF THE SLAG SAMPLES

Introduction

The initial study of the slags identified five groups based on morphology. The XRF analysis showed that one group, the black slags, were a sub-set of either the smithing slags or the copper alloy smithing slags. One group, the white slags, was represented by only a few examples. Examples of the remaining three groups were selected for detailed archaeo-metallurgical analysis. The two major groups were identified as iron-smithing slags and copper-alloy smithing slags. The latter identification was based on morphological details such as copper alloy staining and the XRF analysis, which detected high copper values. However the former group that lacked these details may still have been used for copper-alloy working, but the copper-alloy inclusions were inside the slags rather than at the surface.

The aims of the analysis were to characterise the two main slag groups, the iron-smithing slags and the copper alloy 'smithing' slags, to assess whether they are fundamentally the same slag, and also whether, in the case of those containing copper, they were formed during a period of copper working, or whether they display radically different compositions indicating different genesis. A further aim was to assess whether a deposit of hearth-lining material dated to an early phase of the building was used for copper alloy working.

Summary Results

SFN 1109 36a

Morphologically the slag was atypical smithing slag lump, randomly shaped, black in colour and had an agglomerated texture, ie not flowed. The reflected light microscope study showed that the microstructure comprised fine crystal laths and very fine dendritic phase in a glassy matrix. The SEM examination confirmed the optical study and the bulk analyses showed that the slag is rich in alumina, silica and lime and relatively low in iron oxide compared to archaeological smithing slags. The material attached to the slag is sand (SiO2). The slag is predominantly aluminium lime silicate with some crystals of alumina iron oxide dendrites.

SFN 1411 SSL-Cin

Morphologically similar to SFN 1109 36 a, but considered to have a higher silica content due to its lighter colour. The microstructure is dominated by a silicate or glassy matrix. The bulk analysis shows the slag is dominated by alumina and silica with varying quantities of lime and iron oxide. Overall the slag was an aluminium silicate of varying composition particularly in the levels of lime and iron oxide.

SFN 1255 Cu-SSL

This sample had a varied microstructure ranging from long laths in a silicate matrix to areas with squat light/dark grey crystal in a glassy matrix, containing fine dendrites. Copper alloy prills are common throughout the microstructure, especially in the glassy/dendritic areas. The bulk analysis shows a composition dominated by alumina and silica with lime, and higher levels of iron oxide than were present in the ironworking slags. Copper was present at significant levels, with minor levels of tin, lead and zinc. The phase analyses show that one crystal phase is alumino-silicates solid solutions with varying quantities of lime or iron oxide present, and that others are hercynites (principal oxides alumina and iron oxide). The analysis of one of the oxides metal prills, and one of the large metallic inclusions, show that they are copper, occurring as either oxides or chlorides.

SFN 1324 Cu-SSL

The microstructure comprises fine dendrites in a silicate matrix with copper alloy prills present throughout the section. The bulk analyses show that the slag was dominated by alumina and silica with a significant iron oxide content. The lime content is low, although copper oxide and zinc oxide are present. The phase analyses show that the dendrites are hercynite ($2FeOAl2O3$) in an alumino silicate matrix. The analyses of metallic prills show that they are copper.

SFN 1329 Cu-SSL

The microstructure comprised silicate crystals in a glassy or silicate matrix, with numerous copper alloy prills, dispersed throughout the section. The bulk analyses show that five of the areas (Areas 1–5) are broadly similar and show that the slag is an alumino-silicate slag, with a reasonable level of iron oxide. The analyses distinguish them from other slags by the high zinc oxide content. The phase analyses confirm the presence of zinc oxide in all phases; the two crystalline phases are zinc rich, one being a silicate the other an alumina-iron oxide - zinc oxide mineral silicates. The analyses of the prill and associated corrosion products show that both copper and lead prills are present.

SFN 1369 Lining

Under reflected light microscopy, there was little discrimination of phases. There were no copper alloy prills or evidence of copper corrosion present in the sample. In the SEM, BSE imaging showed that the lining section showed a variable microstructure ranging from an amorphous uniform low Z number (grey) matrix with occasional high Z number (white) crystals to a multiple crystalline microstructure. The black phase is either charcoal or more probably coal, due to the presence of a sulphur peak in the spectrum which is obtained from a similar phase in another part of the slag. The bulk area analyses show that alumina and silica dominate the composition with a significant level of lime. The phase analyses show that the phases identified in the different areas analysed vary in composition, but all are rich in alumina.

Discussion

The two important slag types were morphologically identified as iron smithing slags and smithing slags that include a significant non-ferrous component, indicating they were used to work the copper alloys. The initial optical examination seemed to confirm the phases typical of ironworking slags, eg dendrites of iron oxide and silicate laths were present in the ironworking slags and copper prills were common in the copper working slags. However the SEM x-ray analysis dramatically demonstrated that the dendrites and laths were not typical of ironworking slags. The mean bulk analyses of the iron and copper working slags are given in Table 7.17 and show that all the slags are dominated by alumina and silica. Both of the ironworking slags are low in iron oxide. Two of the copper working slags have higher iron oxide than the ironworking slags. One of the copper working slags (SFN 1324) does not contain significant levels of non-ferrous oxides, but the presence of copper prills in the slag confirms it as a copper working slag. They are present in the other two copper working slags. Not all the copper working prills contain alloying elements, although lead prills were identified in one of them (SFN 1329).

Three of the slags contain dendrites that appeared white (high Z number) under BSE imaging on the SEM. Their composition (Table 7.18) showed that two were similar to hercynite, and the third was an iron rich form of hercynite. Four of the slags and the slagged lining sample contained lath-shaped phases (Table 7.19). In one of the slags (SFN 1329) the phase is fayalitic (*c* 30% $SiO2$, 70% FeO), the remainder are similar being alumina silicates with a significant lime content. The matrix analyses of the slags are given in Table 7.20 and show that the matrix of the slag is alumino-silicate with variable quantities of lime and iron oxide. Most significantly, the matrix in iron slag SFN 1109 36 a, is very similar to that of the slagged lining. This strongly suggests that the slags are one continuum. The overall slag composition is radically different to earlier archaeological iron working slags in which iron oxide and silica dominate to generate an overall fayalitic

TABLE 7.17 MEAN VALUES OF BULK AREA ANALYSES OF ALL SAMPLES

	Iron working slags		Copper working slags			Slagged lining	
	SFN 1109 36 a	SFN 1411	SFN 1255 Cu-SSL	SFN 1324	SFN 1329	SFN 1369	Mean of the slags
Na_2O	2.5	0.6	1.4	0.5	2.7	2.6	1.5
MgO	0.5	6.4	1.7	0.9	1.2	0.6	2.1
Al_2O_3	30.7	26.6	20.5	21.4	15.2	30.6	22.9
SiO_2	44.2	45.4	47.6	49.5	34.8	43.6	44.3
P_2O_5	0.5	0.1	0.4	0.2	0.2	0.5	0.3
SO_3	0.4	0.7	0.2	0.1	0.2	0.4	0.3
K_2O	0.6	1.2	1.8	2.3	1.0	0.6	1.4
CaO	14.4	3.5	6.2	3.0	4.1	14.4	6.2
TiO_2	0.8	1.0	1.1	1.0	0.6	0.9	0.9
V_2O_5	0.1	0.0	0.1	0.0	0.0	0.1	0.1
Cr_2O_3	0.0	0.1	0.0	0.0	0.0	0.0	0.0
MnO	0.0	0.6	0.2	0.1	0.1	0.1	0.2
FeO	4.7	13.9	16.6	19.3	11.0	5.2	13.1
CoO	0.0	0.0	0.1	0.1	0.0	0.1	0.1
NiO	0.1	0.0	0.0	0.1	0.1	0.1	0.1
CuO	0.1	0.0	1.2	0.5	4.9	0.1	1.4
ZnO	0.0	0.0	0.6	0.9	21.6	0.0	4.6
SnO_2	0.3	0.1	0.3	0.1	0.7	0.3	0.3
PbO	0.0	0.2	0.0	0.1	1.5	0.0	0.3

TABLE 7.18 PHASE ANALYSES OF HIGH Z DENDRITES

	Iron working slags		Copper working slags			Slagged lining		
	SFN 1109 36 a	SFN 1411	SFN 1255 Cu-SSL	SFN 1324	SFN 1329	SFN 1369		Classic hercynite
Na_2O	0.5		0.6	0.4				
MgO	2.1		4.5	2.3				
Al_2O_3	15.8		43.4	55.6				59.0
SiO_2	8.3		1.9	0.6				
P_2O_5	0.5		0.0	0.1				
SO_3	0.1		0.0	0.0				
K_2O	0.1		0.0	0.0				
CaO	1.9		0.2	0.1				
TiO_2	4.9		0.6	0.5				
V_2O_5	0.5		0.3	0.4				
Cr_2O_3	0.0		0.0	0.0				
MnO	0.2		0.2	0.0				
FeO	64.2		40.9	36.2				41.0
CoO	0.3		0.3	0.2				

NiO	0.2		0.3	0.0				
CuO	0.0		0.7	0.0				
ZnO	0.3		6.7	3.6				
SnO_2	0.0		0.0	0.0				
PbO	0.1		0.0	0.3				

TABLE 7.19 PHASE ANALYSES OF LATH PHASES

	Iron working slags		Copper working slags			Slagged lining
	SFN 1109 36 a	SFN 1411	SFN 1255 Cu-SSL	SFN 1324	SFN 1329	SFN 1369
Na_2O	1.0		2.4	0.5	4.9	1.2
MgO	0.1		0.0	0.6	0.9	0.0
Al_2O_3	31.5		26.9	15.4	1.3	30.6
SiO_2	49.7		54.4	57.1	26.6	46.8
P_2O_5	0.1		0.1	0.3	0.1	0.1
SO_3	0.1		0.3	0.0	0.1	2.7
K_2O	0.3		1.2	2.5	0.1	0.3
CaO	15.3		11.9	4.5	0.4	16.0
TiO_2	0.1		0.3	1.2	0.2	0.2
V_2O_5	0.0		0.0	0.0	0.0	0.2
Cr_2O_3	0.0		0.1	0.0	0.1	0.0
MnO	0.1		0.0	0.1	0.1	0.0
FeO	1.6		2.0	17.0	4.1	1.4
CoO	0.0		0.0	0.1	0.2	0.1
NiO	0.0		0.0	0.0	0.1	0.0
CuO	0.1		0.1	0.4	0.6	0.1
ZnO	0.1		0.0	0.2	59.9	0.3
SnO_2	0.5		0.2	0.2	0.2	0.2
PbO	0.0		0.1	0.2	0.1	0.0

TABLE 7.20 MATRIX COMPOSITIONS OF THE SLAGS

	Iron working slags		Copper working slags			Slagged lining
	SFN 1109 36 a	SFN 1411	SFN 1255 Cu-SSL	SFN 1324	SFN 1329	SFN 1369
Na_2O	0.9		1.3	0.4	1.8	0.8
MgO	0.4		2.4	0.6	0.7	0.0
Al_2O_3	33.6		19.0	15.8	8.4	34.4
SiO_2	49.5		48.8	56.6	42.0	46.1
P_2O_5	0.9		0.6	0.3	0.2	0.0
SO_3	0.1		0.1	0.1	0.0	0.3
K_2O	0.2		1.7	2.4	1.3	0.1
CaO	10.4		7.3	4.3	8.5	17.2
TiO_2	0.5		1.4	1.2	0.8	0.2
V_2O_5	0.0		0.0	0.0	0.0	0.0
Cr_2O_3	0.0		0.0	0.0	0.0	0.1
MnO	0.1		0.4	0.2	0.1	0.0

FeO	3.3		15.9	17.4	13.5	1.1
CoO	0.0		0.2	0.1	0.1	0.0
NiO	0.0		0.0	0.0	0.1	0.0
CuO	0.1		0.4	0.1	0.2	0.2
ZnO	0.0		0.5	0.3	18.9	0.0
SnO_2	0.1		0.5	0.3	1.0	0.2
PbO	0.0		0.0	0.0	2.5	0.0

composition. These data strongly suggest that the iron slag derived from slagging of the lining, caused by reactions involving coal as a fuel, and that it is not generated by iron working. Thus during the working of the forge slag starts to form by reaction between fuel, fuel ash, sand, and the clay lining of the hearth. As the forging process continued the slag flowed into the hearth and interacted with the copper alloy being worked in the hearth. Hence archaeology has recovered these three stages, the lining and slagged lining, the initial slag, formerly identified as ironworking slag, then the copper alloy working slag. This means that ironworking was not being carried out at any significant level in the Cambridge Street Works.

One slag stands out (SFN 1411). This was described as a smithing slag/cinder, its microstructure was very different - lacking any of the phases present in any of the other slags, and its phase analyses are significantly different in particular free iron oxide is present in the sample. This may derive from iron smithing, but its overall iron oxide content is very low.

The metallic prills in the slags are all copper, but zinc would volatilise off. It was, however, captured in slag SFN 1329, and lead prills were present in the same sample. This supports the theory that the slags derive from the hot working of the brass.

OVERALL CONCLUSIONS

Detailed analysis of the metal residues has been undertaken to examine the chemical composition of the metals produced. Brass is composed of copper, zinc (known as spelter) and a number of trace elements that include tin, cadmium and lead added to create various traits and effects.

A clear methodology for sampling the metallurgical residues, and metalwork was devised at the outset (see Appendices 1.2 and 3.1). Advice was sought as to the potential and best method of recovery, of the samples. Samples were taken, only where there was the best chance of confirming the historical record with regards to the rapid changes of technology during the period. The targeted samples (residues, crucibles and metalwork) were sent for detailed scientific analysis. Analysis included X-radiation fluorescence (XRF), metallographic examination and Scanning Electron Microscopy (SEM) analysis. The conclusions drawn from these analyses, not only provide a benchmark study for a 19th and 20th century brass works and rolling mill, but are also supplement the archaeological and historical records through clear scientific study.

The analysis of the artefacts and the conclusions drawn are based upon the assumption that they were produced on site and not elsewhere. Some of the items of metalwork may have been imported into site, although this is unlikely as the works was known to have been producing artefacts of this nature. There is no such uncertainty with regards to the residues, and crucibles which were stratigraphically secure. Specific conclusions can be drawn from each type of artefact, copper alloy metal, crucible fragments, slags and slagged linings.

The XRF analyses of the non-ferrous metal artefact show that the great majority were leaded tin brass. Other alloys were identified including a lead-antimony-tin alloy ingot, and high copper alloys. The brasses displayed a range of compositions. The effect of corrosion on the XRF spectra detected was examined by the analysis of three artefacts in the as-received condition, followed by removal of the corrosion layer, ie abrasive cleaning down to bright metal. This confirmed that the iron signal derives from the corrosion layer, and that de-zincification occurs during burial/corrosion. The data was inconclusive regarding the degree of de-zincification. Thus, no correction model can be applied to a surface analysis to produce a 'true' metal analysis. The 67 artefacts were divided into three broad groups, cast artefacts, sheet artefacts and wire. The analyses show a wide range of compositions used in the works.

The analyses of the artefacts included in the cast group suggest that the artefacts cast as finished objects contain a higher lead content than the cast ingots. This benefits the casting of the objects and machinability, ie finishing the artefact. High lead contents are detrimental to working brasses, eg to sheet and wire. The sheet artefacts have a wide range of compositions; there is an indication that the washers have a lower zinc content, giving them better malleability than a higher zinc content alloys. The wire artefacts had an overall higher zinc average than the other artefact groups. A significant proportion of the wires had zinc contents that resulted in a duplex microstructure containing the harder beta/phase, strongly indicating that the wire had to be drawn hot, or at least if cold drawn it had to be annealed. The data suggest some alloy manipulation for specific functions.

The metallographic analyses of the Cambridge Street brass artefacts demonstrated that there were three types of microstructure. The lower zinc alloys displayed a remnant dendritic structure, possibly retained due to the presence of lead droplets, sand inclusions and porosity. The microstructure of the second group consisted of alpha grains with finely dispersed lead droplets, some sand inclusions and in some cases the presence of annealing twins. The third group comprised alpha grains with grain boundary beta prime present; this structure deviated from the equilibrium phase diagram indicating rapid cooling. The data suggest that the brass was manufactured with a relatively high zinc content, *c* 30% Zn to avoid the occurrence of the remnant dendritic microstructure, which may have detracted from the malleable working properties of the alloys.

The analyses of the interior surfaces of the crucibles showed that they were dominated by the presence of iron from the ceramic fabric, copper and zinc from the melting of the alloy. In addition lead was present on all the crucibles. Tin was present in six fragments. Antimony was present at minor/trace levels in three crucible fragments. Analysis of the crucible fragments confirm that they were used for melting leaded brasses containing a minor level of tin. The XRF analyses of the crucible fragments confirm that the crucibles were used for melting leaded brasses containing a minor level of tin. This data accords with the analyses of the metalwork.

The XRF analyses of the slags demonstrate that all slags were associated with non-ferrous metalworking. If the slags derived solely from ironworking, and they were contaminated by the zinc vapour prevalent in the works due to the casting of brass, then only zinc would have been detected. These data indicate that the slags derive from the hot-working of brass and iron. The analysis of selected samples has demonstrated that the slag compositions from Cambridge Street are radically different from iron smithing slags encountered on earlier archaeological sites. The initial division of the slags into iron working slags and copper alloy working slags is not supported. The data suggest that the slags represent a continuum that progresses from hearth lining, to slagged lining to slag to copper alloy working slag, as the slag gets incorporated into the working hearth.

Archaeo-metallurgical analysis of specific samples enabled some resolution of whether the slags are predominantly iron working slags with some copper alloy present, or are non-ferrous in origin. The data confirmed that although the slags showed no exterior evidence of copper alloy content they were derived from working in a copper alloy environment and not exclusively for iron.

The XRF data would suggest that these slagged lining fragments were not associated with brassworking. In the case of the slagged lining from 1369, subsequent conclusions were confirmed, ie that it was possibly associated with a water tank overlain by later annealing muffles as opposed to elements of metal-working.

Overall the metallurgical analysis has confirmed that the site was producing a range of leaded brasses. Of these the wrought products, wire and sheet metal tended to be distributed in the vicinity of the rolling and wire drawing mill (Chapter 4). In contrast, crucible fragments and slags were generally associated with crucible furnaces within the bedstead works (Chapter 5), which were used for the hot-working of both iron and copper alloy leading to the formation of smithing slags saturated with non-ferrous metal oxides. The analyses of selected samples, has demonstrated that the slag compositions from Cambridge Street are radically different from iron smithing slags encountered on earlier archaeological sites.

CHAPTER 8: THE CAMBRIDGE STREET WORKS: CONTEXT AND COMPARISON

Chris Hewitson, Will Mitchell and Ray Shill

with Steve Litherland, Leonie Taibi and Christine Winter

To examine the Cambridge Street Works in isolation is to see only part of the story of the works. Too often archaeology is seen in the focus of the site, yet the diverse factors that both influenced and were influenced by the work that occurred on the site are ignored. A national research framework for industrial archaeology has been in place in recent years (Gwyn and Palmer 2005) and this has sought to diversify the scope of research away from prescriptive excavation and recording to a more holistic examination of industry in its setting. It has sought to examine multiple themes that include continuity and change; production and consumption; understanding the workplace; industrial settlement patterns; class, society and identity; social control, paternalism and philanthropy; and the international context of industrialisation (Palmer 2005, 16–17). More recently many of these themes have been placed in a specifically regional, West Midlands setting (Belford 2011, 211–236).

Not all of these themes can be identified in terms of physical archaeological remains, and the use of a multi-disciplinary approach is essential. Archaeology as a means of investigating past societies draws on strands of both below- and above-ground archaeology, using documents, memory, landscape and buildings as equally valuable sources (Belford 2011, 226). The difficulties faced by what is known as 'commercial archaeology' have been highlighted: *'...in the world of developer-funded archaeology it is sometimes difficult to think of a site in its full landscape context, let alone any other form of context such as historical, social or cultural one'* (*ibid*, 225). What follows in the following two chapters is an attempt to contextualise the site, that is, place it in a context of an industry and a city, on the regional, national and global stage, and also to examine the people that were influenced by the site owners, workers, and consumers.

THE GEOGRAPHY OF THE BIRMINGHAM BRASS INDUSTRY

What Manchester is in cotton, Bradford in wool, and Sheffield in steel, Birmingham is in brass; its articles of cabinet and general brass foundry are to be found in every part of the world (Aitken 1866, 229).

The geographic distribution of the metal trades has been the subject of several classic studies, the first by William Aitken in Samuel Timmins *The Midland Hardware District* (1866) followed by those of W H B Court *The Rise of the Midlands Industries* (1938), M J Wise in *Birmingham and its Regional Setting* (1950, 161–223) and Margaret Rowlands *Masters and Men* (1975). Examination of the workforce of the metals industries with reference to the brass industry was undertaken in the 1960s by James Vance in his study *Housing the Worker* (1967). However, no detailed industrial archaeological study has been undertaken of the Birmingham brass industry in the same way that can be said for that of Bristol (Day 1973) or of Swansea (Hughes 2000).

There are a number of reasons for this. The industry stretches between several unitary authorities, unlike the Swansea industry, which is concentrated in a single region. It is not as old as the Bristol industry and historic tendency has been to place greater emphasis on age over individual significance in archaeological terms. The industry is also of a very different scale to the large primary processing and secondary manufacturing industry in Wales and Bristol. Manufacturing has a tendency to proliferate in small workshops and medium-scale factories, widely distributed within a complex, varied industrial landscape. Cambridge Street, as one of the largest works of the mid 19th century, represents the upper end of this scale and as such provides an important case study in examining the brass manufacturing industry.

Studies of the manufactories have been conducted in Birmingham – the most important of these, without a doubt, being the architectural survey of the Jewellery Quarter by English Heritage (Cattell *et al* 2002). In addition, the extensive desk-based assessment of the Digbeth, Deritend, Bordesley area (Litherland 1995; Mould 1999), further desk-based assessment of the adjacent Eastside area (Driver 2005) and the amalgamated desk-based assessment from PPG16 (policy planning guidance 16) work as part of the Birmingham: Life, Work and Death Project (Forster and Rátkai 2008), have extended our knowledge of the industrial remains within the core of the city centre. The study of the remains of buildings around the Icknield Port Loop area in Springhill adjacent to the Edgbaston Reservoir (Driver and Hislop 2004) have provided tangible evidence for the later industry at the turn of the 19th to 20th centuries.

The areas that have not been addressed by these studies are those in the Broad Street fringes, extensively developed in the 1980s and 1990s, with subsequent loss of 19th century industrial architecture. The study along Sheepcote Street, conducted as part of redevelopment in 1999 and 2002, provides a directly comparable site to the remains of the Cambridge Street Works (Litherland and Winter 1999; Hewitson 2002; Conway and Litherland 2003). The results of this study are discussed in detail below.

The 18th-century brass industry

In the first half of the 18th century, manufacturing was concentrated within the medieval core of the town centre, defined by the area from Colmore Row in the northwest to Park Street in the southeast; and along the Digbeth, Deritend, Bordesley corridor. The expansion of the city in the 18th century resulted from the establishment of new estates around the medieval town by 1750 (see Fig. 8.1). The early expansion of the city in the 18th century was into new areas including the Weaman Estate and the Colmore Estate (see Fig. 8.2). This was subsequently followed by the Jennen's and Holte's Estates in the area northeast of the centre. The final area to be expanded in the 18th century was the area southwest and southeast of the centre, stretching from Broad Street to Digbeth and continuing onto the other side of the High Street (McKenna 2005a, 31–32). Specific zones of industry began to develop around the city with a complex of inter-linked industries; the Gun Quarter defined the former area of the Weaman Estate that became closely associated with the gun trade in the later 18th century; The Jewellery Quarter of the Colmore Estate, based around St. Paul's Church became renowned for small highly intricate manufacture, but was also a location for large factories associated with button and steel pen nib manufacture (Vance 1967, 95–127). The Weaman Estate was one of the first areas to be associated with the brass industry and was the location of Turner's

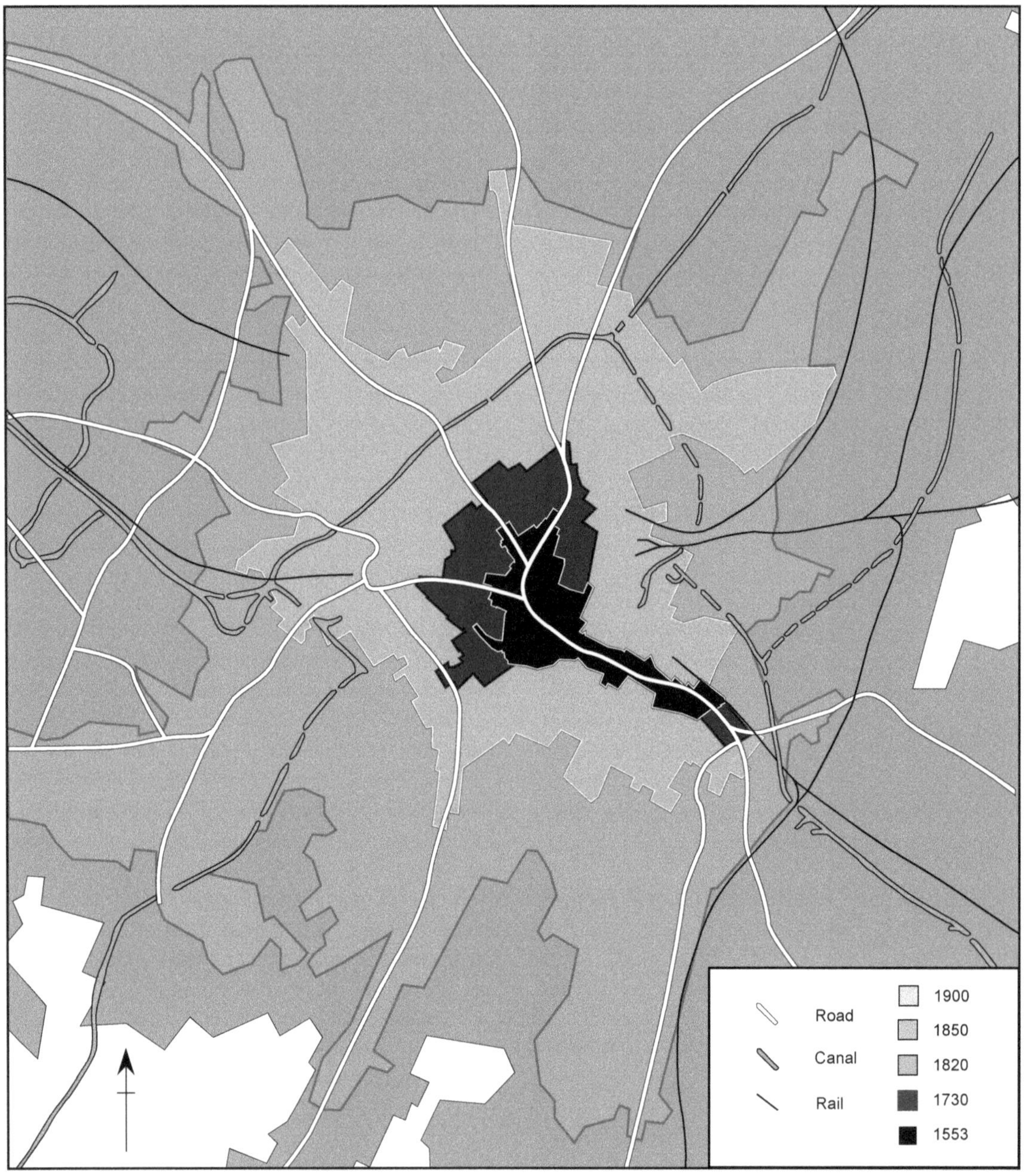

FIGURE 8.1 THE EXPANSION OF BIRMINGHAM, 1553–1900

Brass House in Coleshill, established *c* 1740. This was the first brass foundry in Birmingham. The brassworks were described in Angerstein's Illustrated Diary (Berg and Berg 2001) in some detail;

The brassworks lies the other side of town and belongs to Mr Turner and consists of nine furnaces with three built together in each of the three separate buildings. The furnaces are heated with mineral coal, of which 15 tons is used for each furnace, and melting lasting ten hours. Each furnace holds nine pots, 14 ines high and 9 ines diameter at the top. Each pot is charged with 41 pounds of copper and 50 pounds of calamine, mixed with [char]coal. ... The result of one charge was 75 pounds of brass with a value of £4.10s per cwt. The calamine comes from Derbyshire 40 miles from Birmingham and 12 miles outside Derby, but the copper is brought from Wales.

Suburban industry was a feature of Birmingham during the 17th and 18th centuries, due to a need for water-powered mills for grinding, slitting, rolling and wire drawing (Stephens 1964; Dilworth 1976). A number of locations of brass manufacture developed in the Birmingham hinterland on the tributaries of the rivers Tame and Rea. As metals were rarely differentiated in early records, it is often difficult to establish the specific purpose of a mill, but they were likely to have processed both ferrous and non-ferrous metals. Increasingly, as trades diversified in Birmingham throughout the 18th century, mills were converted to ever-increasing uses including for paper, metal-slitting, gun-barrel boring, thimble, wire and rolling mills along the Tame Valley river system (Fig. 8.3).

Initially in the 17th century, mills were adopted on the River Rea as blade mills but by the mid 18th century water

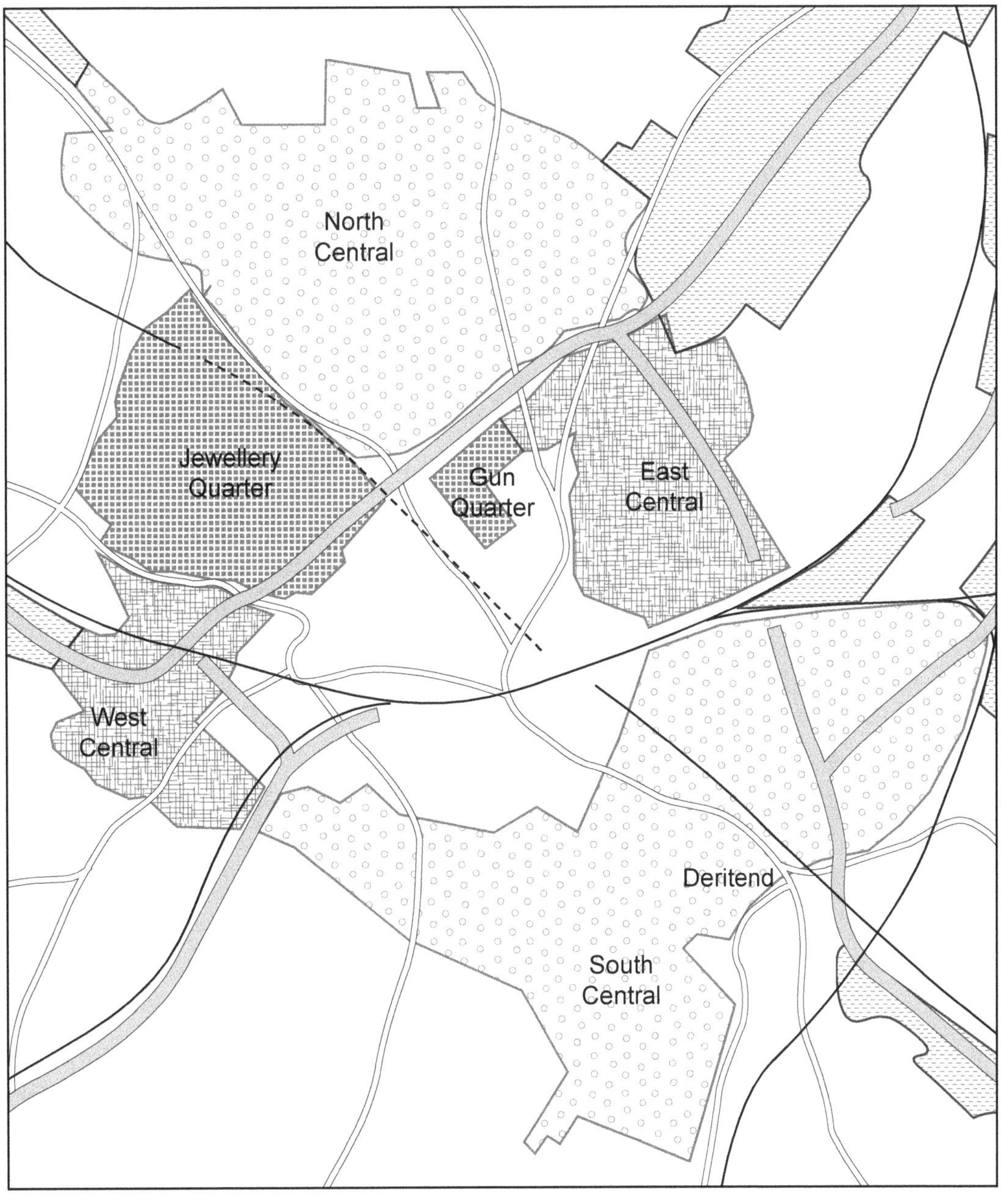

FIGURE 8.2 THE INDUSTRIAL ZONES OF THE INNER SUBURBS OF BIRMINGHAM (ADAPTED FROM WISE 1950, 223, FIG. 44)

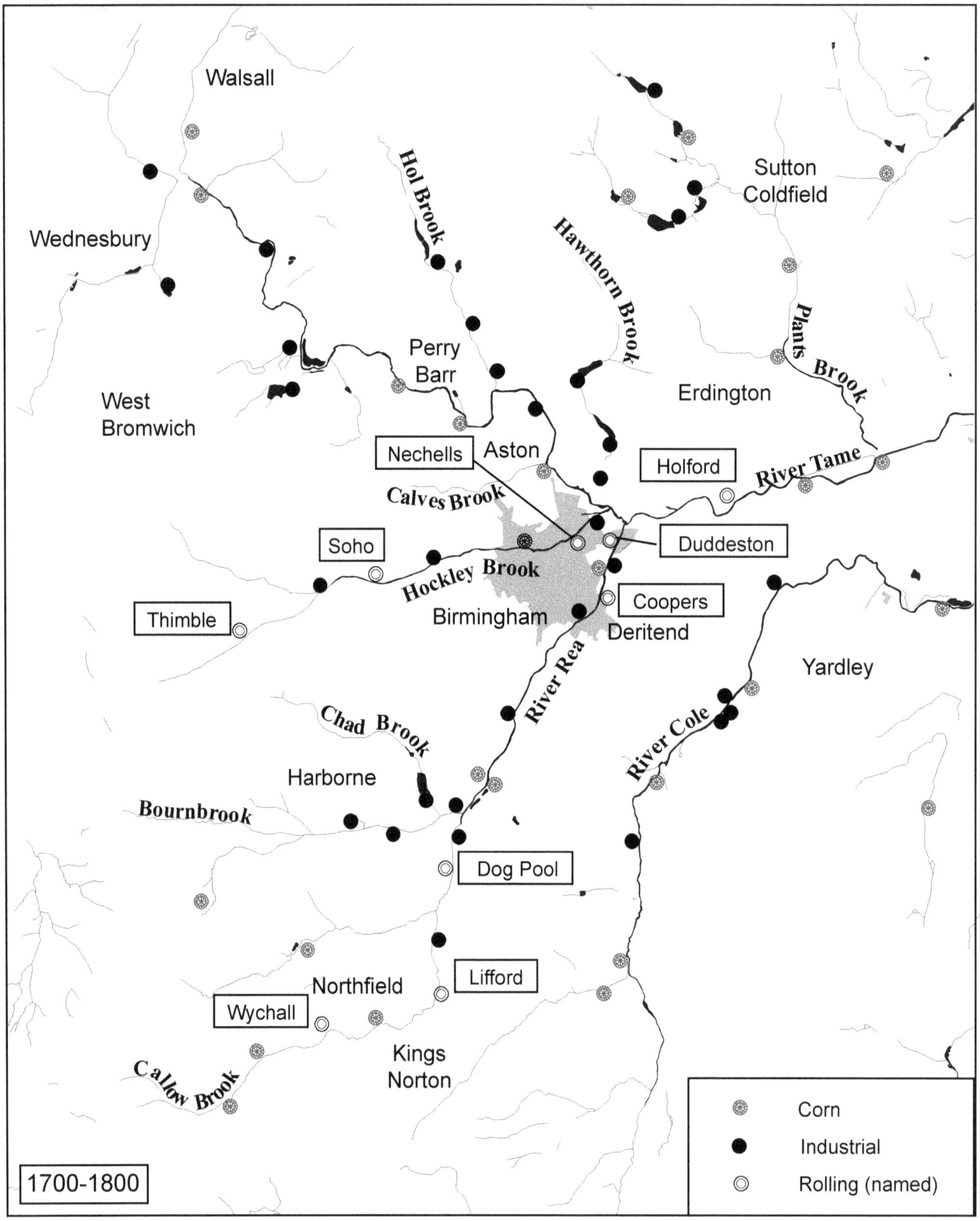

Figure 8.3 Location of rolling mills in Birmingham in the 18th century (after Stephens 1964)

power use had diversified. Willett's Meadow Mill, on the lower River Rea, was converted to a wire mill by 1732. The first mills to be converted to rolling mills appear to have been the Duddeston Mill, owned by James Farmer, ironmonger, in 1744, and visited by historian William Hutton in 1756 who '*went with Will Ryland to Horton's at Duddeston Mill to have some silver rolled*'. The Thimble Mill, on the Hockley Brook, was leased to Samuel Birch of Birmingham, button maker, in 1749. He rebuilt it and converted it into a rolling mill and it was subsequently rebuilt between 1758 and 1833, downstream, as the Nechells Rolling Mill (Stephens 1964).

The most influential suburban manufactory, not only in a regional, but in a national sense, was probably the Soho undertakings of Matthew Boulton and James Watt, located just inside the Smethwick boundary. Between 1761 and 1848 the Soho Manufactory was the most influential business in Birmingham and is credited with many of the initial developments that can be considered to have influenced the development of the modern factory (Gale 1946). In 1757 Edward Ruston and John Eaves, toymakers, of Birmingham, established a mill for rolling metals that was subsequently acquired by Matthew Boulton in 1761–1762. He demolished the mill and subsequently began to build the Soho Manufactory in 1764. The Soho Foundry, the first purpose-built steam engine manufactory in the world, was also established on the adjacent site by Boulton and Watt in 1795 (Belford *et al* 2003).

On the upper River Rea, Lifford Mill in Kings Norton was a medieval mill that was converted by its tenant, Thomas Dobbs, in the late 18th century, to metal rolling. He installed a Boulton and Watt steam engine in 1787. Thomas Dobbs abandoned the old mill on the left hand side of the house and constructed another rolling mill closer to Lifford reservoir. It was revealed by excavations and geophysical survey in 1989 that it to contained three 18th-/19th-century buildings: the watermill itself, the furnace room and the rolling room into which a machine pit had been inserted, possibly to house an auxiliary steam engine (Hannaford 1989; WMAFS 2002).

By the early 19th century many more mills had developed. Wychall's Mill was a metal-rolling mill by the early 19th century (Booth 1978, 63). The Dogpool Mill on the Rea, was a rolling mill in 1836 and the Holford Mill on the River Tame had also converted by the 18th century and was occupied by Thomas Wilmore, in 1815, a '*manufacturer of rolled, plated, gilding and dipping metal, wires*' (Stephens 1964). Suburban industry in the early 19th century was still small-scale and domestic in scope and this would not alter until the later 19th century.

The 19th-century brass industry

The expansion of the town in the 18th century led to development of industry around the city centre. The development of the steam engine meant that there was no longer a need to base industry close to water power. This meant that the brass trade was not highly focused and spread in broad zones in the inner suburbs. A broad geography of the mid 19th century brass industry can be attempted. This was initially done by Wise (1950, 222, fig. 44) who divided the industrial zones in the inner suburbs of Birmingham as North, East, South and West Central. These have been adapted for the following discussion as the Jewellery Quarter and North Central (after Cattell *et al* 2000, 187); the Weaman Estate and Eastside, East Central; Digbeth, Deritend and Bordesley, South Central (after Litherland 1995; Mould 1999) and the Broad Street area, West Central (Fig. 8.2). A more detailed discussion of the Broad Street area will be attempted in order to place Cambridge Street in its local context.

The Jewellery Quarter and North Central

The north central area extends from the edge of Lozells to the northeast to the edge of the Birmingham city centre as defined by the Birmingham and Fazeley Canal. This includes the internationally important Jewellery Quarter, essentially a zone of mixed industrial use of small, medium and large workshops dating from the late 18th to 20th century (Fig. 8.2). The Jewellery moniker is misleading as the area contains a variety of trades that included in particular the button trade, the pen trade and a variety of small-scale fabrication industries in pressed metal (Newman Brothers Coffin Furniture being an example). The use of brass and copper in these fabrication trades meant the area inevitably included elements of the brass trade. A broad zone of heavier industries congregated along the line of the Birmingham and Fazeley canal (Cattell *et al* 2000, 187).

Generally, works in this district specialised in manufacture as opposed to foundry work or rolling. The larger works included the Newhall Works on George Street, a large purpose-built pin and wire factory dated to *c* 1847 (Cattell *et al* 2000, 225–227). More restrained medium-scale foundries include The Wellington Works on Legge Lane, housed Walker and Woodward, brass manufacturers dating from 1860 and worked until the 1970s. The factory appears to have grown organically from a series of buildings throughout the latter part of the 19th century (*ibid*, 248). The Tyndall Works on Sloane Street was built by the brass founder David Mallins in two phases between 1854 and 1887. A later example is Newman Brother's Coffin Factory, 13–15 Fleet Street (*ibid*, 218), built 1892, which specialised in solid brass and plated coffin furniture. In addition there was a great quantity of smaller manufacturers who operated out of converted terraces or organically established workshops at the rear of existing shop frontages that developed throughout the 19th century, using stamped and pressed brass as a staple in the manufacture of a diverse range of products.

However, a number of large manufacturers did exist, and these concerns are directly comparable with the Cambridge Street Works. The Birmingham Mint and Metal Works, Icknield Street, of which only the front range survives, is an example of a factory designed to both produce rolled and cast metal and finished products. It was purpose-built with a quadrangle of workshops around a central courtyard, in which were flat-roofed workshops including casting shops, rolling mills, an engine house and boiler house. An image of the works was shown in the *Illustrated Times* in 1862 (Cattell *et al*, 242–243, fig. 252). Other surviving foundries include the Globe Foundry on Charlotte Street owned by Robert Lloyd Crosbie and Company, manufacturers of brass and iron bedsteads, and built before 1855. The foundry survives today as a converted architect's office (*ibid*, 94–95, 228–229). The Derwent Works Iron foundry on 60–62 Constitution Hill, owned by Taylor and Challen, included a capacity for brassworking (formerly Jennens and Betteridge, *ibid,* 215).

The area was characterised by small- and medium-scale manufactories, many of which survive today. The area east of the Great Hampton Road was somewhat different and larger in scale but has not been discussed in detail here.

The Gun Quarter and Eastside (East Central)

The east central area of Birmingham according to Wise's (1950, 222) definition consists of a large swathe of land from Digbeth and Deritend High Street to the area now defined as Eastside. The area has been adjusted here to include specifically the small area that included Eastside and the Gun Quarter (Weaman Estate). This area lay within the ring formed by the Birmingham and Fazeley Canal that lay on the ridge of land between the Hockley Brook and the River Rea flood plain, east of the city centre. It was similar to the Jewellery Quarter in that it developed from the 18th century, on a domestic scale. Although some still survives, much has now been removed by both 19th-century change, including the establishment of the general hospital, and 20th century development, such as the inner ring road (Fig. 8.2).

The Weaman Estate has largely disappeared, but during most of the 18th and 19th century was the focus of the gun industry in Birmingham and hence became known as the Gun Quarter. Like the Jewellery Quarter discussed above, it was concentrated on small manufactories operating out of workshops that had often been converted from domestic premises. The brass industry was not especially prevalent within the district, but as the gun trade was reliant on specific copper alloys, of which 'gunmetal' was one, it shared much in common with the small manufacturers of the brass industry. The area has been discussed previously in terms of history and architectural development (Wise 1949).

The eastern area of industrial development stretches up towards Coleshill, in what is now known as Eastside. This was the original location of Turner's Brassworks, that later became the location of Elliot's Metal Company, Solid Brass Tube Manufacturer (Aitken 1866, 228). In 1770, the area was characterised by domestic manufacturing and in this respect was not dissimilar to the Weaman Estate (see above). The trade directories in 1767 and 1770 for one street, Chapel Row, list trades that include brass founder, buckle and button maker, plater, sword cutler, watch-maker, chain-making, silverer, cooper and iron maker (Driver 2005, 4). By the mid 19th century the area of Eastside is characterised by larger manufacturers adjacent to smaller domestic industry. Brass founders were located on Penn Street (still extant in 2005), Grosvenor Street and Princes Street (Driver 2005, 4, 12). Patterns of mixed 19th-century domestic manufacturing typified the area of Cardigan Street and Gopsal Street, such as G J Parkinson and Co. at No.14 Cardigan Street, manufacturers of gas regulators. Associated trades included pattern makers and engravers on Howe Street (*ibid,* 12). The location of heavier industry adjacent to the canal is characteristic of the industrial 19th-century patterns seen in the southern part of the Jewellery Quarter, and was also visible in Eastside adjacent to the Birmingham and Fazeley Canal where the Ashted Pump House, Belmont Glass Works and the remains of a copper and brass refinery were located (*ibid,* 18). However, by the 20th century the area was beginning to change and by 1905 court housing was beginning to be demolished and replaced by large industrial units (*ibid,* 9).

Overall, the patterns of development in Eastside are similar to those in the Jewellery Quarter and the Weaman Estate, where medium- to large-scale factories sit side by side with domestic industry. Unlike the Jewellery Quarter, the pace of change in Birmingham resulted in much of the 19th-century fabric and character of this area changing.

Digbeth, Deritend, Bordesley, (South Central)

The south central region as defined by Wise (1950, 223) stretched in a broad zone south of Digbeth and Deritend High Street in the east towards Gas Street Basins in the west. This area still survives as a mix of industrial buildings, now fringed to the north by the entertainment and commercial district of the Chinese Quarter and Hurst Street. For the purposes of this discussion the area between Digbeth and Eastside which retains large swathes of industrial buildings on the Rea valley floor, has also been included as they have similar characteristics (Fig. 8.2).

The development of the area along the River Rea flood plain that formed Digbeth, Deritend and Bordesley had long been associated with the brass industry. The earliest evidence for the brass industry in Birmingham have been recovered from excavations at Park Street (Patrick and Rátkai 2009) and on the banks of the River Rea (Duncan *et al* forthcoming) dating to the period of the late 17th century to the mid 18th century. The pattern of industrial development in the area continued to adopt the line of Digbeth and the Deritend High Street throughout the 18th century.

The slow infill between Eastside and Digbeth began with expansion onto the estates around the Birmingham city centre in the 19th century. This initially saw new businesses being established in traditional locations. The medieval Heath Mill was converted to industrial purposes in the late 18th century, when it was leased to Joseph Cotterill, Wire Maker (Snape's Survey 1796), who converted part of the mill to run on steam power. The mill appears to have housed both a wire drawing business but also the premises of Deritend Forge and a sword cutler at Woolley's Mill. When the mill was advertised for sale in 1819, it was described as being for 'rolling and slitting iron, rolling sheet iron, steel, copper' and included an iron and corn mill. A short-lived triangular mill pond was built at this time to supplement the flow of the river but had gone by the 1850s. The water mill continued to be used until *c*

1841 when it was converted to a steam powered wire, tube and rolling mill and operated during the 1850s as Samuel Walker Ltd, Wire, Tube and Rolling Mill (Demidowicz 1991; Mould 1999, 79).

The arrival of the Digbeth Branch of the canal encouraged development at its terminus. Of these Kempson's Survey of 1794 shows a mill, later called Phipson's Mill, and finally the Fazeley Street Rolling Mill, which marks two steam powered mills. Adjacent to this and established in 1813 were the remains of the Gun Barrel Proof House (Mould 1999, 70–71). The arrival of the railway provided further impetus for industry, and to this day the area is characterised by the tall railway arches across the River Rea flood plain. Organisation had changed from the domestic workshop to larger organised manufactory. Brass foundries, rolling mills and tube works were typical of the area. These included a rolling and tube mill on Floodgate Street and Milk Street by 1901 (*ibid*, 54–55); a brass foundry in 1828, that converted to the Britannia Tube Works by 1889 (*ibid*, 93–95); The Bordesley Mills, Adderley Street, and Kingston Metal Works, Liverpool Street were both tube works in 1885 (*ibid*, 101). There was also a particular focus on bedstead works. On Glover Street, there was a bedstead works built in 1849 to 1855 and another bedstead works was located on Heath Mill Lane by 1905 (*ibid*, 103). This was also the location of large bedsteads works such as Peyton and Peyton and S B Whitfield (Shill 2006b, 48–54). To the west of the Digbeth High Street large factories were laid out on grid patterns and are still visible today with a particular emphasis on the blocks between Bradford Street and Moseley Street. Companies included the William Cooper's Rolling Mill, Bradford Street and the adjacent Deritend Bridge Works, operated by Charles Smith and Sons Ltd that produced cast items in 1888 described as 'medieval metalwork' presumably in the Gothic style (Forster and Rátkai 2008, 57, vol. 2). One of the largest was William Tonks and Sons discussed earlier (see Chapter 5). Established in 1789 they were originally of Hill Street and moved to Moseley Street in 1860 and began making cast and wrought brass products (Strauss *et al* 1864, 51–58).

Overall, the principal characteristic of the area in the later 19th century was large manufactories including amalgamated metal works and subsidiary manufacturers such as bedstead works. The scale of the factories reflected the fact they were established in the mid to late 19th century, when smaller domestic industry had begun to be replaced by amalgamated works.

Broad Street (West Central)

The west central zone, including the area of Broad Street and all the zones on the industrial fringes, has seen the greatest change (Fig. 8.2). The area still contained large areas of industrial buildings, part derelict, in the 1970s and 1980s. In part due to its location, adjacent to a network of canals, it has been the focus for the most redevelopment, and so little of the industrial character of the area now survives. The buildings predominantly date to the last thirty years and few industrial buildings relating to the brass industry now survive. Of these the most significant is the Birmingham Brass House on Broad Street.

The location of elements of the brass industry in the west central area of Birmingham owed much to the location of the Birmingham Brass House. The main buildings are still extant (converted to a restaurant and the Brass House Language Centre) and are located on Broad Street, on what was Brass House Passage, backing onto the canal, now the Sheepcote Street basin. The Brass House was started as a concern in 1781, after instability in the price of copper led local brass founders to group together to form a company to aid the Birmingham industry against the monopoly of metal dealers that had led to increases in prices in the previous years. The Brass House had two furnaces, with eight small and one king pot either side of them. It was taken over by Thomas Pemberton in 1831, but continued to operate until its demise in 1866 (Aitken 1866, 254–255).

In order to place Cambridge Street in its context, discussion below will specifically look at two areas of industrial development around Broad Street. The first is the Baskerville Estate of which the Cambridge Street works was a major part and the second is the northern end of Sheepcote Street. This area provides a direct comparison and contrast to the Cambridge Street site as it contained similar elements of the brass industry. Both provide case studies for the development of the brass industry in Birmingham with a particular emphasis on rolled brass, wire and tubes.

Late 19th-century transition to the suburbs

The later 19th century saw a movement out of the areas around the city centre to new purpose-built sites with better transport links provided by both canal and rail. The industrial zones of Birmingham developed rapidly in the period *c* 1860 to *c* 1910 in areas outside the core of the city centre, specifically at Hay Mills/Tyseley to the southeast, Saltley, Washwood Heath, Aston and Witton to the northeast and an area to the west around the Icknield Port Loop in Ladywood (Wise 1950, 222–224, fig. 44, Fig. 8.4). Into this pattern should be added the development of the Soho area that extended from the 18th century Soho Manufactory and Foundry towards Smethwick and Winson Green (Wise 1950, 222–224, fig. 44). W T Avery of Digbeth (established *c* 1820) took over the Soho Foundry of Boulton and Watt, Birmingham in 1895 (Gale 1946, 32–38; Belford *et al* 2003), the premises retaining original elements of the Soho Foundry, being expanded due to its location outside the core of the built-up city.

Other areas of suburban development included more isolated settlements along the Rea valley, locations that were original locations of industry in the 19th century. These include Selly Oak, Stirchley and Kings Norton that all had mills in the late 18th or early 19th century, but

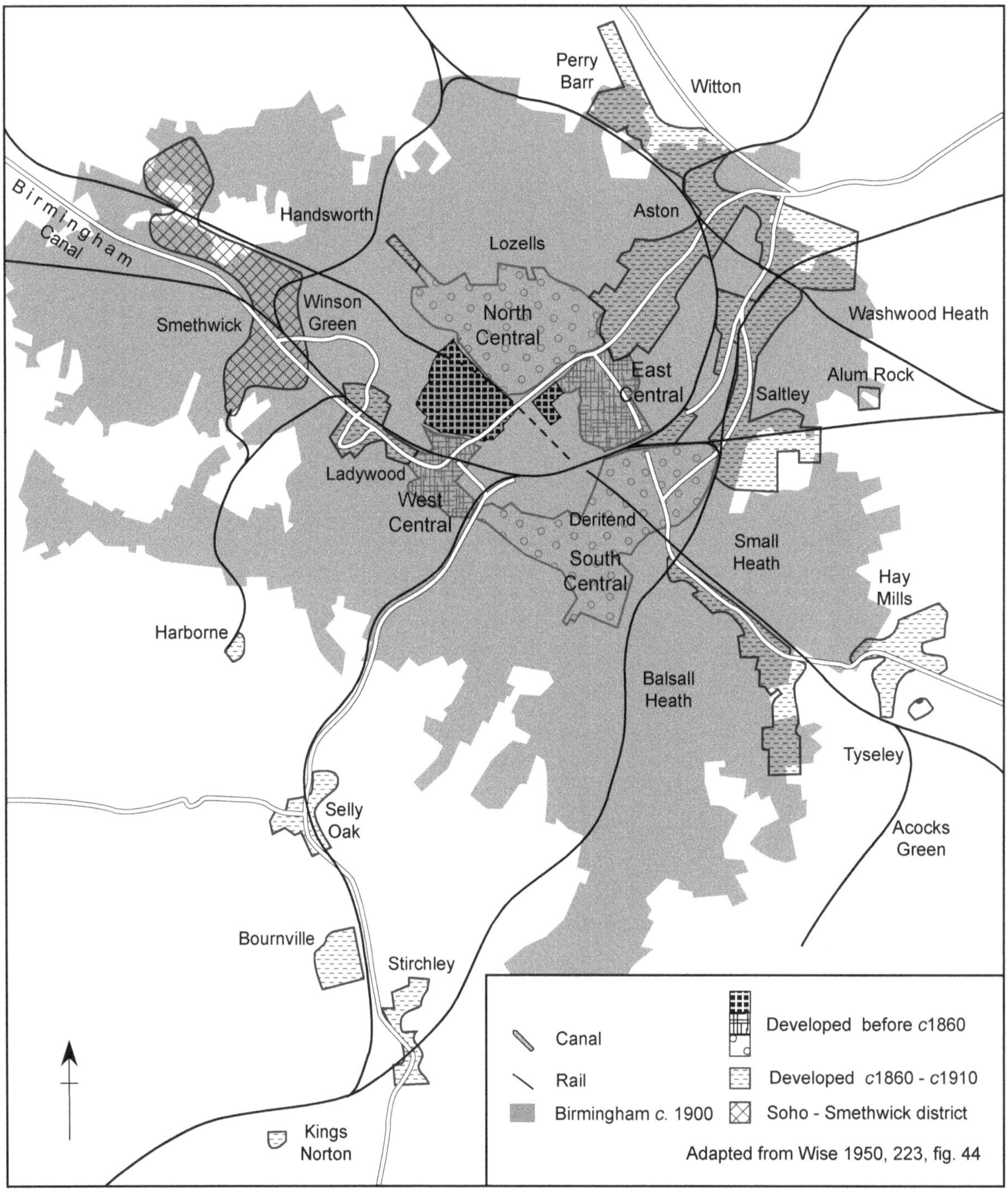

FIGURE 8.4 THE INDUSTRIAL ZONES OF BIRMINGHAM (ADAPTED FROM WISE 1950, 223, FIG. 44)

became more industrialised by the end of the 19th century. Many of the 18th century mills became 19th-century metal manufacturers. By 1863, Dogpool Mill was owned by Tomlinson and Co., tube and wire manufacturers and by 1875 the mill had been taken over by C. Clifford and Sons, metal rollers and tube makers, who were still in occupation in 1957 (Stephens 1964). The buildings at Wychall's Mill, Kings Norton, were by the late 19th century occupied by G Ellis and Sons, metal rollers (Booth 1978, 63).

The relocation of the brass industry to sites outside the inner suburbs will be examined with specific reference to the area that developed around Icknield Port Loop northwest of the city centre in the late 19th century.

THE BASKERVILLE ESTATE – CAMBRIDGE STREET IN CONTEXT

Ray Shill and Chris Hewitson

The development of the surrounding streets around the Cambridge Street works produces a case-study of the 19th-century industrial sectors that sprung up around canal basins in Birmingham. The mix of inter-related industries reflects the brass industry as a whole in Birmingham (Fig. 8.5).

Investment in the transport network made the area of Gibson's and Baskerville Basin a profitable location for development and following the introduction of the canal wharfs, industrial buildings began to occupy the former Baskerville estate. After the establishment of the wharfs, and his withdrawal from the merchant, rolling and slitting trade, Thomas Gibson concentrated his efforts on further development of his property.

The development of the land on the former Baskerville Estate can be traced through the lessees and sub-lessees of several of the buildings. Using rate book searches (rate books Ladywood ward) it has been possible to pick up some of the occupiers, but not all. Trade directories (particularly those that give street listings, post 1860s, eg PO 1864, 1871, 1875; Kelly 1883, 1892) provide a better picture of the occupiers, but they do not necessarily list all of them. What follows is a discussion of the portion of the estate formed by the quadrant of Broad Street, Baskerville Place, Cambridge Street and Easy Row. The larger area around the Cambridge Street works enables us to view a snapshot of the varied development of industry in the inner suburbs. In itself this broader study area represents a slice of the intensely occupied, industrial mass of Birmingham.

Cambridge Street (east to west)

Cambridge Street forms the northern boundary of this study area running east–west. The properties on the south side of the road had transport links with both the street and Gibson's Canal Arm. The Cambridge Street works formed the principal element of the street, but there were also a number of other properties. Robert Winfield's original Cambridge Street factory occupied one of the land plots to the east of Gibson's Lock and became the eventual location of the offices and showroom that faced east onto Easy Hill/Cambridge Street (BCA MS 322/ 1–9).

On Cambridge Street two sections of the land around Gibson's Basin remained separate to the Cambridge Street

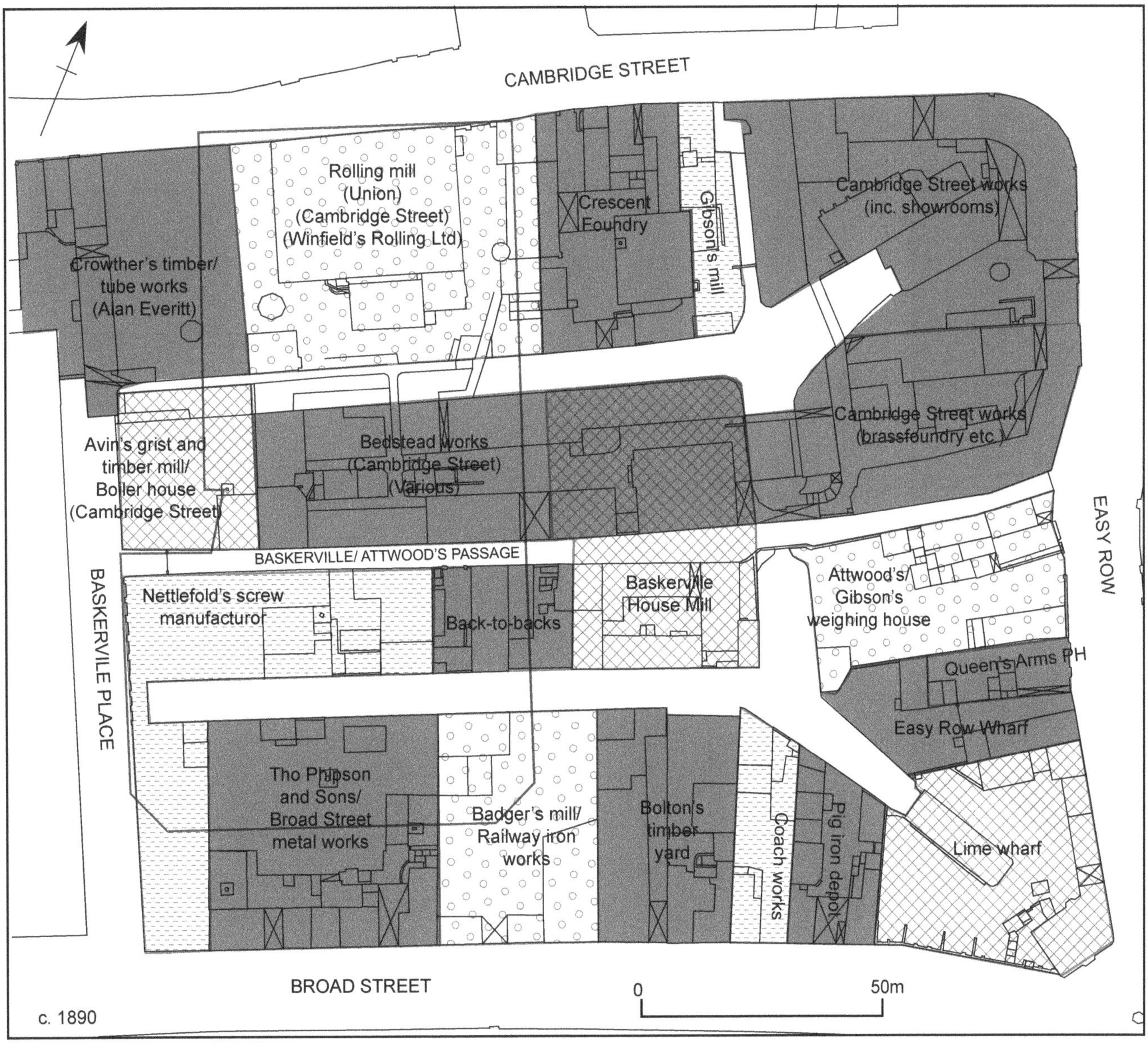

Figure 8.5 The Baskerville estate in the 19th century, land use c 1890

works: the Crescent Foundry, an iron foundry, and Gibson's Mill (Bradford's map 1855–1857; Ordnance Survey 1888). Charles Henry Capper, the Union Rolling Mill's manager, was also an engineer, iron founder and manufacturer of grates, stoves, fences, iron chests, and castings whist at the Crescent (Union) Foundry, Cambridge Street. He was also an agent to Horseley Iron Co, Tipton (Wrightson 1833). He subsequently supplied stationary steam engines from a foundry at Broad Street.

Gibson's Mill was constructed by Thomas Gibson beside the lock. Towards the end of the 19th century the mill tended to be let out as shopping to various tenants including Winfield and Co, but also to various glass cutters, wood turners and steel polishers (Kelly 1883).

The Union Rolling Mill occupied the land to the west of the Crescent Foundry. The evolution and occupation of this land plot has been discussed in detail in Chapter 4 above. There were also buildings used as a casting shop that backed onto the Crescent Foundry. The final plot on the street was the location of the tube works (see Chapter 4).

Easy Row (north to south)

Easy Row (now Paradise Circus) forms the eastern boundary of the study area. The northern end of Easy Row was dominated by the properties of the Cambridge Street works, including the commercial show room (BCA MS 322/ 1–9). These properties had transport links with both the street and short wharves, extending from the canal arm that included Gibson's Lock.

The plot on the corner with Attwood's Passage was part of the Attwood's properties. This was also the location of Gibson's weighing house (Wrightson 1833; Kelly 1883, 1892). The adjacent property to the south was the Easy Row Wharf (PO 1854; Ordnance Survey 1887–1889). This was mostly a coal wharf for the local area. Part of the plot was taken up by the Queen's Arms public house. This was established in the early 19th century and was updated in 1886 by Henry Naden. It eventually closed in 1922 (McKenna 2005a, 121).

The Lime Wharf was at the corner of the Baskerville Basin on Easy Hill. It was adapted for the lime trade during the 1820s. Early occupiers of this wharf were Cooper, Strongitharm and Co who had limestone quarries and mines at Daw End alongside the Wyrley and Essington Canal (Ladywood rate books 1834/19, 1840/36–37). A later long-term tenant was Samuel De La Grange Williams (PO 1864; Kelly 1892).

Attwood's And Baskerville Passage (east to west)

Bisecting the study area was an east–west passage known as Attwood's Passage at its eastern end and Baskerville Passage at its western end. Today this forms the northern edge of Centenary Square. A number of properties developed along here, served by Gibson's Canal Arm on the northern side and the Baskerville Arm on the southern side. The Cambridge Street bedstead works occupied much of the northern side of the passage.

Baskerville House was rebuilt as Baskerville House Mill and was first occupied by Benjamin Cook and Thomas Attwood and later George and Thomas Attwood. The former house was bisected by Attwood's Passage with the main mill property placed on the south side of this passage and the Baskerville Arm of the canal running alongside the southern perimeter of the mill. At the time of the 1841 fire at Bolton's timber yard, the tenants of Baskerville House Mill included; Willmot and Stokes, glass cutters, Mr Morgan, glass and chandelier maker, Barnett and Chambers, nail manufacturers. The remainder was occupied by the Attwoods (BJ 1841). The mill was fortunate to escape the fire, owing to the wind changing direction. These premises occupied only part of the land. Thomas Attwood took out a patent in 1850 to convert used copper rollers into seamless brass tubes and this was presumably undertaken here (Aitken 1866, 328). Later Baskerville House Mill premises were occupied by H and S Phillips and used as a forge for steel and iron. When S Phillips left this partnership in 1866 a stock of steel hammers was included in the sale. H Phillips kept the premises a while longer, but they were eventually let off as shopping occupied by a range of workers including glass cutters, cut tack manufacturers and spoon polishers (PO 1871). Baskerville House Mill survived until 1888. The front wall of the original house survived and remained until its demolition. The demolition of the house was carried out in September 1888 and the materials were sold off in lots (BDP 1888b).

To the west another section of land was leased by Thomas Crowther, timber merchant, box and cask maker (Wrightson 1833). The house buildings north of Attwood's Passage seem to have been absorbed later into Winfield's Cambridge Street works. Adjacent to the mill site were two courts of back-to back houses constructed in Baskerville Passage. These covered a square of land between the passage and Baskerville Basin. The Attwoods also owned the Adelphi Steel Works alongside the Baskerville Basin that fronted Broad Street (Easy Row, BJ 1841).

A square of land at the far end of Gibson's Canal Arm and bordering on Baskerville Place was kept as a timber yard and grist mill that came to be owned by Thomas and then Henry Avins (Pigots 1842; PO 1864). The Baskerville Timber Yard owned by Avins suffered a serious fire in 1862 and was subsequently acquired by George Tarplee and William Gardner. Another fire in 1867 affected Winfield's adjacent bedstead factory. The yard was sold during 1873 and Winfield's then used the land as an addition to their works and used it to house a new boiler (BCA MS 322/12).

Broad Street (east to west)

Present day Broad Street formed the southern edge of the study area and was also the southern perimeter of the

Ryland Estate. Initially there were few premises on this side, but in time, a virtually unbroken façade of different buildings was created from Suffolk Street to Baskerville Place. Property numbering began from Easy Hill/ Suffolk Street Junction and continued as far as the Birmingham Canal Navigations Bridge where the road crossed the canal. This bridge also marked the dividing point between Broad Street (Easy Hill) and Broad Street (Islington). These were separate roads and property numbering began again at the Bridge with the Brasshouse. Much of the maintenance for Broad Street to Hagley and Blakedown Pool was the responsibility of the Turnpike Trustees, but this later fell to the Birmingham Street Commissioners.

Rate books (Ladywood ward, 1860, 1871, 1876, 1881) record the start of the Broad Street properties as beginning as a house Number 1, which was to the west of the Lime Wharf. Number 1 Broad Street had various owners, but was eventually adapted by Southall Brothers and Barclay, who set up a factory for the supply of chemical apparatus. Number 5½ Broad was occupied by T Cox and Co and used as depot for pig iron moving from their Cambridge Street location (PO 1864). Numbers 7 and 8 Broad Street became a coach builder's works owned by William Henry Iliffe and later by George Tye (Post Office 1864, Kelly 1883, 1892). Number 9 Broad Street was developed as a timber yard by William Bolton and there was a serious fire here in 1841 (BJ 1841). This yard later passed to Thomas Dowling who continued the timber and slate business there until 1861 (BDP 1861). The fire that severely damaged Bolton's timber yard also seriously affected Badger's Mill (see below), which was sub-let to various parties. Some £6000 damage was reported, and the structure required rebuilding.

Number 11 Broad Street was leased to Thomas, Isaac and Septimus Badger, nail factors and nail makers from Dudley. The building was known as the Baskerville Mill and had a steam engine to provide mill power. Thomas and Isaac Badger were involved in the iron trade operating ironworks and later smelting furnaces. The Badger brothers were also involved in glass making at Dudley. Baskerville Mill suffered a serious fire in 1844, but the Badgers had the mill rebuilt and let shopping out to different trades people. This mill was advertised for sale in May 1845, when the premises included a range of three-storey shopping and other sheds. The steam engine was 34 HP (ABG 1845).

Number 11 Broad Street was subsequently occupied by Morewood and Co, patent galvanised iron makers until 1862 (BDP 1862), when they transferred this business to Birmingham Heath. During 1869 these premises were rebuilt as a warehouse for Joseph Warden and Sons and became known as the Railway Ironworks. The Railway Ironworks came into use during January 1870. Their previous warehouse in Edgbaston Street was advertised to be let in June 1870 (BDP 1870a/b). The Broad Street premises specialised in the supply iron goods such as bars, angles, T-iron, boiler plates, galvanised and corrugated sheets, telegraph, fence and signal wire, iron telegraph posts, railway spikes, screws, nuts and bolts (PO 1871).

Thomas Phipson and Sons occupied 12 Broad Street as a pin and needle factory controlled by his sons, Samuel Ryland Phipson and Richard Phipson. The Ryland family previously carried on a successful wire drawing and pin making business in Birmingham from the mid 18th century. Samuel Ryland handed over the business for making of pins to his nephew Thomas Phipson about the year 1785 and pin factory on Broad Street was a continuation of this (Phipson 1866, 601–603). Pin making was labour intensive in the period before mechanisation and involved a series of manual processes that started with brass wire and included drawing wire, pin pointing, head cutting, heading, cleaning, whitening and carding. The Broad Street factory employed many children who had jobs in various departments, but there was a high number (80) that worked in the two pin heading shops (Grainger 1840). With increased mechanisation in the trade the Phipsons transferred pin-making to a factory at Nos. 91 and 92 Mott Street by 1849. After this the property (later numbered 14) was owned by Weiss Brothers Merchants for a number of years (PO 1864, Kelly 1892).

Number 13 was a mill and for a time occupied by John Stephens, steam engine maker but was for sale in 1835 (ABG 1835). By 1838 these premises had been taken by Thomas Bolton and Co and became the Broad Street Metal Works. On the corner of Baskerville Place was a factory owned at one time by William Clark, cut nail makers. His business was taken over by Joseph Edwards who continued the making of cut nails and pressed hinges both here and a workshop on Gas Street. These premises were taken over by John Sutton Nettlefold, Screw Maker about the year 1847. The manufacturers Thomas Bolton and Co (Morton 1983) and John Sutton Nettlefold (Chamberlain 1866, 604–609; Jones 1987) have also been the subject of historic research and discussion.

BROAD STREET METAL WORKS AND NETTLEFORD'S SCREW MANUFACTURY – AREA 2 EXCAVATIONS

In many respects the Broad Street Metal Works and Nettlefold's Screw Manufacturers represent closely comparable manufacturers to the Cambridge Street Works. Both specialised in metal products and flourished in a comparable period of the mid to late 19th century (Phase 4) when Birmingham was at the forefront of the metal industries. The technology they adopted and the products they used are closely comparable.

Historical evidence

Ray Shill

In 1838 Thomas Bolton and Co, a partnership of Thomas Bolton (a Birmingham merchant) and Thomas Clowes, took over John Stephen's steam engine mill (Morton

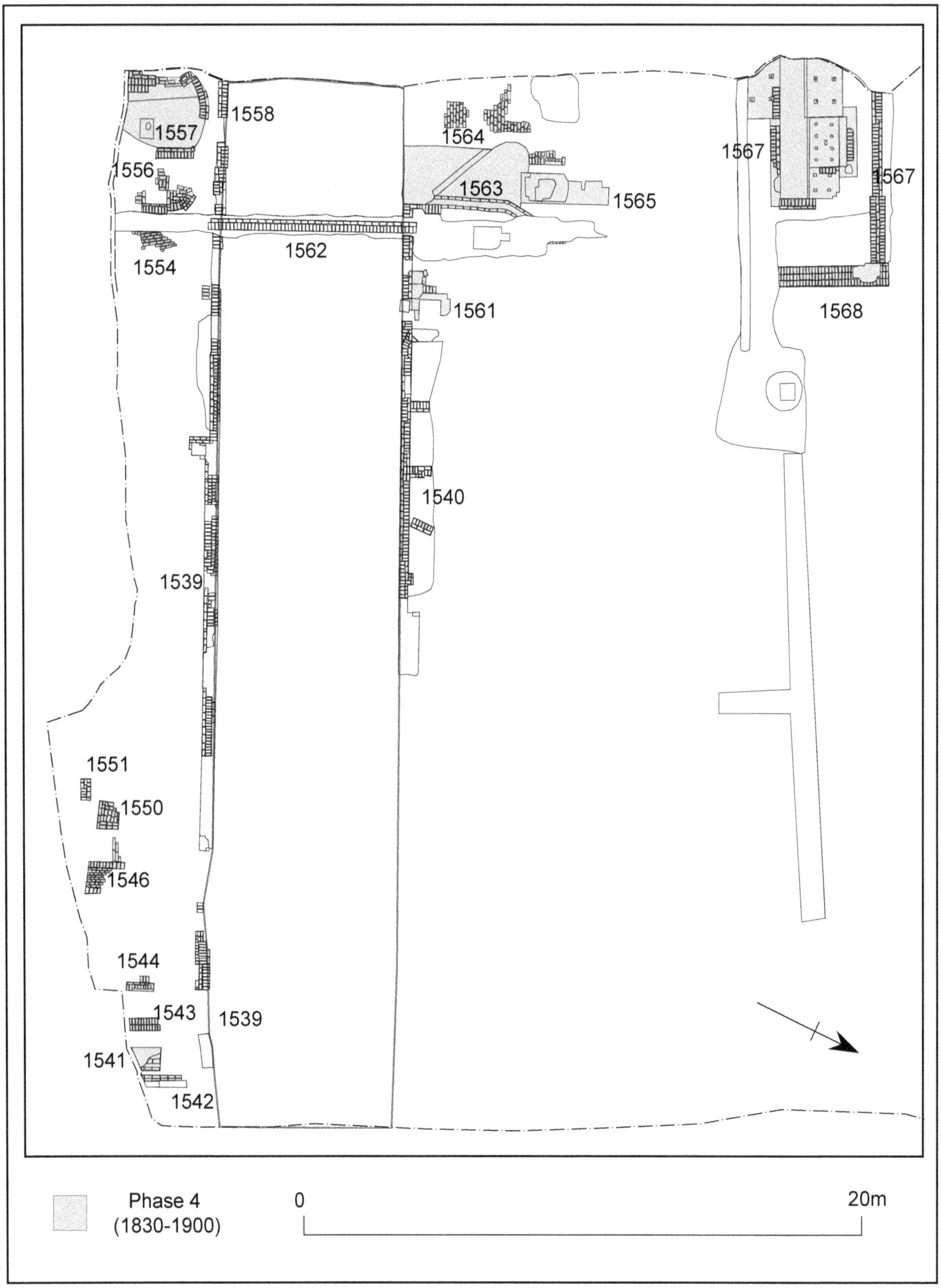

Figure 8.6 Area 2, Phase 4

1983, 35–38). The property included a 30 horsepower steam engine (BJ 1852) that was subsequently replaced by a Boulton and Watt engine. They commenced the manufacture of rolled metal, wire and tubes and were suppliers of copper and cable for telegraph purposes.

As their trade grew, the business was extended to a new works at Oakamoor, but the Broad Street factory was retained. Bolton enlarged their Broad Street premises, which became known as the Broad Street Metal Works. The premises consisted of No. 13 Broad Street later renumbered No. 15 in the 1870s. To the rear and north was a long frontage to the Baskerville Arm that extended as far as the adjacent Railway Ironworks. Copper production moved to their other premises at Oakamoor, Froghall, St Helens and Widnes. Birmingham came to specialise in the making of rolled brass and brass wire, and specialised in fine wire that was woven for use in paper mills. The decision to close the Birmingham mill was made in 1910 and production was gradually run down until closure in 1912 (Morton 1983, 43–106). The premises were subsequently purchased by GKN Engineering (formerly Guest, Keen and Nettlefolds).

Nettlefold had previously owned a water-driven mill at Sunbury-on-Thames. The move to Birmingham was to premises in Baskerville Place and Attwoods/Baskerville Passage that had wharf space alongside the Baskerville Basin. The offices were located at No. 14 Broad Street (later No. 16). A steam engine provided the power for the screw-making machinery.

Nettlefold took various sons into partnership but it was the partnership with the Chamberlain family that proved most successful. They invested around £30,000 in the firm in 1854 in order to purchase the patent for the manufacture by automatic machinery of woodscrews. After this the business was known as Nettlefold and Chamberlain with John Henry Nettlefold as works manager and Joseph Chamberlain junior as the commercial manager (Jones 1987, 133–139). A new factory was built outside the city centre on Heath Street, but the property on Baskerville Place continued to be used as an office, mill and packing house for the business. It included an office of three storeys fronted onto Broad Street, with 20–30 clerks. Adjacent to the office and entered from Baskerville Place was a two-storey mill that manufactured special screws and performed the pattern work. Across the basin was a four-storey warehouse, the top floor contained brass screw stocks, the third floor offices, the second was for japanned screws, and the ground floor was for receiving goods from the main works (*ibid*, 143).

Joseph Chamberlain left the business in 1874 to devote more time to politics. Nettlefolds went from strength to strength, becoming Nettlefolds Ltd in 1880. Fierce competition in the 1880s and 1890s from German and American rivals put pressure on the firms, imperial and domestic markets. This led to horizontal integration with many of the local firms eventually leading to amalgamation as part of Guest, Keen and Nettlefolds (GKN; Jones 1987, 199–232, 361–363). Broad Street retained a factory and offices as part of the group until the compulsory purchase by Birmingham Corporation in the 1920s.

Archaeological evidence

Will Mitchell

Archaeological excavation was conducted in Centenary Square (Area 2) to examine the potential for industrial buildings surviving here. However, the development of Centenary Square in the 1920s appeared to result in large-scale demolition of the buildings in this location that did not occur to the north. However, the truncated remains of industrial buildings were excavated, located on the site of the Broad Street Metal Works and Screw Manufactory. These buildings were situated south of Baskerville Passage and surrounded Baskerville Basin Canal. The historic maps confirmed that this area began to be populated with industrial buildings after the Earl of Dartmouth's Map of 1824–1825 but before Piggott-Smith's Map of 1855–1857, at around the same time as the construction of the Union Rolling Mills. The area would have been as densely covered by buildings as the rest of site, but preservation in this area was poor. Much of the truncation encountered within this area of site was due to the lowered level of the later Centenary Square, and its extensive landscaping (particularly the large concrete fountain foundation pad, planters and service trench elements.

Several brick structures were identified both to the north and south of Baskerville Basin, the majority of which remained without interpretation due to their fragmentary nature (Fig. 8.6). These were likely to have been various floors, walls, drains and other industrial features (1541–1546, 1550–1552, 1554–1558 and 1563–1555).

The most substantial and intact structure was that of a machine base and flywheel pit and associated L-shaped wall (Plate 8.1). The machine base (1567) was made up of a main rectangular pit, 5.4m in length x 1m in width x 1m+ in depth. This was constructed of machine-cut, 9 x 4¼ x 3in red bricks set in an English bond. On both its northern and southern sides were machine bases with built in holding-down pins, its structure being different from the other machine bases identified on site. The L-shaped wall (1566/1568) which surrounded the base on its northern and eastern sides was 7m x 4m, and was likely to have been an external elevation.

During the construction of centenary square Baskerville Basin was filled in. Bridging walls were constructed over the both the eastern (1542) and western (1539 and 1553) ends, presumably to add support to the surfacing above.

Discussion

In Area 2 beneath Centenary Square the poor survival of archaeological remains has highlighted the importance

Plate 8.1 Machine base, Area 2, Nettleford's Screw Manufactory, east facing

of the buildings excavated in Area 1. The archaeological evidence has not succeeded in enhancing our understanding of the industry located in Centenary Square. The survival of very few below-ground structures may emphasise that the majority of the works along Broad Street were not heavy industrial manufacturers but were in fact small-scale manufacturers. This is surprising given that Thomas Bolton and Co were significant rolled and wire brass manufacturers at their Cheadle Works (Morton 1983, 43–106) but it may suggest that the works in Birmingham were on a much smaller scale.

The machine-base and flywheel appear to be part of the mill of Nettlefolds Ltd. The two-storey mill manufactured special screws and performed the pattern work (Jones 1987, 143). The description of the warehouses suggests these special screws were not the iron screws produced by the main Heath Street works but were probably brass and japanned screws (*ibid*, 143). The mill was rebuilt in 1882 to a design by Thomas Clarke a London architect and Nettlefolds son-in-law (*ibid*, 142). The fact that the base was very different to other machine bases on the site supports the fact that it was designed for machining screws, but was also a relatively late mill in comparison with those located at the Cambridge Street works.

SHEEPCOTE STREET – A COMPARTIVE CASE STUDY

Chris Hewitson, Steve Litherland and Christine Winter

The only other brass rolling mill and works to have been extensively investigated in Birmingham was on Sheepcote Street. The Standard Works of Lawson and Holden was a coach works producing finished goods in iron and brass, between 1857 and 1903, and the adjacent Grice, Grice and Booth Works was a tube and rolling mill. This was located on a triangle of land between the New Birmingham Main Line Canal and the Sherborne Wharf Loop of the Old Main Line Canal and Sheepcote Street (Fig. 8.7; NGR SP 0560 8670). The site was made up of large open factory units fronted by stylistic workshops and showrooms that still survive today incorporated into a complex of flats. Number 24, Sheepcote Street was a Grade II listed building. The terracotta façade of the late 19th-century office buildings of 25, Sheepcote Street, and several 19th-century brick buildings behind the street frontage were locally listed Grade A. The site comprised redundant and unoccupied industrial and office premises (Structures A to F) recorded by a previous survey in 1999.

Historic Background

The following section provides a chronological model for the development of the site. The overall development of the study area is depicted on Fig. 8.8.

1769–1853

The Sherborne Wharf Loop formed part of the Birmingham Main Line Canal to Wolverhampton that was completed in the late 1760s. A new main line canal, cut by Thomas Telford, was completed in 1829 and effectively by-passed the Sherborne Wharf Loop, creating a large basin that was subsequently developed for industry. The arrival of the canals began a process of transformation of the western

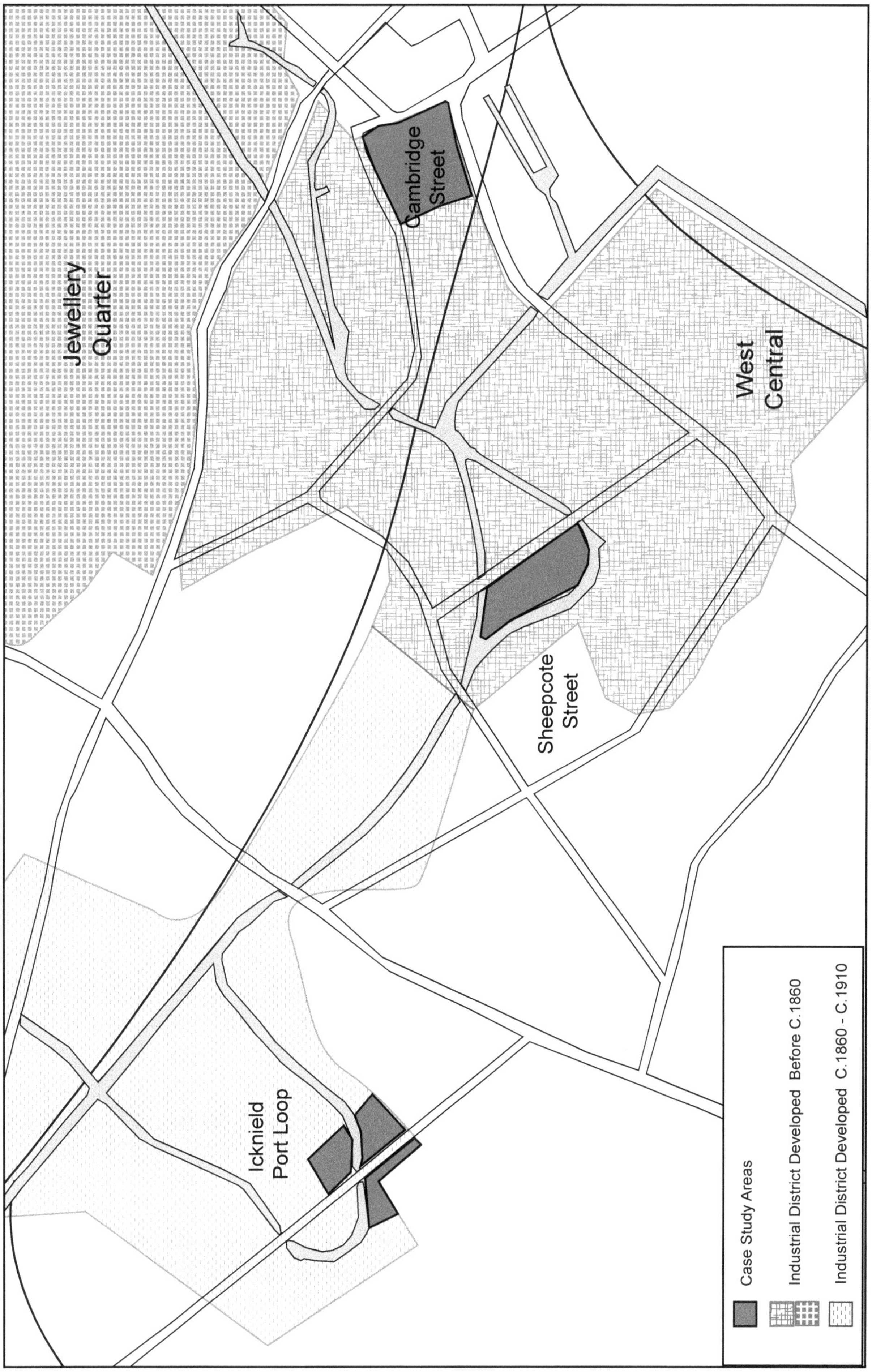

FIGURE 8.7 LOCATION OF CASE STUDY AREAS

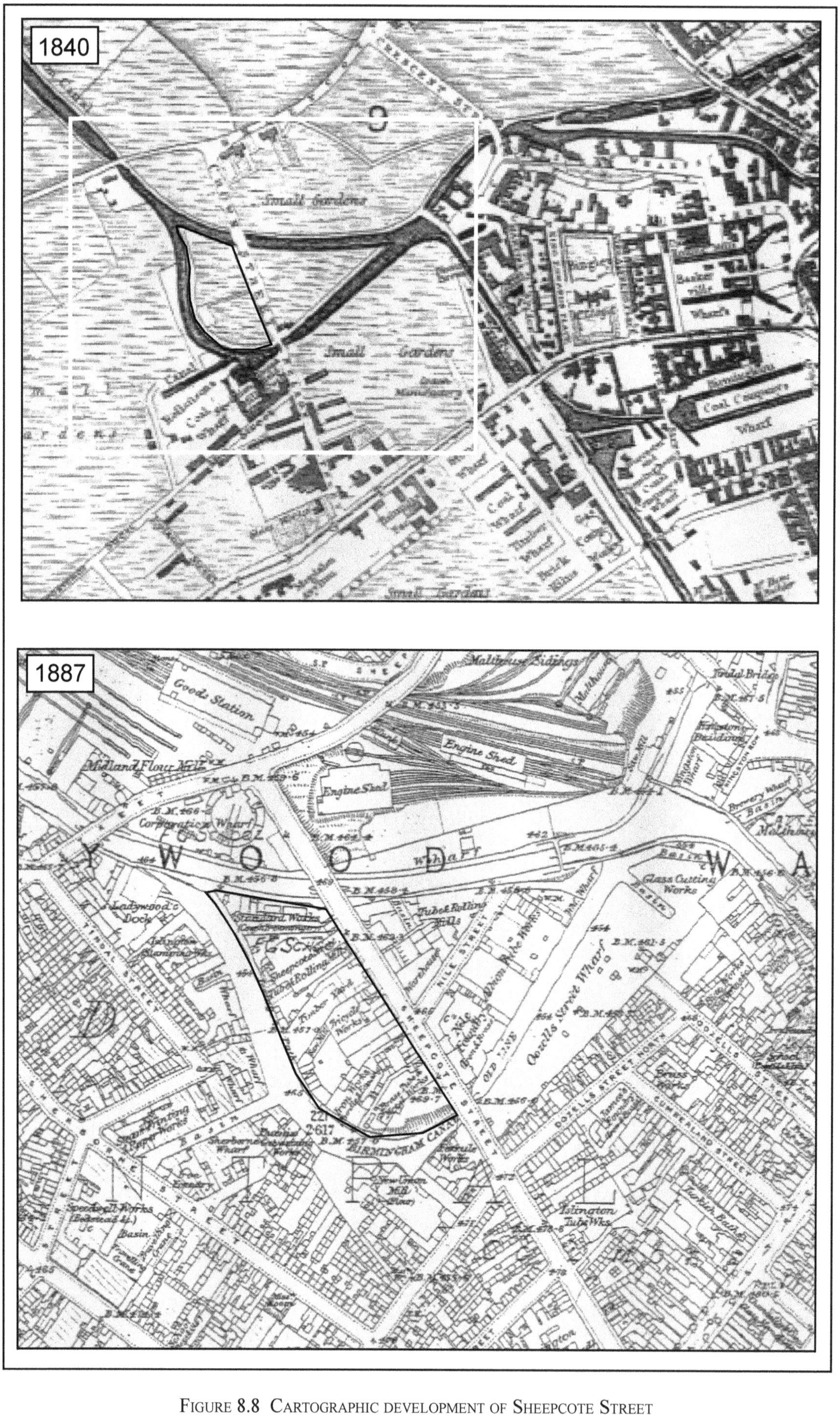

FIGURE 8.8 CARTOGRAPHIC DEVELOPMENT OF SHEEPCOTE STREET

approaches to Birmingham, offering the opportunity to transport raw materials cheaply. The land inside Sherborne Wharf Loop was not developed until 1853. A map produced in 1824/5 by Piggott-Smith for the Earl of Dartmouth shows that Sheepcote Street was called Crown Street. The construction of Crown Street and Nelson Street (to the south) created a nodal link between Broad Street (known as Islington) and the turnpike road to Dudley via Sheepcote Lane. The Manufactory of Johnson Berry and Harris was situated on Nelson Street, and once this industrially-driven expansion had begun more land started to be released for development.

1853–1887

White's Directory of 1855 lists companies between the Nos. 20–30 Sheepcote Street as smiths, wheelwrights and coal merchants, but is not specific about where these businesses were located. The year 1857 saw the establishment of the Standard Works of Lawson and Holden, axeltree and spring makers, at 24 Sheepcote Street. Lawson and Holden previously worked from 147 Great Charles Street, but were listed at 24 Sheepcote Street up to 1875. The Post Office Directories for this period also list a coal merchant called George Stafford at 24 Sheepcote Street. In 1875 the

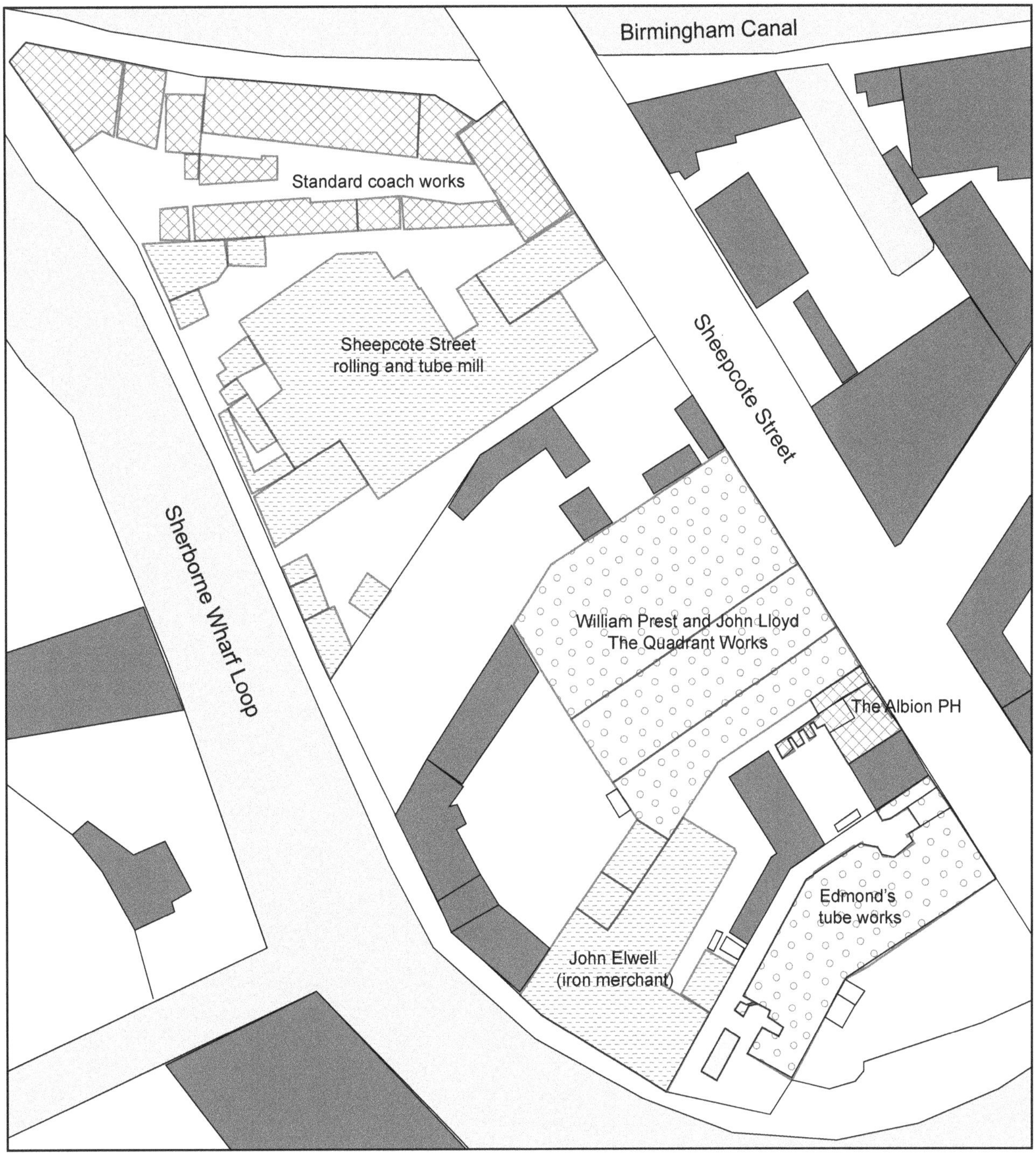

Figure 8.9 The Sherbourne Wharf Loop and Sheepcote Street c 1887

coal business was taken over by William Bottrell but had ceased by 1881. The Standard Works is still recorded on the First Edition Ordnance Survey in 1887, but as making coach ironmongery, suggesting that the company may have changed ownership (Fig. 8.8). In 1871 Grice, Grice and Booth, the owners of the copper tube and rolling mills, first appear in the trade directories and are visible on the First Edition Ordnance Survey map in 1887 (Fig. 8.8).

1887–1903

The First Edition Ordnance Survey maps published in 1887 depict an industrialised enclave, largely contained within the Sherborne Wharf Loop, which was dominated by metal working (Fig. 8.9). At this time the company was known as Grice, Grice and Booth, but by 1892 Kelly's Directory lists the company as James Booth and Co. In 1896 they took over the premises of the Standard Works at No. 24 Sheepcote Street. Building plans for the construction of the present frontage to No. 25 Sheepcote Street in 1890, have survived (BCA MS 3375/411595). The building plans denote the proposed functions of each room within the new building and also show the extent of the new work and the way in which existing walls were adapted into the build.

The expansion of James Booth and Co in 1896 resulted in the takeover of the adjacent Standard Works and the alteration of the buildings on the site. Building work created a series of plans, three of which have survived (not illustrated) - the change in the plot is revealed in Fig. 8.10. The first plan is titled 'Sketch Plan of Restored Shopping', dated June 1896. The plan is colour-coded and seems to refer to alterations to Structure F (i) which involved the strengthening of the main elevations and the insertion of a new roof which is similar to the one currently in position. This work must have involved the demolition of the two, two-storied, structures denoted inside Structure F (i) on the 1:500 Ordnance Survey map of 1887 (reproduced as Fig. 8.9). However, the south-facing elevation of F (i) depicted on the plan differs from that in existence today, as it follows the pillar-and-panel arrangement of the north wall. Two more bays on the west end of Structure F (i) were shown on the 1896 plan divided from the main shopping by a partition wall with one opening, which was labelled 'the engine house'. No trace of this building survives today.

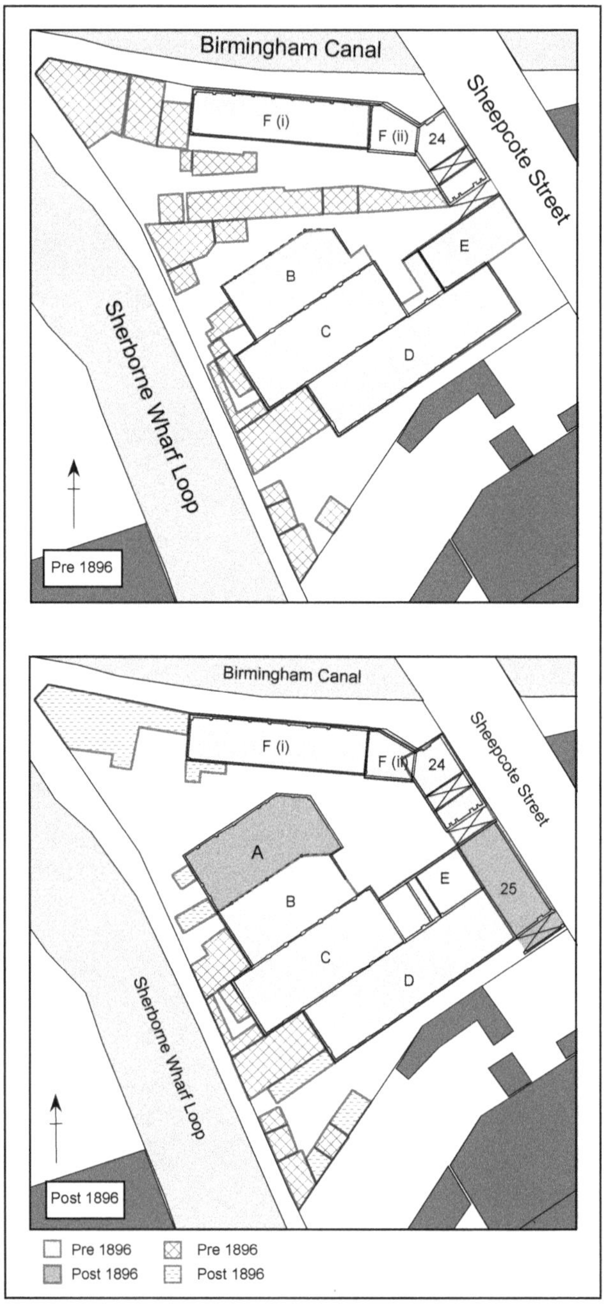

FIGURE 8.10 DEVELOPMENT OF BOOTH AND CO TUBE AND ROLLING MILL IN THE 1890S

The second plan, dated July 18th 1896, depicts the construction of a casting house with a similar roof to Structure F (i). The north wall contained a range of ten flues arranged in pairs. No trace of this structure, which was located to the west of Structure F (i), has survived. The third plan, dated August 26th 1896 depicts the construction of Structure A, its' amalgamation with Structure B, and the demolition of the southern range of workshops which used to form part of the Standard Works. At this point Structure F (i) is called 'the tube drawing and finishing shop'. It seems likely that the south-facing elevation of F (i) was altered to allow greater access to the tube drawing and finishing shop at this stage. It is unclear whether the arcaded south-facing wall of F (i) was built at the same time as the new roof depicted on the first plan.

The second edition Ordnance Survey of 1903 confirmed the completion of work outlined in the building plans above (not illustrated). The boundary between the two businesses, which was visible on the first edition map, is no longer present, and the gateway to the former Standard Works has been filled-in. With the exception of a number of relatively minor additions and the demolition of a number of small buildings to the rear of Structures A, B, C and D, the basic ground plan of the tube works remained unchanged.

PLATE 8.2 NOS 24 AND 25 SHEEPCOTE STREET, NORTHWEST FACING

The Buildings

Both works shall be described here together as they were in their final manifestation. The works existed on two levels. The level along Sheepcote Street provided access to the main ranges of offices and shopping at the front of the plots. At the rear, at basement level, workshops stretched back towards the canal of the Sherborne Wharf Loop. The two offices of the Standard Works (No. 24) and those of James Booth and Co. (No. 25) are described as they were numbered in 1999, whilst the workshops have been numbered as they were in the 1999 survey.

Number 24 Sheepcote Street (Fig. 8.10)

This was an imposing three storey mid Victorian building with basement, built *c* 1857 (Plate 8.2). The slate roof was hipped at the north gable end. The main elevation five bays in length, built in Flemish bond, machine-cut blue/black engineering brick on the ground floor, machine-cut red brick above. Decoration was with engineering and red brick voussoirs above the window and moulded string courses between floors. The arched support of the cart track through the bricked-up gateway was still *in situ* identified in the cellar. The cellar had been broken through

PLATE 8.3 BUILDING F (I), NO. 24 SHEEPCOTE STREET, SOUTHEAST FACING

to workshops at the rear. This was presumably blocked with the amalgamation of No. 24 into No. 25, when James Booth and Co. took over the Standard Coach Works. The ground to second floor of the interior had been altered but suggested an interior plan of shopping and show rooms on the ground floor with offices on the upper floors.

The workshops

Building F (i) was a single-storey workshop, nine bays in length that ran adjacent to the canal behind No. 24 Sheepcote Street. It probably dated to the original build of the Standard Works in 1857 (Plate 8.3). It was built in hand-made red brick in the English bond with a pitched roof with four evenly spaced ventilation hoods. Originally the nine bays to the north, facing the canal, had cast-iron framed windows (according to 1896 building plan) but had subsequently been bricked-up. The southern nine bays were all arcaded and open to the remainder of the works. Building F (i) connected directly to a second, Building F (ii) to the east, that linked it to No. 24 Sheepcote Street. It likewise dated to 1857. This was originally three-storey shopping, but subsequently the upper storey had been removed. It was built in hand-made engineering brick and was four bays in length. The ground floor was open with a single cast-iron column centrally placed and it continued straight through to the basement of No. 24 Sheepcote Street. The first floor had been refurbished as an office.

The workshops of the Standard Works appeared to have been altered as part of the amalgamation with the Rolling and Tube Works. Originally Building F (i) was probably an assembly shops adjacent to the canals. A plan dated June 1896 shows the changes to the building. Notably at the western end of the building were a further two bays described as the engine house, that shows the location of power for the works. A two-storey element of the works (depicted on the 1:500 Ordnance Survey map of 1887) was demolished and the southern arcaded element was built to allow easier access. A second plan dated July 18th 1896 shows the adjacent buildings in the western corner of the land plot (now demolished), which have a series of ten flues arranged in pairs along the canal and is called the 'casting house'. A third plan dated August 26th 1896 shows the entirety of all the southern workshops between the Standard Works and the tube works and rolling mill have been demolished and replaced by Building A. Building F (i) is now called the 'the tube drawing and finishing shop'.

Number 25 Sheepcote Street (Fig. 8.10)

This was a purpose-built late Victorian office building built *c* 1890. There was an elaborate front façade in Neo-Gothic style, with terracotta mouldings. The use of riveted I-section girders, cast-iron pillars, steel floor-joists and vaulted ceilings were typical of this period. The main elevation comprised fourteen bays, in Flemish bond machine-cut red brick. Two sets of semi-circular arched double-doorways, with end bays one either end formed by double-bay gateways. Windows on the ground and first floor were balanced-sashes, the head formed by a moulded terracotta sill. The second floor windows were likewise balanced sashes with semi-circular arches. The interior contained many original features, including the tiled floor of the entrance hall and the wooden ceiling of the counting room. The 1896 building plans reveal that the basement was a workshop, whilst the ground floor had the metal store and the packing room. The first floor was split into two with the counting room on the northern side and an office and living space on the south. The second floor was likewise bedrooms and stores.

The workshops

Building A was constructed between 1896 and 1903 as a single-storey extension to Building B with which it formed a large open workshop. The walls of the structure were of pier and pillar design in machine-cut red brick in English bond. The central division between the two buildings was formed by an arcade of cast-iron pillars that supported an internal valley between the Building A and B. The roof was pitched with ventilation hoods above.

Buildings B and C were two long single-storey workshops of contemporary build in machine-cut red brick in the English bond and dated to the earliest period of the works *c* 1871. The walls were generally of pier and panel construction except the shared wall between the structures, where open bays were formed by wooden lintels between the piers. Both had pitched roofs supported on composite wooden trusses with iron tie-beams equivalent to the queen posts. Originally it would appear that the structure was one large open space that was subsequently altered upon the construction of Building A; this corresponded and post-dated the expansion into the adjacent Standard Works. Building C had a separate room in the southwest bay.

Building D was a long single-storey factory workshop built of machine-cut polychromatic red and engineering brick in the English bond (Plate 8.4). It extended directly back from No. 25 Sheepcote Street adjacent to Building B. The walls were again pier and panel and shared the wall with both Buildings C and E demonstrating it was slightly later and dated to the period 1871 to 1888. The shared wall with Building C again consisted of pier and panel construction wall with wooden beams between the piers to create an open arcade, subsequently bricked-up. Bricked-up openings also passed to Building E. Openings at first floor level suggested that walkways probably existed at this level above the machine floor. The pitched roof was again supported by trusses of composite wood and iron-tie construction with ventilation hoods above.

Building E was slightly enigmatic, being much smaller than those around it. It was a two-storey building built in machine-cut red brick in the English bond, with a pitched roof. It dated to the period 1871 to 1888 and originally the plan extended all the way to Sheepcote Street, before it was subsequently curtailed by the construction of 25 Sheepcote Street in *c* 1890. The ground floor plan consisted of two

PLATE 8.4 INTERIOR BUILDING D, NO. 25 SHEEPCOTE STREET, EAST FACING

rooms with a central cross-passage. The first floor was a single open room that had been altered to allow direct access to the packing room of No. 25 Sheepcote Street.

Discussion of the works

Land use in the surrounding area in the 19th century was sharply defined (see Fig. 8.8). Immediately to the north was a transportation corridor defined by the New Main Line Canal and the Birmingham to Dudley railway, which provided ready access to the local and national transportation systems. Much of the surrounding area, beyond the transport corridor, was devoted to high density housing, including courts and back-to-backs. Within the industrialised zone many, but not all, of the works along the south-west side of Sheepcote Street were arranged behind a frontage which presented a 'respectable' face to the street (Fig. 8.9). Access to the works behind was provided by several gateways. The scale of these works was not domestic, but neither were the works particularly large-scale concerns. Several of the plot fronts conform to a *c* 15m width, which then open out into an irregularly-shaped space between the street and the canal behind. The scale of industrial activity is still largely consistent with that of family-owned businesses. Some types of metal working, such as tube making, were more heavily mechanised, with production taking place within large factory units as opposed to skilled craftsmen working in 'shoppings'.

The two types of business illustrate the contrasting requirements of works devoted to the production of processed metal (Sheepcote Street Tube and Rolling Mills, 25 Sheepcote Street) and those devoted to the production of finished goods (Standard Works, 24 Sheepcote Street). The different scale of the two works is immediately obvious. The Standard Works consisted of two long, but narrow, ranges of workshops set either side of yard with access through the 'respectable' frontage of 24 Sheepcote Street. None of the workshops was particularly large. Production would typically have focused around small hearths for casting and forming, supplemented by 'fast and loose' belt driven lathes and polishers for finishing the products. The small-scale buildings could be readily modified or extended within the confines of the site to adapt to new techniques. The wharf to the rear of the plot may have been used for importation of raw materials, including metal sheets and coal. The narrow front gateway controlled access to the site. Finished goods probably left the works on carts through this gateway for local and national distribution rather than via the canal.

In contrast, the Sheepcote Street Tube and Rolling Mills comprised large open workshops. These would have represented a substantial capital investment. It is not surprising that the works was set up first, to be followed in 1890 by the construction of a 'respectable' office block at the front of the site. Overall the workshops represented a series of covered spaces that represented the machine floor of the rolling and tube mill. Given the understanding gained from the excavation of the Cambridge Street works it may be suggested that Buildings B and C, and subsequently Building A represent the floor of the rolling mill. The separate room in Building C may have housed the engine for power. These would provide direct linear linkage to these buildings with wheel pits running parallel and drive shafts perpendicular to the building. Building D would appear to be the most ideally located for the tube works. Building E could be interpreted as a suitable location for the engine house but it seems more likely, given its location close to the street, that this was originally some form of shopping and offices that were subsequently replaced by 25 Sheepcote Street. The location of ancillary structures like the casting house and muffles do not appear to have formed part of the recorded buildings. A number of buildings were located between the canal and Buildings A–D, the rolling mill and tube works. These could be the remains of the boiler houses, casting shop and muffles. The boiler houses in particular may be represented by a series of small buildings adjacent to the original workshops, Buildings B and C. Two small buildings in the southwest corner of the plot may represent the muffles and casting shop.

When the works were reorganised in the 1890s there appears to have been an emphasis on improving the open-plan nature of the site (Fig 8.10). Internal walls were arcaded, and the new workshop Building A was built with cast-iron columns as opposed to a solid brick wall. This would allow the movement of metals around the workshops more fluidly, and would create greater levels of efficiency. The acquisition of the Standard Works also allowed further capacity for casting metal with the new casting shops constructed, and the rationalisation of the works, with the Standard Works converted to tube manufacture and possibly the new Building A also involved in this work. It appears to have allowed the works to expand.

The re-organisation and expansion of the 1890s were in contrast to the Cambridge Street works and could not be more telling. The construction of new offices and the takeover of the Standard Works by James Booth and Co are in direct opposition to the decline of the Cambridge Street works. Concentration on large-scale manufacturing processes of tubes and rolled brass was clearly more suitable for the market at this time. The newer works adopted sites with the potential for expansion as opposed to the now limited sites within the inner suburbs. Nos 24–25 Sheepcote Stre*et a*lso had much better communication routes via the canal, without the necessity of using Gibson's Lock.

ICKNIELD PORT LOOP – A CASE STUDY OF SUBURBAN MOVEMENT

Leonie Taibi and Chris Hewitson

During the 19th century there was a rapid increase in the number of industrial sites around the Icknield Port Loop, mainly metalworking industries (Skipp 1980, 53). By 1887 around 20 separate companies had established themselves around the Icknield Port Loop. In particular the area became quickly adopted for new tube, rolling and wire works. The area provided extensive new land for purpose built factory sites. The survey as part of a study of the area in 2005 (Driver and Hislop 2004) revealed the extent of surviving or partially surviving metal works in the area.

Icknield Port Rolling and Wire Mills, Icknield Port Road (Fig. 8.11, site 1; Plate 8.5)

This site on the corner of Osler Street and Icknield Port Road has been occupied by industrial premises since the 19th century. Presently used by the Hermetic Rubber Company, the site was originally the Icknield Port Rolling and Wire Mills. Trade directories dating from 1863 list Vivian Henry Hussey and Co as the manufacturers. Amongst the goods they produced were brass and copper tubes, German silver and brass and copper wire. The site became occupied by the Winfield Rolling Mills in the

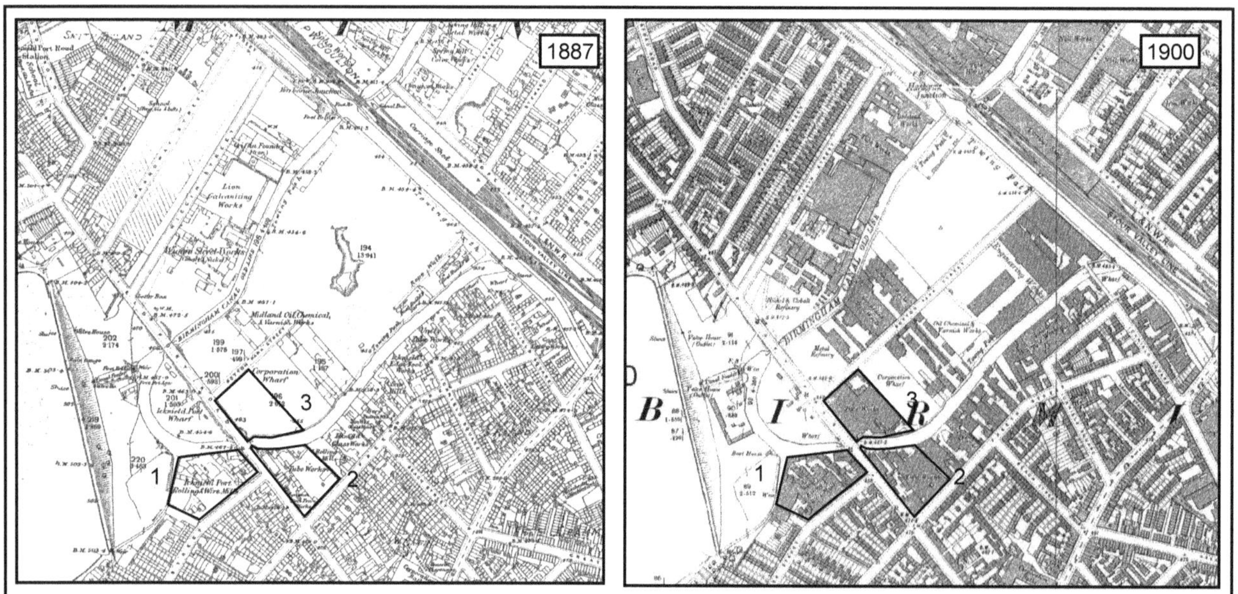

1. Icknield Port rolling and wire mills 2. William Morris rolling mill 3. Weldless Steel Tube Co Ltd

FIGURE 8.11 CARTOGRAPHIC DEVELOPMENT OF TUBE AND ROLLING MILLS AT THE ICKNIELD PORT LOOP, 1887 TO 1900

PLATE 8.5 ICKNIELD PORT ROLLING AND WIRE MILLS, ICKNIELD PORT ROAD, NORTHWEST FACING

1930s when they became the Icknield Port Rolling and Wire Mills. The front of the site is today occupied by a large 1930s brick-built office building with hipped plain tile roofs. Flanking gateways give access to the works, which is largely accommodated in the original later 19th-century red brick industrial buildings. These are aligned northwest to southeast and comprise a double-pile structure with prominent coped gables.

William Morris Rolling Mill, Freeth Street (Fig. 8.11, site 2; Plate 8.6)

William Morris Rolling Mill at the Freeth Street site was listed in trade directories since 1878. The works occupied a large site on the corner of Freeth Street and Icknield Port Road and were further extended in 1900 to occupy the corner of Rotton Park Street (this part of the tube works still survives). The Freeth Street frontage appears to date from the mid 20th century. An inspection of the interior of the building revealed that although the front of the site is occupied by a comparatively recent office block, a number of older rolling mill buildings survive to the rear. These industrial buildings date from the late 19th century through to the early 20th century. The principal block comprised a double-pile construction aligned northwest–southeast, the panelled side walls being articulated by pilasters carrying I-beams. Steel spinal framework carrying the central valley and steel roof trusses. This roof structure dates from the later 20th century following a fire at the factory. Extension to the rear, taking the structure as far as the canal, part of which retains its timber roof trusses. To the southwest of this building, another structure is roofed with wooden Belfast trusses. Another building incorporates the lower part of a large, square, brick chimney (Plate 8.6). A chimney in the same position is marked on the 3rd edition OS map (1914, not illustrated).

Weldless Steel Tube Co Ltd, Icknield Port Road (Fig. 8.11, site 3; Plate 8.7)

The premises of the Weldless Steel Tube Co Ltd were part of the complex that occupied the corner of Rotton Park Street and Icknield Port Road site. They existed from 1878 until the construction of a large cinema in the 1930s (now Bill Landon and Sons). This tube works is an extension of the demolished tube works on Freeth Street (see above). No. 298 and 300 Icknield Port Road, date from the early 20th century and are single-storey factory buildings, accessed via the back of the rolling mill. The remainder of the tube works is known from trade directories to have been established by 1900. The main building fronting onto Icknield Port Road and Rotton Park Street is red brick with terracotta dressings.

Birmingham Tube Works and Rolling Mills, Heath Street South (not illustrated)

This manufactory was constructed early in the 1890s and was run by the tube manufacturers Earle, Bourne and Co (Kelly 1892). The partnership of John Earl and George Bourne started in 1874 but from 1896 it was run by the Earle family. The works occupy a large area with access

PLATE 8.6 WILLIAM MORRIS ROLLING MILL, FREETH STREET, EAST FACING

PLATE 8.7 WELDLESS STEEL TUBE CO. LTD, ICKNIELD PORT ROAD, SOUTHWEST FACING

from Heath Street South and the east side of the main factory building was constructed against the canal. Below this was a further tube works and rolling mill, which took up the rest of the Mason College site. This manufactory began operating around 1898, in which year Kelly's Directory lists Grice, Grice and Son as tube manufacturers (formerly of the partnership at Sheepcote Street). This again was a large structure spanning between Heath Street South and the canal. The south end of this factory was built against the Spring Hill Coach Iron Works (not named on this map), which retains its original arrangement of factory buildings.

Discussion of the Icknield Port Loop area

The area as a whole was specifically adopted by new metal industries seeking larger sites outside the city centre. It had several advantages: available land, good transport links by both canal and rail – goods sheds lay on the Stour Valley Line to the northeast accessed via the canal, it lay in an area which although removed from the city centre was sufficiently close to take advantage of pre-existing supply and trade links, the area was surrounded by newly built terraces in the late 19th century that would house the necessary workforce for the factories. The scale of the works at these sites is closer to that of the Sheepcote Street works than that of the Cambridge Street complex. The buildings were large, laid out with specific industries in mind and did not adopt the 'all industries under one roof' approach that had characterised the mid 19th century at the Cambridge Street Works. They in fact display the characteristic change towards larger manufacturing works in the industry towards the later 19th and early 20th century. It is notable that even in these new locations expansion on the site was restricted and most businesses remained medium in scale and did not expand after the Second World War.

CHAPTER 9: THE RISE AND FALL OF THE CAMBRIDGE STREET WORKS

Chris Hewitson

The Cambridge Street site is a snapshot of the development of the British brass and metal manufacturing industry in the 19th and early 20th century. They provide an example of the advances in technology, construction and production of brass as a product as well as later mirroring its downturn as a commodity. The fortunes of the companies that ran the Cambridge Street site throughout its lifetime reflect economic and social changes in the industry as a whole: the early innovations of the Union Rolling Mill and its founders, the massive expansion and success of the Winfield's works, wholesale change in ownership, fragmentation of the works, and the utilisation of the works for the War effort.

The Cambridge Street works was intrinsically linked with brass production in the 19th century, in semi-processed brass sheet, wire and tube, large manufactured goods including metallic bedsteads, smaller household goods and intricate craft design. The excavations have allowed the examination of two differing aspects of the Cambridge Street works trade: the north of the canal was the heavy industrial base of brass rolling, wire drawing and tube manufacture whilst to the south was the four-storey bedstead works, the tertiary element of the industry involving manufacture. The following discussion seeks to bring together the factors that influenced the rise and fall of the Cambridge Street works over its lifespan between the 1820s and 1930s. These have been defined as economic changes, technological changes and social changes that affected the industry.

THE ECONOMIC CHANGES – GROWTH AND DECLINE

The development of the Cambridge Street works and its eventual demise reflects changes in the patterns of industry. The expansion and contraction of Winfield's must be viewed within the wider context of the 19th-century economy. At the start of the 19th century, Cambridge Street was located on the rural fringe of Birmingham. Over a century rapid expansion of the city occurred to the north and west of the centre (McKenna 2005a; Fig. 8.1), which gradually became part of the core. The movement of the brass industry away from the city centre was typical of the development of industry in the city as a whole. Initially the brass industry was domestic in scope. Areas including the Jewellery Quarter, the Weaman Estate and Eastside, developed out of an 18th-century industry that was small in scale and based predominantly in a domestic setting or in small manufactories. By the early to mid 19th century, larger firms had begun to develop as a result of new technology and the advent of the principals of the factory system. The Cambridge Street works represents one of a number of works that developed in this context in areas around the city centre in areas peripheral to the Jewellery Quarter, in new estates south of the city on the River Rea flood plain and west of the city around Broad Street. The areas were associated with canal links - the Birmingham and Fazeley Canal, the Digbeth Branch, the Birmingham Main and Old lines to Wolverhampton and the Birmingham and Worcester. The key to 19th-century development in these areas was available land and transport links provided by the canals.

The Cambridge Street works was one of a series of firms that developed on the Baskerville Estate. Although archaeological evidence for the works of other firms was poor due to later developments (such as the construction of the civic complex), the evidence from documentary sources allows us to see the scope of industry in the area. Cambridge Street came to be the most dominant manufacturer in the local area but it was one of a series of factories that produced a range of products, both part-finished, (eg The Crescent Foundry) and finished, (eg The Broad Street Metal Works, Nettlefold's Screw Manufactory). The pre-dominance of the metal industries within this sector of the town was also apparent. In particular a relationship can be seen with the copper alloy manufacturers (eg steam engines, chemical apparatus, pin, needles and screws) that suggested possible adjacent markets for the semi-manufactured products of the Cambridge Street rolling mill. Wire could be made into needles, pins and screws, tubes and rolled brass would be used in the manufacture of steam engines and chemical apparatus. Iron manufacturers such as the Crescent Foundry provided wrought iron for manufacturers such as the railway and nail works. The Cambridge Street bedstead works was another business which used iron as well as brass in its manufacturing process. The patterns revealed are of inter-dependent metals industries where industrial linkage on a localised level can be assumed. Obviously this block of buildings was not inter-dependent at the expense of the rest of Birmingham and this provides an example of how related industries thrived in close proximity due to the interchange of both products and knowledge. The patterns of development that formed the Jewellery and Gun Quarter have been acknowledged for a number of years (Wise 1949, 57–72) and similar patterns have been identified for the cloth trade in London (Hall 1960, 155–178). These patterns have also been examined in the 20th century (Birmingham, Taylor and Wood 1972;

Sheffield, Watts *et al* 2003, 615–630). The concept easily transfers to the 19th century Baskerville Estate - Amin has stressed how, within clusters of traditional industries, *'economic interdependency, social familiarity, and face-to-face contact... helped firms to reduce the cost of their transactions ... [and]... facilitate the flow of information and knowledge'*(Amin 2000, 153). This is clearly how early industrial estates developed in Birmingham: exchanging trade and knowledge in a confined area.

Yet as transport networks improved, the necessity for an immediate geographic relationship between different industries decreased. The progression of firms from the inner suburbs to the outer suburbs from 1870 to 1930 has demonstrated that businesses moved to new sites when they had good transport links. The development of new sites along the corridor of the canal and railway is apparent from the investigations at Sheepcote Street and Icknield Port Loop. The Sheepcote Street site in particular has demonstrated the advantage of scale in the new sites that allowed the business to expand and adapt new technology. These sites had an innate advantage compared to constricted site at Cambridge Street in the inner suburbs. The sites at Icknield Port Loop demonstrate once again how the primary brass manufacturing industry expanded from the core of Birmingham and adopted new industrial sites away from inner suburbs, allowing greater scope for expansion in the later 19th and early 20th centuries. The transition of firms such as Grice, Grice and Sons from Sheepcote Street and Winfield's Rolling from Cambridge Street to the Icknield Port Loop demonstrate the benefit of these moves.

These movements should be set against the wider economic climate of the 19th and early 20th centuries. After the post-Napoleonic War depression of 1817–1820, Britain's export trade in textiles, iron and machinery grew in the early 19th century due to improved technology, attracting foreign custom with low cost. Further economic decline was to follow periodically in the late 1830s to early 1840s and the 1870s (McCloskey 1994, 249–253). The reduction in profits as a result of the 'Great Depression' from 1873–1876 was felt throughout the heavy metals industries. Kenrick's of West Bromwich saw sales double between 1870 and 1892, yet net profit fell from 26% in 1876 to 6% in 1886 (Staples and Staples 2001, 41). From the 1870s the increase in foreign competition, from the continent and North America, appears to have been one reason for the decline in the profitability of metal producing trade. Yet over the entire period there was growth in productivity in a series of industrial sectors including the brass industry (McCloskey 1994, 249–253). What is apparent is that individual firms, including the Cambridge Street works, experienced a continual rise in productivity between 1820 and 1890, despite fluctuations in economic growth. In the late 19th century the Cambridge Street works business collapsed and this was not an isolated occurrence. The collapse in traditionally-run firms was common in the late 19th and early 20th centuries and has been defined as the collapse of structures of paternalism (Staples and Staples 2001, 57–92). Suggested reasons include incompetence, a conservative outlook, or successful performance that did not ultimately aid the good of the business, the use of outdated technology as a cost saving being a prime example (Pollard 1994, 72–78). The collapse of the Cambridge Street works as a firm may have resulted from foreign competition from both North American and continental markets into traditional areas of their production. Yet the archaeological and historical evidence suggests a different reality. Production continued on the site in the form of a number of smaller firms, some continuing traditional products whilst others adopting new lines. The period before the First World War has long been seen as a period of decline in Britain during which traditional industries, including iron and steel, receded in economic importance. The inter-war economic downturn accelerated the process (Harley 1994, 330). At the Cambridge Street site the demise of production could be seen as a result of a period of economic downturn, yet production of both brass and aluminium continued within the city in different locations. A change in location was designed to improve efficiency in the face of economic pressure. In the case of Cambridge Street it is arguable whether a change in location was inevitable or merely the result of the construction of Centenary Square.

THE TECHNOLOGICAL CHANGES – PRODUCTION AND PRODUCTS

The development of the Cambridge Street site allows examination of advances in technology, construction and production. It also shows the changing fortune of brass as a product. This has been examined in Chapters 4, 5 and 7 in terms of what products were being produced, how, and whereabouts in the works. Here we shall bring these separate strands of the argument together to discuss the influence of technology on the Cambridge Street works. Technological changes can be neatly divided into two aspects, production and products.

As suggested in Chapter 4, the Union Rolling Mill developed because of competitive design of technology. Rolling mills in Swansea based on steam technology appeared at first the Hafod Works in 1819 (Toomey 1985, 361–363) and later the Morfa Works in 1828 (Hughes 2000, 48–49). This may have encouraged the building of an integrated rolling mill with the latest technology applied. The Union Rolling Mill cannot claim to be the earliest rolling mill in Birmingham as the application of James Watt's reciprocating steam engine to rolling had been undertaken since the 1780s. The principal advantage of the stationary steam engine was its mobility, which allowed rolling mills to be set up away from the traditional locations adjacent to the water power of streams and rivers. In the early part of the 19th century, long-term contracts between raw product supplier and manufacturer were adopted for the purchase of raw materials. These could be risky and open to change in market circumstances. By the 1820s middlemen, agents who acted on commission and owned rolling mills, began to buy cake copper and undersell established prices (Toomey 1985, 312–325). The Union Rolling Mill therefore took advantage of economic change to implement technological change to the advantage of the Birmingham market. Again the developments of the 1840s - the improvement of the Cambridge Street Rolling and Wire Mill should be examined within the national context of the industry. The 1840s were a period of instability in the copper markets with low prices, but by 1847 shortages in copper supply due to increased demand from the patent (ship) sheathing, railway engine manufacture and the brass industry itself meant export of foreign markets had to be restricted in order to satisfy the domestic market (Toomey 1985, 331–334). The construction of a new steam engine on the site may reflect the influence of economic change on raw product prices. Investment in new technology could be used to combat reduced profit margins. This is an example of what has been suggested as 'not an Age of Cotton or of Railways or even of Steam entirely; it was an age of improvement' (Floud and McCloskey 1994a, 118). The incremental improvement of technology led to improvements in production.

The 1870s and 1880s was a period when technological investment was low in the Cambridge Street works, in part due to economic (the 1870s depression; the fire at the bedstead works in 1888) and social (the rise of unions) difficulties but also due to increased investment in other areas of development (eg electric lighting venture in the 1880s). Yet other metal firms in the West Midlands were able to invest in new technology or improve their works. In West Bromwich, hollow-ware manufacturer, Kenrick, reacted to economic decline in the metals trades with an aggressive investment in new plant and technology. They were successful and in part this may be explained by the financing arrangements - the loans were secure as Kenrick was a member of the board of Directors of Lloyds Bank (Staples and Staples 2001, 41). In contrast during the 1880s Winfield's and Co. were in legal proceedings over reclaimed costs from their unsuccessful lighting venture with Crompton and Co. Change in management and flotation as a limited company in the late 1880s cannot have helped stability. The hammer blow of the fire to the bedstead's works in 1888 was also detrimental, but ultimately the absence of investment in new machine stock must have placed them at a severe economic disadvantage in contrast to a company like Kenrick's.

Perhaps a better parallel is provided by the adjacent land plot of the Birmingham Metal Works. Here a company built on the same lines as the Cambridge Street works, specialising in brass wire, sheet and tubes, also saw demise. Yet in contrast the overall business survived. This in no small part must be due to the variance in enterprise between the two brothers who ran the firm. Alfred and Francis Bolton took over the firm at a young age. Alfred, 26, dealt exclusively with the Oakamoor, Cheadle side of the business whilst the younger Francis, 25, took over the Birmingham side based on Broad Street at the Birmingham Metal Works (Morton 1983, 44). Alfred expanded the business rapidly into new markets of electrical cable and wire, whilst continuing to invest in new machinery in the traditional rolling and tube making. In contrast Francis allowed the Birmingham business to stagnate whilst he concentrated investment on unsuccessful property acquisitions (*ibid*, 43–63). The reality of poor management of the firms may be true but the Birmingham firm's confined location cannot have helped (*ibid*, 63). In contrast, the Oakamoor site had ample space for expansion. At Birmingham old stock and machinery could not be easily replaced, as it would have to be done within the confines of the site. Replacing old machinery would therefore involve decommissioning it and replacing in the same location and no work would be undertaken during this process.

This may provide a more realistic model for the Winfield's Ltd period of the Cambridge Street works. Poor managerial decisions affected the business throughout the 1880s and an inability to finance and replace machinery that in many cases was over 50 years old put it at a distinct economic disadvantage. The improved rolling machinery of the 20th century is clearly visible in the archaeological evidence. Winfield's Rolling Ltd invested in new machinery and was able to survive for another 30 years. Yet ultimately the example of Oakamoor holds fast and in the 1930s the Winfield's works migrated to a site with more ample space for expansion at the Icknield Port Loop, away from its confined city centre location. This shows that the Cambridge Street rolling and wire drawing mill was not necessarily uneconomic and there were markets that could sustain the mill. Again the advent of new technology was visible as attempts were made to keep the rolling mill competitive: a new electricity plant that provided power to the rolling mill and new gas-fired technology to power the muffles.

Economic downturn may be a reflection of changing demand for products. . Unlike other manufacturers who specialised in either large heavy-duty products or small high-quality articles' Winfield's were very diverse. Consumption of heavy rolled metal and wire remained high because it was easily adaptable to changing markets, where general brass wire could be replaced by copper wire for electrical purposes. Sheet brass and metal could be adapted to a variety of trades that would have included munitions casing during the First World War. The same could be said for tube and engineering plumber's brass foundry departments that could adapt from steam and water based trade to the expanding gas market. Yet in comparison, the more traditional brass products, such as fine metalwork and bedstead's were subject to changes in fashion. In the case of the bedstead works this was a highly profitable line of production that became unfashionable.

Even the raw material was subject to fluctuations in consumption. The metal used in the 19th century at the works remained predominantly brass. The start of the century saw the expansion of the brass trade reaching its height in the 1860s. Although new metal compositions such as Muntz's Yellow Metal (a high-zinc brass derivative; Aitken 1866, 313–314) were developed in Birmingham, the metal used at the Cambridge Street works appears to have remained the same. The examination of the brass from the site suggests that from the 19th to the 20th century the composition of brass metal appears to have remained within the typical range of compositions from 80/20% to 60/40% (copper/zinc ratio). There tended to be a correlation towards the higher end of the zinc ratio at 65%/35% (copper/zinc ratio; see McDonnell, Chapter 7) to allow for zinc loss during processing. This may suggest that the Cambridge Street works may have been developing new lines of products, but was conservative in the raw material it used.

The continuation of separate departments of the Cambridge Street works as individual businesses in the years immediately prior to the First World War reflected the continuing viability of the product range. But they were heavily reliant on cast brass technology. This was being replaced in the early 20th century by new processes such as extrusion and the use of the power press to shape brass in a die. In addition new materials were being used. The ultimate use of the southern side of the site for the manufacture of aluminium castings highlights the change of material technology towards lighter more flexible materials, which corresponded with the decline of brass. The advent of the motor car and the future development of aircraft technology meant lighter materials were more suitable. The sale and re-organisation of the works allowed the development of this new technology at the site. But ultimately even the Birmingham Aluminium Casting Company moved from the site to take advantage of a larger site where new technology could be employed (see Shill, Chapter 5).

SOCIAL CHANGE – MASTERS AND WORKERS

The industrial transformation of Birmingham that occurred over the course of the 19th century led to the social transformation of the working environment, both in respect to the relationship between the owners and the workforce but also to the workforce and their workplace. The Cambridge Street works bridges the transition from the domestic industrial works to the large modern manufacturer.

The manufacture of brass in Birmingham in the 18th century operated from small workshops and enterprises making 'toys' that is small items such as trinkets, candlesticks, buttons, pins, snuff boxes and buckles, as well as larger items such as swords and guns. The workers were used to adapting their skills and moving easily between trades. Yet in other trades, a particular examples in Birmingham being the gun trade (Wise 1949, 57–72; Behagg 1998, 1–15) the sub-division of labour meant that the workers became highly skilled and could command high wages. What followed by the end of the 18th century was flexible working patterns, often working long hours at the end of the week in order to have Monday, as well as Sunday, as free time. The tradition known as 'St Monday' or 'piece work' was the norm and although there were some large factories, notably the brass foundries, most of the work continued to be done in small enterprises. The practice of garret masters employing their own families and thus undercutting established firms originated in 18th-century practices. Aitken (1866, 223) comments that, '*the peculiar characteristic of industrial Birmingham during the last century was the large number of small masters employing a few workmen in the various trades*'. The start of the 19th century saw more organised works develop but the role of the garret master employing cheap labour in the form of women, children and juveniles continued. Robert Winfield began his work as one of these small tradesmen but the path of the Cambridge Street works highlights the rapid change that small businesses undertook in the first half of the 19th century.

The transition to a more complex model of management was one of the achievements of the 19th century. The Cambridge Street works was a good example of a large-scale employer of the mid 19th century. Most brass houses in Birmingham employed just over 100 men but Winfield's were employing 800 people with a weekly pay roll of nearly £3,000, an average of nearly £4 a worker by 1866 (Stephens 1964, 81–139). Winfield's in its greatest period of success can be seen be a model for what Staples and Staples term 'Paternalistic Despotism' (2001, 52–56). The paternalistic management style of the English family firm has been argued to have been a more successful model for economic success in these years than European (in particular German and French) military or bureaucratic models (Joyce 1980, 136). Robert Winfield had certainly developed paternalistic instincts by the middle part of the century. Certainly as a large employer, his practices were at the forefront of the paternalism in the West Midlands before model firms such as Cadbury developed.

Clearly there was a transition from the earliest period of Winfield's work in the 1820s, when the business was small scale, to the business at its height in the 1860s. One of the themes apparent from the excavation work is the separation of departments within the works. Although the Cambridge Street works was a single business, to understand the management of the Winfield's works it may be better to see each department as an individual entity. Interconnectivity between the rolling and wire mill was kept separate from the bedstead works, with only walkways providing any link between the two. There would appear to be sufficient allowance for each individual department to operate separately. Yet cooperation was still possible where scale made sharing of facilities more economic. An example would be annealing hearths serving both the bedstead works and rolling and wire mill. In this way a layered managerial structure was present at the works that bridged the gap between the small-master system of the 18th century and amalgamated businesses that operated by the end of the 19th century.

A paternalistic management system was interrupted at the Cambridge Street works by the death of Robert Winfield and from this period a change in structure is apparent.

Winfield's works became firstly a cooperative and then a limited firm in the period from 1870 to 1897, before its eventual demise and sale. The success and failure of this final system of management should not be viewed in isolation. Technological advances and failures, economic pressure due to fierce economic times following the depression, and external social change were all factors.

The high proportions of young males and females in the brass trade highlighted the division of labour apparent in the industry. In 1861 of 8,334 people employed, 28% (2,379) were males under 20 years of age and 25% (2,119) females (Aitken 1866, 361). The heavier elements of the industry were the preserve of adult males. Yet the vast majority of work, piercing, stamping, laquering, finishing and wrapping was repetitive, strenuous but essentially light work. The bedstead trade in particular had a tendency towards large factories were the economies of scale and division of labour could maximise profits. The make-up of the employees of the bedstead trade varies in comparison to the brass trade: there were a high proportion of boys, around 50%, whilst women and girls made up a further 16% (Peynton 1866, 626). No apprenticeship was required and much of the work was unskilled, involving filing, assembly and finishing, and this may explain the low proportion of men. By the turn of the 20th century when Edward Cadbury, examined women's work and wages, some 31 different types of brass work were listed with wages varying from 30s (£1.50p) per week down to 3s 6d (17p) for the female employees (Cadbury *et al* 1907, 39–40). The increased mechanisation of work meant fewer and fewer jobs required men. Women and children were substituted as cheap sources of labour and Birmingham centres were less reliant on male income than their Sheffield counterparts in this respect (Smith 1982, 70). That this was a prevalent trend by the end of the 19th century is apparent as Frederick Ryland presented a report to the Association of Cast Hollow Ware Manufacturers in 1894 on '*the substitution of machine-turning for hand turning in hollow-ware and replacement of men by women*'.

A particular facet of the rise of women and children's labour in the brass trade was the resultant rise of trade unionism. The division of the labour between the two genders became evident during the 1880s when strikes were led by W J Davis against 'the frequent attempts to displace the labour of men with that of women and in departments...that are only fit for the male sex' (quoted in Morgan 2001, 85). The supposition is that there was a restrictive male monopoly on the skilled areas of the brass trade. A number of areas traditionally deemed 'men's work' were without question capable of being performed by females, as suggested by Arthur Chamberlain of Smith and Chamberlain gas-fitters, who instigated the strike by employing women at the lathe, finishing brass items. The motivation appeared to be monetary, not benevolence, as women were clearly cheaper and were less difficult employees as they were less likely to form unions. Yet how this affected the Cambridge Street works is debatable. Some departments such as the rolling mill and wire drawing mill were the preserves of the traditional male dominated brass trades. However, as Peyton (1866, 627) suggests the absence of skilled trades in bedstead manufacture meant that there were no trade unions.

There was a return to sweated labour in the market by the 1890s. Within the foundries, the workforce was divided into gangs, charge hands sub-contracted their own group. A head brass-caster would usually have between eight and ten people working for him, and they would produce castings at an agreed price. The contractor paid the charge hand who, in turn, paid his workers, often in the form of day wages. Blind piecework for fixed price work was not abolished until an Act of Parliament in 1901. It did not come into force in the brass trade until 1907. The increase in abusive labour practices reflects the change in economic climate that characterised the late 19th century. Whether or not these practices were carried out at the Cambridge Street works is arguable. Women were certainly employed in the bedstead works in the light department in the 1890s (IMechE 1897, 403–404). Yet the pressure of highly competitive work practices and the rise of trade unionism must have put an economic strain on the Cambridge Street works by the end of the 19th century.

The expansion of Birmingham's industrial trades led to an expansion of the city's housing. In the 18th century the prevalence of smaller businesses meant that workers lived and worked in close proximity. Early industry had its origins in the domestic setting in the late 18th and early 19th century, slowly converting terraced housing with shopping to the rear. Housing and shopping arrangements of this type are still visible in the Jewellery Quarter (Cattell *et al* 2000, 8–19). The development of more substantial purpose-built works, of which Cambridge Street is an example, through the middle part of the 19th century led to the development of back-to-back and court housing that housed workers in the vicinity of the works. Examples are again visible in the Jewellery Quarter (*ibid*, 19-23) but also at the National Trust run back-to-backs at 55–63 Hurst Street and 50–54 Inge Street. Examples of these were visible in Attwood's Passage opposite the bedstead works from the mid 19th century. The presence of a small number of back-to-back houses within this landscape of industrial premises is an enigma. To the modern eye the idea of residential and industrial areas co-existing seems strange. However, it was the norm from the medieval and post-medieval period and this can be seen as a continuation of this pattern.

Standards of living appear to have remained high in this period. According to 'The Society for the Diffusion of Useful Knowledge', Birmingham in 1833 was

... watched and paved, and lighted with gas under the provision of a recent Act of Parliament. The flagging of the footpaths is gradually proceeding and the edging of the flags (flagstones) with scored iron curbs, the invention of a native of the town, adds both to its neatness and durability. Although forges and furnaces so much abound, the air of Birmingham is deemed pure and salubrious which is

possibly owing to the dryness of its red sandy soil. The vicinity (ie the Edgbaston area) *abounds with the many pleasant villas and retreats of its opulent manufacturers* (1833, 444).

In 1842 the Poor Law Commission Report on Sanitary Conditions (Chadwick 1842), upheld the surprising revelation that Birmingham was a comparatively healthy place to live despite the undoubted squalor of the slum areas. The natural drainage, availability of building land and a good water supply meant that housing and living standards were better than in other industrial towns and cities.

In this case the presence of a row of back-to-back houses is not unusual but a reflection of local residential patterns. The pattern for the area was the expansion of worker accommodation in the form of back-to-back and court housing over much of the inner suburbs particularly west of the city centre. Studies of residential and work location in 1850 in Birmingham suggest a much closer relationship between the two than today with most people living within walking distance of their place of work (Vance 1967, 95–123).

The brass industry however, represents an industry in transition from the close relationship of the workplace and workforce and the disassociation that prevails in late modern society. '*No longer were separate industry quarters in evidence; rather there were a number of manufacturing districts. These manufacturing districts represent successive increments to the industrial land use located in open land at the edge of the city*' (Vance 1967, 123). This was certainly the case with the Cambridge Street works and its location in the Baskerville Estate.

Birmingham's growth expanded most rapidly in two distinct periods, the early 18th century when it became the third largest town in England and Wales by 1750, and subsequently in the later 19th century when it grew to rival and eventually surpass Manchester in size to become the second largest city in Britain (Berg 1991, 180). Patterns of habitation clearly shift further over the later part of the 19th century as clear industrial zones form along the transport routes. The industrial suburban development of the later 19th century occupied linear routes following the lines of transport, the Birmingham Canal, northwest, the Warwick and Birmingham Canal, southeast and the Tame valley floor both east and west (Wise and Thorpe 1950, 222–226). Housing developed around these areas in organised grid patterns, of continuous rows of speculative terraces. Areas such as Bordesley, Balsall Heath, Leasowes and Aston developed around the new industrial regions (McKenna 2005a, 69–72). The patterns along the Birmingham New Canal towards Icknield Port were not dissimilar. The Cambridge Street site, once on the edge of the suburbs, now lay in the core. The social association of the worker with his workplace was lost as large domestic suburbs of terrace and court housing expanded. New factories could establish on the edge of these suburbs away from cramped confined city-centre sites and yet still have an available workforce.

What becomes apparent is that social conditions of the workforce reflected a social change in the city. Whereas in the formative years of the Cambridge Street works the relationship between the worker and workplace was close and domestic in nature, by the early 20th century domestic and industrial suburbs had become zoned. This can be compared with the physical relationship of master and worker. Responsibility and paternalism, personified by Robert Winfield, was replaced by a more distant relationship between the board of directors and their staff. The rise of unionism was a reflection of the social change that overtook relations between master and worker. Owners increasingly saw the workforce as a commodity and took advantage of the cheapest and most productive elements: women, young men and children.

CONCLUSIONS

In many ways the fortunes of the Cambridge Street works mirrored those of the British Empire. Its rise and the height of its success followed a close chronological sequence. It was formed in the period that followed the Napoleonic Wars and economic success was at the height of the British Empire in the mid 19th century. The beginning of its demise was apparent before the First World War. In the aftermath of this event it was forever changed.

Viewed as an individual site it is just bricks and mortar, a series of walls and waste rubble. Yet in the individual areas of the site the technological changes that shaped the brass industry are visible. The advent of steam applied to the manipulation of metals, to the importance of reworked metal, brass sheet, wire and tube in the supply of the manufacturing trades of Birmingham, to the continued ability of men to cast brass in traditional fashion. New technologies can be seen across the site, concrete replaces brick, electricity replaces steam, gas replaces coal and ultimately aluminium replaces brass. Consumption, not technology, led to change to the markets in brass and copper, and to the materials of the 19th century being replaced by aluminium. The bedstead, once the height of fashion, became yesterday's fad.

The individuals that worked the site are now gone, traces of their passing are left in the tar pitched shoes, the broken bottles and newspaper cuttings. Yet were these the people who made the brass bed knobs, the gas fittings and candelabras, or merely the remains of those who destroyed what came before? The scraps of metal, strips of wire and the blank of pressed brass washers perhaps represent them better, the remnants of work conducted on the site.

Why did the Cambridge Street works rise and fall? Was one man alone, Robert Winfield, capable of sustaining such growth and did it begin to fail with his death? Can a simple correlation between singular management and success and diverse management and failure truly be seen? The reality is perhaps a more complex blend of technology, changing markets, geographical location, a socially changing society and an inability to react to an increasingly competitive market. The sale of the works at the end of the 19th century did not radically change its function. Elements of the works were still productive, but consisted of smaller companies operating within individual spheres of production. If one should fail, it failed alone.

Of all the factors the location of the site may have been the one most responsible for the end of production at Cambridge Street. The site that was prime development land in the 1820s, by the 1920s lay in a zone of congested industry on the edge of the urban centre. The transport links that had been its selling point a century before were now what held it back; railways and roads replaced canals and the passages around the site were confined and narrow. Space to expand was restricted by the density of development, factory upon factory covering every in of the inner city. This can be compared to the site that Winfield's Rolling Ltd eventually moved to by the Icknield Port Loop, which was located close to both rail and canal links. The pattern of industrial migration from the city centre to suburban locations that had begun in the late 19th century continued.

The homes of our metal manufactures were once thriving industrial workshops, spewing steam and smoke, where men, women and children toiled. These inner-city land plots have been transformed into open civic spaces. Industry has been replaced by commerce, entertainment and learning. The new Library of Birmingham represents the next phase in the development of the site.

APPENDIX 1: DETAILED EXCAVATION AIMS AND METHODOLOGY

APPENDIX 1.1: DETAILED AIMS

Industrial Function

- Power mechanism - examination of the stationary steam engine, boilers, chimney and wheel pits. Three-dimensional visualisation of the operating sequence.
- Rolling/wire drawing process - examination of the machine bases, annealing furnaces and pickling vats. Process analysis of how different areas of the site functioned, how did material enter and move around the works. Three-dimensional visualisation of the operating sequence.
- Transport - examination of the role of the canal, how the finished product was transferred to the canals. How the canal was maintained in water.
- Workshop functionality - examination of the bedstead works in detail to establish individual functionality of the rooms. Examination of floor scars to ascribed machine placements in each room. Understanding how the rooms containing hearths operated.
- Product of the brass works - examination of historical design catalogues, scientific analysis of the collected brass samples.
- Tinning/enamelling - examination of 20th–century processes, can these specific industries be related to the 1st World War and change in industrial function.

Economic Function

- The Brass economy – examination of the industry generally. Historic examination of the national development but also examining the brass works as part of the regional economy of Birmingham and the Black Country. Specific geographic work to place the works in the development of the area.
- Growth - examination of the excavation evidence and mapping evidence to establish changing patterns of the brass works geographically.
- Decline - examination of the historic reasons for the decline of the works, and comparison with geographical/archaeological reasons, eg outmoded technology, poor transport links, absence of space for expansion, fragmentation of the business.
- Production - examination and identification of specific products produced by the works. Attribution to different areas of the work. How these areas of the works can be seen to have altered both through historic work, mapping and examination of the excavation results. Cross-examination of which areas grow and when.
- Consumption - examine the historical consumption of products of the brass works and their change over time. Examine where and to whom the products are sold. Tie these themes into growth and decline.
- Social Function
- Patterns of human movement - examination of specific areas of the works and identifying how humans would operate within these areas, eg hot areas, loading areas, workshop areas, managerial/ office areas. Examine the functional flow of people from one area to another within the site and immediate area.
- Social hierarchy – examine the historic, geographical and archaeological record for specific areas of social hierarchy. Where do the management work? Is there a separation between the heavier production and artisan craft?
- Identity and domesticity – who was working in the brass works, where did they live and what did they do? What was the life of the average Winfield's worker like? Examination of the census records may be able to establish the geographic spread of brass workers in the area and allow an understanding of the influence of the industry on this sector of Birmingham.

APPENDIX 1.2: EXCAVATION AND POST EXCAVATION METHODOLOGY

Methodology

The proposed development covered two areas (1 and 2) that were excavated in three stages.

- Area 1 involved an area of *c* 3500m².
- Area 2 involved the excavation of an area of 1350m².

The first two stages were excavated in Area 1 between July and October 2009. Area 2 was excavated in November and December 2009 (Fig. 1.1).

Excavation Technique

All topsoil and modern overburden was removed using a 360° tracked mechanical excavator. The particular excavator used had a rotating bucket, which proved invaluable at removing the overburden and demolition rubble. The material which could not be removed by machine was removed by hand, which in many cases, due to the scale of the site, was very time consuming, as is often the case on industrial sites. Subjective decisions were then undertaken in consultation with the City Archaeologist to establish areas for more intensive excavation by hand. Areas for specific hand excavation were targeted depending on the information required. These decisions were made specifically to:

- clarify relationships between structures to establish structural phases of construction,
- establish specific purposes where they were not immediately obvious,
- collect physical dating evidence in the form of artefacts
- or collect industrial residues.

Once it was realised that earlier structural phases of the site were obscured, then the later structural phases were removed by machine. This occurred on three separate occasions in Area 1, in each case the phase being recorded then removed. This continued until the earliest phase or the natural subsoil was encountered.

Surveyed, Drawn and Photographic Record

Due to the scale of the site, machining carried on for several weeks and the survey process followed. Initial site survey was prepared using an EDM and located on the Ordnance Survey National Grid. A grid was laid out across the site, which was used for both the EDM survey and traditional drawn planning. Once the survey was complete then a 1:50 scale plan was produced and used as the basis for a more detailed 1:50 scale hand-drawn survey of the site. This was done on a regular basis as parts of the site were exposed. All details were added to this drawn plan, such as changes in brickwork, flooring and structure. Details of specific features could be drawn when needed and added to the plans at the post-excavation phase. Preliminary ideas about phasing and layout could also be attached to these plans. As the excavation progressed further surveys were done. These were treated as overlay plans. Information on small find and sample locations were added as spot data to the hand-drawn plan.

The EDM survey plans were able to fit directly onto Ordnance Survey plans (both historic and modern). By importing these plans into a GIS program, the site could be placed accurately within its modern and historic environment. Further historic maps could be added, the benefit being some of these plans had details about location of machinery and the role of the buildings, the locations of which could be directly compared to the archaeological evidence. The drawn site plans were imported into GIS and illustrated within the program rather than in Adobe Illustrator or other illustrator package. This provided detail adequate enough to be used in interpretation and phasing.

Elevations or profiles were drawn as necessary for the interpretation of the site of significant vertical stratigraphy at a scale of 1:20.

Scale plans were supplemented by photographs using black and white monochrome, colour slide and digital photography. These included feature and group specific photos as well as overall photographs of the site.

The good preservation of the archaeological and documentary material has allowed for an accurately researched and interpreted site, where it has been possible to use supplementary techniques other than those most commonly used. The comprehensive survey work done in the field, along with the use of GIS and extensive historical research, has provided a usable model for 3D analysis. 3D models have been put together which show the site as it may have looked at certain periods in its history. Site survey, historic plans, illustrations and photographs were all used to generate these models. The resulting models allow the general reader/viewer, a clearer and instantly recognisable image of the site and aid their understanding of its interpretation.

The Written Record

A comprehensive written record was maintained using a continuous numbered context system on pro-forma cards. The traditional single context recording system, normal on most archaeological sites was adapted as interpretation was undertaken on site combining archaeological information with historical sources.

This was done in a two-fold manner. An historian was involved at an early stage to provide specific background historical evidence in addition to that provided by the desk-based assessment. This included:

- Geographical and chronological development of the site,
- Function and use of specific geographical areas,
- Understanding of the processes at work in the works.

With all the background research and supplementary material it was possible from an early stage to begin the process of interpretation. This was done by means of a specific interpretive level of recording - the group sheet. Groups could be whole buildings, but could also be rooms within buildings or concentrations of machine footings. Walls and features were linked into associated groups according to stratigraphic and historical association as opposed to recording them as individual contexts. One record sheet was required for each group allowing interpretation of a number of related contexts. Where it was unclear which context belonged to which group, these features were given their own individual context numbers. This aided the context process from the beginning and allowed interpretation to be undertaken during the excavation process.

Finds and Samples

Due to the modern nature of the site (much dated to the 20th century) a selective recovery policy was adopted for the finds with active discard of material undertaken on site.

Not all finds were deemed suitable for retention but were specifically selected on the basis of criteria that included:

- finds suitable for metallurgical analysis,
- finds suitable for dating evidence,
- specific machine parts,
- artefacts relevant to the life cycle of the works in contrast to its demolition.

In addition some artefacts were recorded on site and subsequently disposed of due to petro-chemical and heavy metal contamination, or size and weight that prevented their removal from site and subsequent storage.

Recovered finds were cleaned, marked and remedial conservation work undertaken as necessary. Treatment of all finds conformed to guidance contained within the Birmingham Archaeology Fieldwork Manual and First Aid for Finds (Watkinson and Neal 1998).

Samples were taken to identify macroscopic technological residues in accordance with Archaeometallurgy (English Heritage 2001) and Science for Historic Industries (English Heritage 2006). Advice was sought at early stage in the excavations as to the potential and recovery methodology of the metallurgical samples. Waste products, slags and resides featured in the archaeological assemblage but it was clear that a structured recovery process was necessary from the outset. A clear sample strategy was put into place, which the site excavators were made aware of. The purpose of the recovery of this material was to confirm the historical record with regards to the rapid changes in technology during the period. Specific alloy variations may have been used for specific purposes, in cast and wrought products, recovery of the metallurgical samples may have confirmed these.

Samples were recovered where information on their origin was clear (for example, the slags from the bases of the crucible furnaces). Samples were also recovered where their appearance was unusual (in comparison to elsewhere from the site), or unique (an example of each type of slag was sampled). In these cases, the location in which they were found was not necessarily the position in which they were created. In the case of large assemblages of residues or broken crucibles, only a certain percentage was kept; this was agreed at the outset of the project with the specialists and the excavators. Evidence from the crucibles shows distinct alloy compositions, and the slags were identified as smithing and hearth bottoms from hot-working wrought copper alloys. Copper alloy and iron (alloy) artefacts gave an idea of the types of processes undertaken and products discarded of.

Post-Excavation Assessment and Research

Initial post-excavation assessment of the site focused on the establishment of a detailed phase chronology for the site. A detailed examination of the stratigraphic relationships was needed to confirm the phasing. In addition, interpretation continued to be revised as part of the post-excavation process. Initial assessment of the finds and samples were undertaken and recommendations made for further work. These formed the initial results of the excavation (see Mitchell and Hewitson 2010). Revised aims and objectives were formulated based on these initial results, and the report below forms the basis of these revised aim and objectives.

The Archive

The full site archive includes all artefactual remains recovered from the site. The site archive has been prepared according to guidelines set down in Appendix 3 of The Management of Archaeology Projects (English Heritage, 1991), the Guidelines for the Preparation of Excavation Archives for Long-term Storage (UKIC, 1990) and Standards in the Museum Care of Archaeological Collections (Museum and Art Galleries Commission, 1992). The paper archive will be deposited with the Birmingham Museum and Arts Gallery subject to permission from the landowner.

APPENDIX 2: ARTEFACT ASSEMBALGE SUMMARIES

APPENDIX 2.1: BRICK AND CERAMIC BUILDING MATERIAL CATALOGUE

Context	Type	Initial description	Structure/context description	Dimensions (inches)	Wgt(g)	Fabric
1064	Eng.		Bedsteads works	9¼ x 4½ x 3¼	4450	1
1068	Eng.	Burnt on orange?	Canal wall	9 x 4¼ x 3¼	4850	4
1068	Eng.		Canal wall	8½ x 4 x 3½	4500	6
1079	Fire	Sooted on stamped side. Stamp: 'BEST STOURBRIDGE'	Floor brass furnaces -bedstead works	9 x 4½ x 2½	2850	5
1163	Red	one bed sooted, other covered in mortar	Lancaster boilers	9 x 4¼ x 2¾	3475	2
1165	Red	one edge is very burnt	Lancaster Boilers	9 x 4¼ x 3	4050	3
1165	Fire	Baked red sand over most of the surface	Lancaster Boilers	9 x 4¼ x 2½	3550	10
1183	Fire	Stamped: 'KING BROTHERS STOURBRIDGE'	Chimney culvert/ flue	8¾ x 4 x 2	2400	9
1186	Fire	One edge partial burnt/eroded	Annealing hearth base	8¾ x 4¼ x 2½	3015	10
1189?	Red	One edge much more orange	Early phase of boiler base	9 x 4⅛ x 2 ⅞	3700	3
1206	Eng.		Boiler feed pump	4 x 3½ x 2¾	3600	4
1209	Eng.	Very black, looks handmade	Canal wall	8¾ x 4 x 2½	2850	5
1209	Eng.	Stained	Canal wall	9½ x 4½ x 3	4550	4
1252	Red		Large wheel pit	8¾ x 4¼ x 2½	3200	2
1283	Red	One edge burnt, other covered with mortar	Steam engine base	9 x 4¼ x 2¾	3275	2
1345	Red?	All surfaces except one burnt. One that is not is mortar covered.	Wheel pit	9 x 4 x 3	4200	7
1362	Eng.		Early boilers	9 x 4½ x 3⅛	5000	7
1380	Red	Sooted on one end and vitrified on the other	Earlier muffles/furnace	8¾ x 4¼ x 3	4025	3
1381	Fire	Vitrified/encrusted down one edge	Earlier muffles/furnace	9 x 4½ x 2½	3275	8
1384	Red	Yellow residue on a lot of it	Flue/early boiler base	8¾ x 4¼ x 2¾	2925	1
1386	Eng.		Early boiler base	9¼ x 4¼ x 3¼	4850	5
1388	Red	Very fragmentary	Flue/early boiler base	9 x 4¼ x 2¾	3175	2
1390	Eng./red?	Plinth brick	Early culvert	9 x 4¼ x 2¼–3	3450	5
1539	Red		Canal wall (Area 2)		4000	2
1540	Red		Canal wall (Area 2)		4300	2
U/S	Eng.	Stamped with 'TIMMIS & Co LTD'		? x 6 x 2¼	4850	5
U/S	Eng.			? x 4½ x 2½	1700	5
U/S	Red	Brick stamped with '1839'		9½ x 4½ x 3	3775	2
U/S	Fire	Seven fragments with lots of slag adhered			2330	10
U/S	Red	Brick fragment, 'L' shaped		?x ? x 3	1175	3
U/S	Eng.	Bull nosed			4725	5

APPENDIX 3: ARCHAEO-METALLURGY METHODOLOGY AND DETAILED STATISTICAL ANALYSIS

APPENDIX 3.1. METHODOLOGY OF X-RAY FLUORESCENCE ANALYSIS OF THE COPPER ALLOY METAL

The instrument used was a Bruker S1 Turbosdr hand-held XRF instrument operating at 40kV. Samples were analysed for 30 live seconds, the spectrum is stored and a normalised composition determined using a bespoke Bruker Fundamental Parameters Programme (FP). All elements heavier than calcium (Ca, Z=20) can be detected. The calculated two-sigma error on each element was calculated and, overall, shows values of the order of +/- 0.2%. The data was generated in a comma delimited file and then exported to an Excel spreadsheet, where the data was examined and relevant tables were generated. One sample was analysed five times to assess the consistency of the data generated. Three samples were analysed in the as-received condition, then an area was cleaned back using fine grit (1200 grade) grinding papers to produce a clean metallic surface. The samples were then re-analysed and the data interrogated to assess the difference between corroded and fresh metal surfaces.

The same method of analysis was applied to the crucibles. A reading was taken from the interior surface of a crucible and a second one from the exterior surface. The metal was held within the crucible, but at the high temperatures required to melt the alloy (>900°C) the metals diffuse into and through the crucible wall. Due to the low volatilisation temperature of zinc (910°C), this element readily diffuses into the crucible fabric. It would be expected that a higher concentration of metals would be found on the inside of the crucibles compared to the exterior surfaces.

The same method of analysis was applied to the slag samples. A reading was taken from the surface of each slag sample; analysis was undertaken only for iron and the non-ferrous elements. The metals would be present mostly as oxides in the slag, although some metallic prills might be present and the slag composition would be dominated by silica and iron oxide. The data are presented as elemental percentages of iron and the non-ferrous metals, so that they represent the relative abundance of these elements.

The same method of analysis was applied to the hearth linings. A reading was taken from the surface of each hearth lining sample; analysis was conducted only for iron and the non-ferrous elements. The metals would be present mostly as oxides in the slag, although some metallic prills might be present and the clay hearth lining composition would be dominated by alumina, silica, lime, and iron oxide. The data are presented as elemental percentages of iron and the non-ferrous metals so that they represent relative abundance of these elements.

APPENDIX 3.2. ASSESSMENT OF THE INFLUENCE OF CORROSION ON DETERMINING THE BULK ANALYSIS OF A SAMPLE BY X-RAY FLUORESCENCE

One sample was analysed five times to assess the consistency of the results (Table A3.2.1). The data show that there is slight variation in the iron and the zinc content. These variations reflect subtle variations in the corrosion.

Table A3.2.2 presents the data for a small bar presumed to be in the as-cast condition, which for Cu-Zn alloys up to *c* 35% will produce a granular microstructure. The processed data shows that in the as-received condition the analysis indicates a low percentage of iron (Fe) in the corrosion layer, although the data is presented as elemental values, the corroded layer will in an oxide/hydroxide/chloride compound form. The analysis shows that the bar was manufactured as a leaded brass, with circa 0.5% Pb present in the alloy. The corroded surface is enhanced in Fe, Zn and Pb compared to the cleaned surface data. The data of the cleaned surface shows that the alloy is approximately a 70/30 Cu/Zn alloy. The Cu/Zn ratios are shown in Table A3.3.3 demonstrating that the Cu/Zn ratio is 2.3, which equates to a 70/30 (Cu/Zn) composition.

Two samples of sheet alloy were also examined in the as-received and cleaned condition; a rectangular off-cut (20mm *20mm, 2mm thick) and a circular disc (15mm diameter, 2mm thick). Both samples are leaded brasses. The samples were analysed in the as-received condition (ie corroded). Firstly, both sides of the samples were analysed (Table A3.2.4). The results from the rectangular off-cut show no difference between the two surfaces. However, Side 2 of the disc appears to be more corroded than Side 1, since the analysis shows enhancement in Fe% and Zn%. This probably reflects the position of the off-cut in its burial environment. Table A3.2.5 presents the data from the cleaning of one surface and removal of the iron content from the analyses. In both cases, as expected the corroded surface is depleted in zinc. The ratios (Table A3.2.6) show a clear shift between the as-received and the cleaned surface of both samples but are not so marked in the case of the disc.

The differences in readings between the as-received corroded surface and the cleaned surface are shown in Table A3.2.7. There is a difference between the two sheet samples, which show depletion of zinc and enhancement of copper in the corrosion layer, and the bar (assumed to be cast), which shows the reverse, ie, enhancement of zinc and depletion of copper. The lead is consistent, ie the corroded layer is enhanced in lead, which is to be expected, as the lead forms particles at the grain boundaries in the microstructure and is readily available to be removed from the bulk metal by the corrosion process. This data are insufficient to argue the case that as-cast objects corrode in a different manner to the worked artefacts, ie the sheet. However, they highlight the question, which should be pursued. The percentage difference calculation shows wide variation of element behaviour. These results mean that a correction factor cannot be applied to the data to model the true bulk composition from the corroded surface analysis. However, using the corroded surface data, removing the iron content, and then normalising the data, gives a good approximation to the bulk cleaned metal analysis.

TABLE A3.2.1 FIVE ANALYSES OF THE SAME SAMPLE (WEIGHT %)

Context	Analysis	Fe	Cu	Zn	As	Sn	Sb	Pb
1406	Offcut square corroded 1	2.9	66.2	28.8	0.0	0.1	0.0	1.6
1406	Offcut square corroded 2	2.9	66.3	28.8	0.1	0.0	0.0	1.6
1406	Offcut square corroded 3	0.1	65.3	33.2	0.0	0.0	0.0	1.0
1406	Offcut square corroded 4	2.9	66.3	28.9	0.0	0.0	0.0	1.5
1406	Offcut square corroded 5	0.1	65.3	33.3	0.0	0.0	0.0	1.0
	Average	1.8	65.9	30.6	0.0	0.0	0.0	1.3
	SD	1.5	0.5	2.4	0.0	0.0	0.0	0.3
	Max	2.9	66.3	33.3	0.1	0.1	0.0	1.6
	Min	0.1	65.3	28.8	0.0	0.0	0.0	1.0

TABLE A3.2.2 XRF DATA FOR BAR SFN 58 SAMPLE 12 IN THE AS RECEIVED CONDITION AND AFTER REMOVAL OF CORROSION LAYER.

FP	SFN 58	Fe	Cu	Zn	Sn	Pb	Total	Total less Fe%
159	Bar	1.2	64.9	32.2	0.0	1.2	99.5	98.3
164	Bar cleaned	0.1	69.2	29.8	0.0	0.6	99.7	99.6
Corrected, by removal of Fe%								
159	Bar		66.0	32.7	0.0	1.3	100.0	
164	Bar cleaned		69.5	29.9	0.0	0.6	100.0	

TABLE A3.2.3 RATIOS OF ELEMENTS

FP	SFN 58	Cu/Zn	Cu+Pb/Zn	Cu/Zn+Pb
159	Bar	2.0	2.1	1.9
164	Bar cleaned	2.3	2.3	2.3
Corrected, by removal of Fe%				
159	Bar	2.0	2.1	1.9
164	Bar cleaned	2.3	2.3	2.3

TABLE A3.2.4 XRF DATA OF BOTH SIDES OF SHEET EXAMPLES AND CLEANED SURFACE. SAMPLES FROM CONTEXT 1406

FP	Sample	Fe	Cu	Zn	Sn	Pb	Total
69	Offcut 3 side 1	3.8	65.5	28.7	0.0	1.6	99.6
70	Offcut 3 side 2	3.9	65.0	29.0	0.0	1.7	99.6
71	Disc 9 side 1	1.2	64.6	33.4	0.0	0.4	99.5
72	Disc 9 side 2	2.4	60.9	35.6	0.0	0.5	99.4
74	Offcut 3 side 1 cleaned	0.1	64.1	34.1	0.0	1.1	99.4
75	Disc 9 side 1 cleaned	0.2	63.6	35.5	0.0	0.3	99.6

TABLE A3.2.5 XRF DATA FOR TWO SHEET SAMPLES FROM CONTEXT 1406

FP	Sample	Fe	Cu	Zn	Sn	Pb	Total	Total less Fe
69	Offcut 3 side 1	3.8	65.5	28.7	0.0	1.6	99.6	95.8
74	Offcut 3 side 1 cleaned	0.1	64.1	34.1	0.0	1.1	99.4	99.3
71	Disc 9 side 1	1.2	64.6	33.4	0.0	0.4	99.5	98.4
75	Disc 9 side 1 cleaned	0.2	63.6	35.5	0.0	0.3	99.6	99.4
Corrected by removal of Fe								
FP	Sample	Fe	Cu	Zn	Sn	Pb	Total	
69	Offcut 3 side 1		68.4	30.0	0.0	1.6	100	
74	Offcut 3 side 1 cleaned		64.5	34.3	0.0	1.1	100	
71	Disc 9 side 1		65.7	33.9	0.0	0.4	100	
75	Disc 9 side 1 cleaned		64.0	35.7	0.0	0.3	100	

TABLE A3.2.6 RATIOS OF ELEMENTS BETWEEN THE AS-RECEIVED AND CLEANED SURFACES (SHEET SAMPLES FROM CONTEXT 1406)

FP	Sample	Cu/Zn	Cu+Pb/Zn	Cu/Zn+Pb
69	Offcut 3 side 1	2.3	2.3	2.2
74	Offcut 3 side 1 cleaned	1.9	1.9	1.8
71	Disc 9 side 1	1.9	1.9	1.9
75	Disc 9 side 1 cleaned	1.8	1.8	1.8

TABLE A3.2.7 DIFFERENCES BETWEEN THE CORRODED SURFACES (CORR) AND THE CLEANED SURFACES (CLEAN)

	Offcut 3	Disc 9	Bar	Average
Cu-Corr	65.5	64.6	64.9	65.0
Cu-Clean	64.5	64.0	69.5	66.0
Cu-Difference	1.0	0.6	-4.6	-1.0
%-Difference	1.5	0.9	-7.1	-1.5
Zn-Corr	28.7	33.4	32.2	31.4
Zn-Clean	34.3	35.7	29.9	33.3
Zn-Difference	-5.6	-2.3	2.3	-1.9
%-Difference	-19.5	-6.9	7.1	-5.9
Pb-Corr	1.6	0.4	1.2	1.1
Pb-Clean	1.1	0.3	0.6	0.7
Pb-Difference	0.5	0.1	0.6	0.4
%-Difference	31.3	25.0	50.0	37.5

APPENDIX 3.3. DATA TABLES FROM X-RAY FLUORESCENCE ANALYSIS

TABLE A3.3.1 XRF ANALYSES OF THE CAST ARTEFACTS (FE% INCLUDED)

Context/SFN	Description	Fe	Cu	Zn	As	Zr	Sn	Sb	Pb	Total
1406	Lump 1	24.4	52.8	18.9	0.0	0.1	0.6	0.3	2.2	99.4
1406	Fleur de lis	6.8	68.6	22.7	0.1	0.0	0.1	0.0	1.3	99.6
1035 SFN12b	Hook	1.5	83.4	13.2	0.1	0.0	0.3	0.0	1.4	99.9
1017 SFN 5a sample 3	Shield boss	2.5	73.1	21.6	0.0	0.1	0.0	0.0	2.5	99.8
1017 SFN 5a sample 3	Swivel	2.5	15.7	77.9	0.0	0.0	0.0	0.0	3.4	99.5
1009 SFN 3	Ingot	0.6	0.7	0.4	0.3	0.1	3.2	15.1	77.5	98.0
SFN58 sample 12	Ingot	1.2	64.9	32.2	0.0	0.0	0.0	0.0	1.2	99.6
1245	Ingot 1	2.2	69.7	23.6	0.0	0.0	0.2	0.0	4.1	99.8
1245	Ingot 2	1.9	88.0	7.9	0.2	0.0	0.0	0.0	1.8	99.9
1245	Dumb-bell	1.2	62.8	30.8	0.0	0.0	0.0	0.0	5.0	99.9
1245	Stopper 1	1.0	77.1	19.8	0.0	0.0	0.0	0.0	1.8	99.8
1245	Stopper 2	5.8	67.5	21.1	0.0	0.0	0.3	0.0	4.9	99.7
1245	Stopper 3 round head	1.2	73.1	21.9	0.0	0.0	0.4	0.0	2.6	99.1
1245	Tube	2.3	57.1	36.6	0.0	0.0	0.2	0.0	3.3	99.5
1245	Angle section	2.0	54.1	38.3	0.0	0.1	0.2	0.0	5.0	99.6
	Average	3.8	60.6	25.8	0.1	0.0	0.4	1.0	7.9	99.5
	SD	6.0	23.6	17.6	0.1	0.0	0.8	3.9	19.3	0.5
	Max	24.4	88.0	77.9	0.3	0.1	3.2	15.1	77.5	99.9
	Min	0.6	0.7	0.4	0.0	0.0	0.0	0.0	1.2	98.0

TABLE A3.3.2 XRF ANALYSES OF THE CAST ARTEFACTS (FE EXCLUDED)

Context/SFN	Description	Cu	Zn	As	Zr	Sn	Sb	Pb	Total
1406	Lump 1	70.4	25.2	0.0	0.1	0.8	0.4	2.9	100
1406	Fleur de lis	73.9	24.5	0.1	0.0	0.1	0.0	1.4	100
1035 SFN12b	Hook	84.7	13.4	0.1	0.0	0.3	0.0	1.4	100
1017 SFN 5a sample 3	Shield boss	75.2	22.2	0.0	0.1	0.0	0.0	2.6	100
1017 SFN 5a sample 3	Swivel	16.2	80.3	0.0	0.0	0.0	0.0	3.5	100
1009 SFN 3 sample	Ingot	0.7	0.5	0.3	0.1	3.3	15.5	79.5	100
SFN 58 sample 12	Ingot	66.0	32.7	0.0	0.0	0.0	0.0	1.3	100
1245	Ingot 1	71.4	24.2	0.0	0.0	0.2	0.0	4.2	100
1245	Ingot 2	89.8	8.1	0.2	0.0	0.0	0.0	1.9	100
1245	Dumb-bell	63.7	31.2	0.0	0.0	0.0	0.0	5.0	100
1245	Stopper 1	78.0	20.0	0.0	0.0	0.0	0.0	1.9	100
1245	Stopper 2	71.9	22.5	0.0	0.0	0.3	0.0	5.3	100
1245	Stopper 3 round head	74.6	22.3	0.0	0.0	0.4	0.0	2.6	100
1245	Tube	58.7	37.6	0.0	0.0	0.2	0.0	3.4	100
1245	Angle section	55.4	39.2	0.0	0.1	0.2	0.0	5.1	100
	Average	63.4	26.9	0.1	0.0	0.4	1.1	8.1	
	SD	24.2	18.0	0.1	0.0	0.8	4.0	19.8	
	Max	89.8	80.3	0.3	0.1	3.3	15.5	79.5	
	Min	0.7	0.5	0.0	0.0	0.0	0.0	1.3	

TABLE A3.3.3 XRF ANALYSES OF THE CAST BRASSES, FE EXCLUDED, ORDERED BY ZN CONTENT

Context/SFN	Description	Cu	Zn	As	Zr	Sn	Sb	Pb	Total
1017 SFN 5a sample 3	Cu alloy swivel	16.2	80.3	0.0	0.0	0.0	0.0	3.5	100.0
1245	Angle section	55.4	39.2	0.0	0.1	0.2	0.0	5.1	100.0
1245	Tube	58.7	37.6	0.0	0.0	0.2	0.0	3.4	100.0
SFN 58 sample 12	Bar	66.0	32.7	0.0	0.0	0.0	0.0	1.3	100.0
1245	Dumb-bell	63.7	31.2	0.0	0.0	0.0	0.0	5.0	100.0
1406	Lump 1	70.4	25.2	0.0	0.1	0.8	0.4	2.9	100.0
1406	Fleur de lis	73.9	24.5	0.1	0.0	0.1	0.0	1.4	100.0
1245	Bar 1	71.4	24.2	0.0	0.0	0.2	0.0	4.2	100.0
1245	Stopper 2	71.9	22.5	0.0	0.0	0.3	0.0	5.3	100.0
1245	Stopper 3 round head	74.6	22.3	0.0	0.0	0.4	0.0	2.6	100.0
1017 SFN 5a sample 3	Cu alloy shield boss	75.2	22.2	0.0	0.1	0.0	0.0	2.6	100.0
1245	Stopper 1	78.0	20.0	0.0	0.0	0.0	0.0	1.9	100.0
1035 SFN 12b	Cu alloy hook	84.7	13.4	0.1	0.0	0.3	0.0	1.4	100.0
	Average	66.2	30.4	0.0	0.0	0.2	0.0	3.1	
	SD	17.0	16.6	0.0	0.0	0.2	0.1	1.4	
	Max	84.7	80.3	0.1	0.1	0.8	0.4	5.3	
	Min	16.2	13.4	0.0	0.0	0.0	0.0	1.3	

TABLE A3.3.4 XRF ANALYSES OF SHEET METAL ARTEFACTS INCLUDING FE CONTENT

Context/SFN	Description	Fe	Cu	Zn	As	Sn	Sb	Pb	Total
1406	Strip 1	11.0	36.9	49.0	0.0	0.2	0.0	1.9	99.0
1406	Strip 2	2.9	68.9	27.5	0.0	0.0	0.0	0.5	96.9
1406	Strip 3	5.2	74.5	18.9	0.0	0.0	0.0	1.1	94.6
1406	Strip 4	0.7	67.5	30.7	0.0	0.0	0.0	0.7	98.9
1406	Strip 5	1.1	67.3	30.5	0.0	0.0	0.0	0.7	98.6
1406	Strip 5	1.4	69.7	27.5	0.0	0.0	0.0	1.0	98.3
1406	Strip 6	3.1	73.5	22.4	0.1	0.0	0.0	0.7	96.7
1406	Disc 1	5.6	63.0	29.9	0.0	0.2	0.0	0.9	94.1
1406	Disc 2	1.2	79.4	18.4	0.0	0.0	0.0	0.6	98.5
1406	Disc 3	1.5	69.1	28.1	0.0	0.0	0.0	0.8	98.1
1406	Disc 4	0.6	73.6	24.2	0.0	0.1	0.0	0.9	98.8
1406	Disc 5	0.6	72.0	26.4	0.0	0.0	0.0	0.6	99.1
1406	Disc 6	0.6	68.3	30.4	0.0	0.0	0.0	0.4	99.1
1406	Disc 7	4.2	66.7	27.8	0.1	0.1	0.0	0.7	95.4
1406	Disc 8	6.7	65.7	25.7	0.1	0.1	0.0	1.3	92.9
1406	Disc 9	2.7	60.5	35.4	0.0	0.0	0.0	0.6	96.5
1406	Disc 10	12.8	71.0	12.7	0.2	0.1	0.0	2.8	86.8
1406	Disc 11	1.9	70.4	26.5	0.0	0.0	0.0	0.7	97.6
1406	Offcut 1	1.4	73.5	23.6	0.0	0.0	0.0	0.9	98.1
1406	Offcut 2	2.5	68.7	27.6	0.0	0.0	0.0	0.7	97.0
1406	Offcut 3	3.8	65.6	28.5	0.0	0.0	0.0	1.6	95.7
1406	Offcut 4	3.0	68.1	27.6	0.0	0.1	0.0	0.8	96.6
1406	Offcut 6	5.8	67.6	25.5	0.0	0.0	0.0	0.7	93.8
1406	Offcut 8	4.6	52.9	23.4	0.0	0.0	0.0	0.9	77.2
1406	Strip 7	12.7	80.9	2.7	0.4	0.8	0.0	1.6	86.4
1324	Cu alloy strip 1	1.9	84.4	4.4	0.4	1.0	0.0	7.8	98.0

1324	Cu alloy strip 2	0.6	90.7	5.9	0.0	0.8	0.0	1.9	99.2
1324	Cu alloy strip 3	0.7	67.4	24.8	0.0	1.3	0.0	5.7	99.2
1406	Bowl	3.5	63.7	32.3	0.0	0.0	0.0	0.3	96.4
1406	Strip flat end	7.2	38.5	53.6	0.0	0.0	0.0	0.1	92.2
1406	Offcut	0.5	69.9	28.8	0.0	0.0	0.0	0.3	99.1
Us 99.99	Disc	2.9	72.6	21.7	0.0	0.2	0.0	2.0	96.5
Us 99.99	Cylinder	3.4	82.6	12.2	0.1	0.3	0.0	0.9	96.1
SFN 49	Strip 1	0.7	66.1	32.4	0.0	0.0	0.0	0.5	99.0
SFN 25 sample 8	Cu alloy offcut	0.7	95.9	2.5	0.1	0.0	0.0	0.2	98.7
1016 SFN1	Cu alloy washer	5.9	92.4	0.0	1.1	0.0	0.0	0.3	93.8
1006 SFN 2 sample 1	Cu alloy frags	1.8	59.3	34.9	0.0	0.5	0.0	3.4	98.1
SFN 58 sample 12	Strip	10.4	42.8	41.3	0.0	0.0	0.0	5.2	89.3
1245	Washer 1	0.6	72.1	23.9	0.1	0.0	0.0	3.0	99.1
1245	Washer 2	0.9	80.6	16.3	0.0	0.0	0.0	2.0	98.9
1245	Sheet off cut	1.4	76.8	17.9	0.0	0.3	0.0	3.5	98.4
	Average	3.4	69.5	24.5	0.1	0.1	0.0	1.5	
	SD	3.3	12.2	11.3	0.2	0.3	0.0	1.6	
	Max	12.8	95.9	53.6	1.1	1.3	0.0	7.8	
	Min	0.5	36.9	0.0	0.0	0.0	0.0	0.1	

TABLE A3.3.5 XRF ANALYSES OF SHEET ARTEFACTS, DATA EXCLUDES FE%, AND NORMALISED

Context/SFN	Description	Cu	Zn	As	Sn	Sb	Pb	Total
1406	Strip 1	41.9	55.7	0.0	0.3	0.0	2.2	100.0
1406	Strip 2	71.1	28.4	0.0	0.0	0.0	0.5	100.0
1406	Strip 3	78.8	20.0	0.0	0.0	0.0	1.2	100.0
1406	Strip 4	68.2	31.0	0.0	0.0	0.0	0.7	100.0
1406	Strip 5	68.3	30.9	0.0	0.0	0.0	0.7	100.0
1406	Strip 5	70.9	28.0	0.0	0.0	0.0	1.0	100.0
1406	Strip 6	76.0	23.2	0.1	0.0	0.0	0.8	100.0
1406	Disc 1	67.0	31.8	0.0	0.2	0.0	1.0	100.0
1406	Disc 2	80.6	18.7	0.0	0.0	0.0	0.6	100.0
1406	Disc 3	70.5	28.7	0.0	0.0	0.0	0.8	100.0
1406	Disc 4	74.5	24.5	0.0	0.1	0.0	0.9	100.0
1406	Disc 5	72.7	26.6	0.0	0.0	0.0	0.6	100.0
1406	Disc 6	68.9	30.7	0.0	0.0	0.0	0.4	100.0
1406	Disc 7	69.9	29.1	0.1	0.1	0.0	0.8	100.0
1406	Disc 8	70.7	27.7	0.1	0.1	0.0	1.4	100.0
1406	Disc 9	62.7	36.7	0.0	0.0	0.0	0.7	100.0
1406	Disc 10	81.8	14.6	0.2	0.1	0.0	3.2	100.0
1406	Disc 11	72.1	27.2	0.0	0.0	0.0	0.7	100.0
1406	Offcut 1	74.9	24.1	0.0	0.0	0.0	1.0	100.0
1406	Offcut 2	70.8	28.4	0.0	0.0	0.0	0.7	100.0
1406	Offcut 3	68.5	29.8	0.0	0.0	0.0	1.7	100.0
1406	Offcut 4	70.5	28.6	0.0	0.1	0.0	0.8	100.0
1406	Offcut 6	72.0	27.2	0.0	0.0	0.0	0.7	100.0
1406	Offcut 8	68.5	30.3	0.0	0.0	0.0	1.1	100.0
1406	Strip 7	93.6	3.1	0.5	0.9	0.0	1.8	100.0
1324	Cu alloy strip 1	86.1	4.4	0.4	1.1	0.0	8.0	100.0

1324	Cu alloy strip 2	91.5	5.9	0.0	0.8	0.0	1.9	100.0
1324	Cu alloy strip 3	68.0	25.0	0.0	1.3	0.0	5.8	100.0
1406	Bowl	66.1	33.5	0.0	0.0	0.0	0.4	100.0
1406	Strip flat end	41.7	58.1	0.0	0.0	0.0	0.2	100.0
1406	Offcut	70.6	29.1	0.0	0.0	0.0	0.3	100.0
Us 99.99	Disc	75.2	22.5	0.0	0.2	0.0	2.0	100.0
Us 99.99	Cylinder	86.0	12.7	0.1	0.3	0.0	0.9	100.0
SFN 49	Strip 1	66.8	32.7	0.0	0.0	0.0	0.5	100.0
SFN 25 sample 8	Cu alloy offcut	97.2	2.6	0.1	0.0	0.0	0.2	100.0
1016 SFN1	Cu alloy washer	98.5	0.0	1.2	0.0	0.0	0.3	100.0
1006 SFN 2 sample 1	Cu alloy frags	60.5	35.6	0.0	0.5	0.0	3.4	100.0
SFN58 sample 12	Strip	47.9	46.3	0.0	0.0	0.0	5.8	100.0
1245	Washer 1	72.8	24.1	0.1	0.0	0.0	3.0	100.0
1245	Washer 2	81.5	16.5	0.0	0.0	0.0	2.0	100.0
1245	Sheet off cut	78.0	18.2	0.0	0.3	0.0	3.5	100.0
	Average	72.5	25.7	0.1	0.2	0.0	1.6	
	SD	12.0	12.3	0.2	0.3	0.0	1.7	
	Max	98.5	58.1	1.2	1.3	0.0	8.0	
	Min	41.7	0.0	0.0	0.0	0.0	0.2	

TABLE A3.3.6 XRF ANALYSES OF WIRE SAMPLES, INCLUDING FE CONTENT

Context/SFN	Description	Class	Fe	Cu	Zn	As	Sn	Sb	Pb	Total
1406	Wire 1	w	0.5	75.4	23.0	0.1	0.0	0.0	1.0	99.9
1406	Wire 2	w	2.7	55.8	41.1	0.0	0.0	0.0	0.5	100.1
1406	Wire 3	w	1.9	59.0	38.6	0.0	0.0	0.0	0.2	99.7
1406	Wire 4	w	0.5	75.5	23.0	0.2	0.0	0.0	0.6	99.8
1406	Wire 5	w	6.4	80.5	11.3	0.1	0.2	0.0	1.6	100.0
1406	Lump 1 wire	w	1.0	98.2	0.0	0.0	0.0	0.0	0.1	99.3
1406	Strip wire end	w	7.2	59.4	32.0	0.0	0.0	0.0	1.1	99.6
SFN 49	Wire 1	w	2.0	70.6	25.3	0.1	0.0	0.0	1.9	99.8
SFN 49	Wire 2	w	1.8	50.4	46.0	0.0	0.2	0.0	1.5	99.9
SFN 49	Wire 3	w	1.6	54.6	41.8	0.0	0.0	0.0	2.0	99.9
SFN 49	Wire 4	w	1.9	59.8	36.1	0.0	0.2	0.0	1.9	99.9
SFN 58	Wire	w	4.9	36.8	55.9	0.0	0.0	0.0	2.2	99.7
SFN 58	Wire2	w	1.5	60.3	36.8	0.0	0.0	0.0	1.2	99.8
	Average		2.6	64.3	31.6	0.0	0.0	0.0	1.2	
	SD		2.2	15.6	15.0	0.1	0.1	0.0	0.7	
	Max		7.2	98.2	55.9	0.2	0.2	0.0	2.2	
	Min		0.5	36.8	0.0	0.0	0.0	0.0	0.1	

TABLE A3.3.7 XRF ANALYSES OF WIRE ARTEFACTS, FE REMOVED AND DATA NORMALISED

Context/SFN	Description	Class	Cu	Zn	As	Sn	Sb	Pb	Total
1406	Wire 1	w	75.8	23.1	0.1	0.0	0.0	1.0	100.0
1406	Wire 2	w	57.3	42.2	0.0	0.0	0.0	0.5	100.0
1406	Wire 3	w	60.3	39.5	0.0	0.0	0.0	0.2	100.0
1406	Wire 4	w	76.0	23.2	0.2	0.0	0.0	0.7	100.0
1406	Wire 5	w	85.9	12.1	0.2	0.2	0.0	1.7	100.0
1406	Lump 1 wire	w	99.8	0.0	0.0	0.0	0.0	0.1	100.0
1406	Strip wire end	w	64.2	34.6	0.0	0.0	0.0	1.1	100.0

SFN 49	Wire 1	w	72.2	25.9	0.1	0.0	0.0	1.9	100.0
SFN 49	Wire 2	w	51.4	46.9	0.0	0.2	0.0	1.5	100.0
SFN 49	Wire 3	w	55.5	42.5	0.0	0.0	0.0	2.0	100.0
SFN 49	Wire 4	w	61.0	36.9	0.0	0.2	0.0	1.9	100.0
SFN 58	Wire	w	38.8	58.9	0.0	0.0	0.0	2.3	100.0
SFN 58	Wire2	w	61.3	37.4	0.0	0.0	0.0	1.2	100.0
	Average		66.1	32.5	0.0	0.0	0.0	1.2	
	SD		15.8	15.5	0.1	0.1	0.0	0.7	
	Max		99.8	58.9	0.2	0.2	0.0	2.3	
	Min		38.8	0.0	0.0	0.0	0.0	0.1	

TABLE A3.3.8 INTERIOR ANALYSES

Context/SFN	Description	Fe	Cu	Zn	Sn	Sb	Pb
1113	Wall 1	20.3	43.3	30.4	0.0	0.0	1.5
1113	Wall 2	4.8	78.7	13.5	0.3	0.4	0.9
1547	Wall	0.9	43.8	53.5	0.0	0.0	1.3
1042 SFN 86e	Wall	6.4	48.5	39.8	0.0	0.0	5.2
1255	Base	14.1	34.0	41.3	1.8	0.2	6.6
1255	Wall 1	10.5	23.3	62.4	0.3	0.0	3.3
1255	Wall 2	17.0	25.4	34.8	16.0	0.9	4.6
1255	Wall 3	7.0	18.5	55.1	1.7	0.0	17.3
1109	Base	10.6	68.9	12.9	1.3	0.0	5.0
1109	Base	2.1	91.2	5.6	0.0	0.0	0.3
1109	Wall 1	19.4	9.8	61.2	0.0	0.0	1.0
1109	Wall 2	2.8	53.3	42.8	0.0	0.0	0.4
	Average	9.7	44.9	37.8	1.8	0.1	3.9
	SD	6.8	24.9	19.2	4.5	0.3	4.7
	Max	20.3	91.2	62.4	16.0	0.9	17.3
	Min	4.8	43.3	13.5	0.0	0.0	0.9

TABLE A3.3.9 INTERIOR DATA ORDERED BY INCREASING COPPER CONTENT

Context/SFN	Description	Fe	Cu	Zn	Sn	Sb	Pb
1109	Wall 1	19.4	9.8	61.2	0.0	0.0	1.0
1255	Wall 3	7.0	18.5	55.1	1.7	0.0	17.3
1255	Wall 1	10.5	23.3	62.4	0.3	0.0	3.3
1255	Wall 2	17.0	25.4	34.8	16.0	0.9	4.6
1255	Base	14.1	34.0	41.3	1.8	0.2	6.6
1113	Wall 1	20.3	43.3	30.4	0.0	0.0	1.5
1547	Wall	0.9	43.8	53.5	0.0	0.0	1.3
1042 SFN 86e	Wall	6.4	48.5	39.8	0.0	0.0	5.2
1109	Wall 2	2.8	53.3	42.8	0.0	0.0	0.4
1109	Base	10.6	68.9	12.9	1.3	0.0	5.0
1113	Wall 2	4.8	78.7	13.5	0.3	0.4	0.9
1109	Base	2.1	91.2	5.6	0.0	0.0	0.3
	Average	9.7	44.9	37.8	1.8	0.1	3.9
	SD	6.8	24.9	19.2	4.5	0.3	4.7
	Max	20.3	91.2	62.4	16.0	0.9	17.3
	Min	7.0	9.8	55.1	0.0	0.0	1.0

TABLE A3.3.10 EXTERIOR ANALYSES

Context/SFN	Description	Fe	Cu	Zn	Sn	Sb	Pb
1113	Wall 1	42.5	3.7	46.3	1.9	2.3	0.3
1113	Wall 2	60.5	8.5	20.0	0.8	1.6	0.6
1547	Wall	35.8	26.4	32.7	1.2	1.5	0.2
1042 SFN 86e	Wall	8.2	7.1	77.1	0.0	0.0	7.1
1255	Base	40.9	16.0	35.2	1.0	0.9	5.4
1255	Wall 1	23.0	32.8	32.2	1.6	1.9	6.8
1255	Wall 2	43.7	20.7	27.3	2.0	1.9	2.8
1255	Wall 3	10.0	60.8	23.3	1.4	0.5	3.8
1109	Base	42.4	13.6	31.8	1.3	2.0	3.2
1109	Base	45.4	2.9	39.9	2.6	3.3	0.2
1109	Wall 1	7.9	37.0	50.9	1.0	1.2	0.4
1109	Wall 2	26.1	18.9	49.9	1.3	1.6	0.3
	Average	32.2	20.7	38.9	1.3	1.6	2.6
	SD	17.0	16.7	15.6	0.7	0.9	2.7
	Max	60.5	60.8	77.1	2.6	3.3	7.1
	Min	42.5	3.7	20.0	0.8	1.6	0.3

TABLE A3.3.11 COMPARISON OF THE INTERNAL AND EXTERNAL SURFACES

Context/SFN	Description	Cu-int	Cu-ext	Diff-Cu	Zn-int	Zn-ext	Diff-Zn
1109	Base	91.2	2.9	88.3	5.6	39.9	-34.3
1109	Base	68.9	13.6	55.3	12.9	31.8	-18.9
1113	Wall 2	78.7	8.5	70.2	13.5	20.0	-6.5
1113	Wall 1	43.3	3.7	39.6	30.4	46.3	-15.9
1255	Wall 2	25.4	20.7	4.7	34.8	27.3	7.5
1042 SFN 86e	Wall	48.5	7.1	41.5	39.8	77.1	-37.3
1255	Base	34.0	16.0	18.0	41.3	35.2	6.1
1109	Wall 2	53.3	18.9	34.4	42.8	49.9	-7.1
1547	Wall	43.8	26.4	17.4	53.5	32.7	20.8
1255	Wall 3	18.5	60.8	-42.3	55.1	23.3	31.8
1109	Wall 1	9.8	37.0	-27.2	61.2	50.9	10.3
1255	Wall 1	23.3	32.8	-9.5	62.4	32.2	30.2
	Average	44.9	20.7	24.2	37.8	38.9	-1.1
	SD	24.9	16.7	38.7	19.2	15.6	23.0
	Max	91.2	60.8	88.3	62.4	77.1	31.8
	Min	68.9	2.9	55.3	5.6	31.8	-34.3

TABLE A3.3.12 GROUP 1 X-RAY FLUORESCENCE ANALYSIS OF SMITHING SLAG

Context	SFN	Sample	Description	Fe	Cu	Zn	Sn	Sb	Pb
1009	3		SSL	30.5	31.6	13.0	2.5	3.0	15.8
1051	18	6	Glassy SSL slag	79.6	8.6	3.9	1.4	1.8	0.0
1051	20b	7	Bubbly SSL slag	88.7	4.1	1.9	0.8	1.2	0.1
1051		5	Glassy SSL slag	67.0	3.3	10.1	0.1	1.9	0.2
1109	36a		SSL	87.0	5.4	3.2	0.5	0.9	0.0
1109	36a		SSL	0.1	65.2	33.3	0.0	0.0	1.0
1255			SSL	86.2	8.8	3.3	0.5	0.3	0.6

1279		15	SSL	75.8	13.6	1.2	0.0	0.4	0.0
1325			SSL cindery	96.1	2.1	1.4	0.0	0.0	0.0
1396	89		SSL	81.2	1.9	4.7	1.1	3.5	0.0
1406			SSL /cinder	68.3	9.4	11.7	1.4	2.4	0.1
1406			Small SSL lump	35.4	28.6	27.4	2.1	2.3	1.5
1406			SSL	36.7	22.7	34.8	1.6	1.7	1.2
	26	9	SSL high fe	98.6	0.2	0.2	0.0	0.0	0.0
	29		SSL1	53.3	21.0	21.2	1.1	1.2	1.1
	29		SSL 2	25.8	56.1	16.4	0.3	0.0	1.1
	29		SSL 3	31.2	65.2	1.7	0.3	0.4	0.5
	58	12	SSL	6.0	31.9	58.8	0.2	0.0	2.6
Average				58.2	21.1	13.8	0.8	1.2	1.4
SD				31.4	21.6	15.9	0.8	1.1	3.7
Min				0.1	0.2	0.2	0.0	0.0	0.0
Max				98.6	65.2	58.8	2.5	3.5	15.8

TABLE A3.3.13 GROUP 2 X-RAY FLUORESCENCE ANALYSIS OF SMITHING SLAGS AND HEARTH BOTTOMS

Context	SFN	Sample	Description	Fe	Cu	Zn	Sn	Sb	Pb
1317			Cu-SSL 1	19.1	76.5	1.7	0.8	0.0	1.7
1317			Cu-SSL 2	30.0	44.7	11.5	6.4	0.0	6.9
1325			Cu-SSL -severe staining	3.8	92.1	3.1	0.0	0.0	0.7
1255			Cu-SSL -flowed 1	17.3	74.2	2.8	0.6	0.0	4.8
1255			Cu-SSL -flowed 2	7.9	83.5	5.5	0.5	0.0	2.1
1255			Cu-SSL -flowed 3	53.4	42.9	3.2	0.0	0.0	0.3
1323			HB1 Cu	53.6	42.6	3.1	0.2	0.0	0.2
1323			HB2 Cu	59.2	35.4	1.7	0.9	0.0	2.6
1329			Cu-SSL 1	16.1	35.2	43.0	0.5	0.0	4.9
1329			Cu-SSL 2	15.6	68.9	12.1	0.6	0.0	2.6
1324			Cu-SSL 1	18.6	71.2	5.5	0.5	0.0	3.6
1324			Cu-SSL 2	27.7	50.1	10.4	0.9	0.0	10.4
1018	4	4	Cu-SSL	1.3	32.2	66.5	0.0	0.0	0.0
1411	91		Cu-SSL	92.1	2.1	4.7	0.3	0.3	0.0
Average				29.7	53.7	12.5	0.9	0.0	2.9
SD				25.8	24.8	18.8	1.6	0.1	3.0
Min				1.3	2.1	1.7	0.0	0.0	0.0
Max				92.1	92.1	66.5	6.4	0.3	10.4

TABLE A3.3.14 GROUP 3 X-RAY FLUORESCENCE ANALYSIS OF BLACK COLOURED SLAGS

Context	Description	Fe	Cu	Zn	Sn	Sb	Pb
1406	Black slag	30.9	43.3	18.5	3.9	0.5	2.1
1317	Cindery black slag	45.9	38.9	12.4	0.4	0.0	1.3
Average		38.4	41.1	15.5	2.2	0.3	1.7
SD		10.6	3.1	4.3	2.5	0.4	0.6
Min		30.9	38.9	12.4	0.4	0.0	1.3
Max		45.9	43.3	18.5	3.9	0.5	2.1

TABLE A3.3.15 GROUP 4 X-RAY FLUORESCENCE ANALYSIS OF SLAGGED STONEWORK AND SLAGGED CLAY STRUCTURE

Context	SFN	Sample	Description	Fe	Cu	Zn	Sn	Sb	Pb
1406			Small slagged stone	81.4	6.0	6.3	1.1	2.2	0.1
1017	5b	4	Slagged structure	0.6	47.6	46.0	0.0	0.0	5.5
Average				41.0	26.8	26.1	0.6	1.1	2.8
SD				57.1	29.5	28.1	0.8	1.6	3.8
Min				0.6	6.0	6.3	0.0	0.0	0.1
Max				81.4	47.6	46.0	1.1	2.2	5.5

TABLE A3.3.16 GROUP 5 X-RAY FLUORESCENCE ANALYSIS OF WHITE COLOURED SLAGS

Context	SFN	Description	Fe	Cu	Zn	Cd	Sn	Sb	Pb
1334		White slag FAS?	65.0	7.3	5.8	8.5	0.0	0.0	1.7
1388		White slag	19.8	3.0	68.4	4.1	0.0	0.0	0.0
1101	40b	White cinder	83.6	1.2	2.0	2.9	0.0	2.4	0.2
		Average	56.1	3.8	25.4	5.1	0.0	0.8	0.7
		SD	32.8	3.1	37.3	2.9	0.0	1.4	0.9
		Min	19.8	1.2	2.0	2.9	0.0	0.0	0.0
		Max	83.6	7.3	68.4	8.5	0.0	2.4	1.7

TABLE A3.3.17 XRF ANALYSES OF THE HEARTH LINING SAMPLES (WEIGHT %, N.D. – NOT DETECTED).

Context/Sample Number	Description	class	Fe	Ni	Cu	Zn	Sn	Sb	Pb
Cont 1018 sfn 4 sample 2	Hearth structure	HL	16.6	0.0	23.6	48.0	1.6	1.3	6.5
1406 GMcD 2	Hearth lining	HL	42.6	0.3	5.1	42.3	1.9	2.2	0.4
1406 GMcD 2	Small hearth lining	HL	23.1	0.0	4.5	61.5	1.7	2.1	0.6
1406 GMcD 2	Small hearth lining	HL	23.1	0.0	4.5	61.5	1.7	2.1	0.6
1369 GMcD 9	Slagged lining 1	HL	85.5	2.7	0.7	1.7	0.0	0.0	0.0
1369 GMcD 9	Slagged lining 2	HL	84.6	2.3	2.3	1.4	0.0	1.7	0.0
1369 GMcD 9	Slagged lining 3	HL	77.6	0.8	4.7	5.4	2.1	3.4	0.0

TABLE A3.3.18 OVERALL XRF ANALYSES OF THE THREE MAIN PRODUCT TYPES AND SPECIFIC RATIOS (WEIGHT %)

	Cu	Zn	As	Sn	Sb	Pb	Cu/Zn	Cu/Pb	Zn/Pb
Cast	66.2	30.4	0	0.2	0	3.1	2.2	21.4	9.8
Sheet	72.5	25.7	0.1	0.2	0	1.6	2.8	45.3	16.1
Wire	66.1	35.3	0	0	0	1.2	1.9	55.1	29.4
Overall	70.1	27.9	0.1	0.1	0	1.8	2.5	38.9	15.5

TABLE A3.3.19 RATIOS OF HEARTH LINING XRF ANALYSES

Context/Sample Number	Cu	Zn	Sn	Pb	Cu/Zn	Cu/Pb	Zn/Pb
Cont 1018 sfn 4 sample 2	23.6	48.0	1.6	6.5	0.5	3.6	7.4
1406 GMcD 2	5.1	42.3	1.9	0.4	0.1	14.5	120.5
1406 GMcD 2	4.5	61.5	1.7	0.6	0.1	8.0	107.9
1406 GMcD 2	4.5	61.5	1.7	0.6	0.1	8.0	107.9
1369 GMcD 9	0.7	1.7	0.0	0.0	0.4		
1369 GMcD 9	2.3	1.4	0.0	0.0	1.7		
1369 GMcD 9	4.7	5.4	2.1	0.0	0.9	122.7	141.8

APPENDIX 3.4. METHODOLOGY OF METALLOGRAPHIC ANALYSIS OF THE COPPER ALLOY METAL AND SLAGS

A sample was cut from the artefacts and mounted in cold setting resin, ground and polished to a ¼ micron finish. The sections were examined using a reflected light microscope in the unetched condition to reveal the occurrence of casting defects (porosity), non-metallic inclusions, and corrosion penetration. The sections were then etched in alcoholic ferric chloride and re-examined to identify the microstructure. Digital images were recorded of both the unetched and etched condition at various magnifications. The mounted specimens were dagged and placed in the FEI Quanta 400 SEM under high vacuum (operating conditions accelerating voltage 20kV, Filament at saturation, emission current *c*100 Micro-Amps. Back Scattered Electron imaging was used (termed BSE image in the text). Oxford Instruments' Inca Software was utilised to obtain quantitative data of both bulk area and phase analyses. Three areas were analysed on each sample to generate a mean bulk value. The areas were analysed at various magnifications commonly between x300-x800, although magnifications as high as x2500 were utilised. Individual phases were analysed in spot mode. Analysis data were measured to *c* 0.1%. See Appendix 1 for details of minimum detectable levels.

The copper-zinc phase diagram is characterised by a narrow liquidus-solidus field, which is in contrast to the other important copper alloy phase diagrams relevant for historical and archaeological copper alloys, eg copper-tin and copper-arsenic. The narrow solidus-liquidus field means that the dendrites tend to be fine and that no coring occurs. The maximum solubility of zinc in copper is very high - *c* 32.5% at the solidus temperature. This means that all alloys containing 32.5% or less of zinc will result in a granular microstructure, termed alpha grains. These alloys are conventionally divided into two types relating to their colour - red brass with zinc contents of *c* 15% (or less), and yellow brass containing *c* 30% zinc. Alloys with between 32.5 and 37% zinc will initially freeze at 900oC as alpha dendrites surrounded by beta phase, but if cooled slowly below 400oC then it transforms to alpha grains. Between 37–39% zinc the alloy initially freezes out as grains of beta, which again below 250oC transforms to alpha grains. This means that all alloys up to 39% zinc have the same microstructure of alpha grains but will differ in colour, the higher the zinc the less red/whiter the metal, and differ in hardness, and working properties. Between 39–45.5% zinc the alloy initially freezes out as β phase which transforms to grains of alpha with β' (Beta Prime) present at the grain boundaries. Between 45.5 and 50% zinc the microstructure at ambient temperature is grains of β'. Between 50 and 59% zinc the microstructure is grains of β' with γ (gamma). The addition of lead to brass enhances the machinability and some casting properties of the alloy. The lead is insoluble in copper in the solid state, and occurs as droplets throughout the microstructure at the inter-dendritic or inter-granular locations. It has slight detrimental effects on wrought brasses, hence, it is unlikely to be used for alloys used in wire manufacture. The addition of tin enhances corrosion resistance and would alter the casting properties slightly.

For the slags a thick section was removed from the selected slags, mounted in cold setting resin and ground and polished to ¼ micron finish. The samples were examined using a metallurgical reflected light microscope and digital images recorded. The samples were coated in carbon and were dagged and placed in a FEI Quanta 400 SEM under high vacuum (operating conditions accelerating voltage 20kV, Filament at saturation, emission current *c* 100 Micro-Amps. Back Scattered Electron imaging was used (termed BSE image in the text). Oxford Instruments Inca Software was utilised to obtain quantitative data of both bulk area and phase analyses. Five areas were analysed on each sample to generate a mean bulk value. The area were analysed at various magnifications commonly between x300-x800. Individual phases were analysed in spot mode. Analysis data was measured to *c* 0.1% see Appendix 1 for details of minimum detectable levels.

APPENDIX 3.5. DETAILED RESULTS FROM METALLOGRAPHIC ANALYSIS OF THE COPPER ALLOY ITEMS

The metallographic analyses are ordered into as-cast artefacts, sheet artefacts, and wire artefacts.

SFN 58 Bar Ingot

In the unetched condition there was little porosity and some non-metallic inclusions. When etched the microstructure comprised alpha grains with annealing twins and finely dispersed lead inclusions. This demonstrated that the bar had been subject to some hot-working and was not in the as-cast condition. It is probable that it had undergone some hot rolling possibly with the intention to reduce it strip or wire. The XRF analyses (Table A3.5.1) and the SEM analyses (Table A3.5.2) are very similar (Table A3.5.3) and place the alloy at the high end of alpha brasses. The XRF data presents both the data obtained in the as-received condition (ie including the corroded surface), and as cleaned. It appears that there is closer correspondence between the as-received XRF data and the SEM data for zinc. It would initially freeze as dendrites of alpha and interdendritic beta, which on cooling would transform to alpha grains. The SEM analyses showed the presence of silica and the back-scattered image demonstrated that there were small particles less than 10 microns in diameter distributed throughout the microstructure. Analysis of other samples showed that these are sand particles presumably derived from the sand moulds used in casting. This ingot is cast from a high zinc alloy, which would allow for subsequent zinc loss as it was worked to a final product eg sheet or wire.

SFN 1245 Bar 1

In the unetched condition there was clear inter-dendritic porosity concentrated in the centre of the bar, with some non-metallic inclusions present throughout the microstructure. When etched it revealed a distinct dendritic microstructure possibly due to the lead, alpha dendrites in solid solution with lead droplets and porosity. Under normal casting conditions the microstructure should have been alpha grains, however, rapid cooling may have caused the formation of the dendrites. The XRF data (Table A3.5.4) and the SEM data (Table A3.5.5) are compared in Table A3.5.6. The SEM data give a higher zinc value than the XRF, which can be explained by de-zincification during corrosion, and places the alloy composition in the second group of alloys, which initially cool as dendrites of alpha surrounded by beta phase. Normally, on cooling this microstructure would transform to grains of alpha below 250°C, however, the presence of the lead droplets (Table A3.5.7, Plate 5), the porosity, and the sand particles, will fossilize the dendritic structure. The BSE image emphasises the finely distributed lead particles (white) and the porosity and sand particles (black), but also shows that there is no coring of the dendrites and that effectively the composition is uniform across the sample. The analysis of a lead droplet is given in Table A3.5.7 and shows beam penetration resulting in values for copper and zinc. This is due to the small size of the droplets *c* 4 microns in diameter. The bar was in the as-cast condition and again is high in zinc, which would allow for zinc loss during further hot-working. The as-cast dendritic microstructure was retained due to the lead, porosity and sand particles, and may indicate rapid cooling (quenching) from high temperature.

SFN 1245 Bar 2

The unetched specimen showed little porosity concentrated in the centre of the bar, with some non-metallic inclusions dispersed throughout the microstructure. There is corrosion penetration into the sample from the surface utilising the porosity as corrosion pathways. Etching revealed an as-cast microstructure of alpha dendrites with the fine lead droplets dispersed throughout the microstructure. The lead droplets are heavily concentrated towards the centre of the casting; hence the surface XRF analysis may underestimate the lead content. The structure shows coarse grains with dendritic microstructure within these grains, possibly due to rapid cooling. The XRF data (Table A3.5.8) and SEM data (Table A3.5.9) are compared in Table A3.5.10 and show that the XRF severely underestimated the zinc content, presumably due to sever de-zincification during the burial and corrosion of the artefact. Analyses of examples of two dendrites (Table A3.5.11, show that they do not differ from the overall composition. Hence, indicating that the as-cast microstructure was preserved by the presence of the lead droplets, porosity and sand particles.

SFN 1245 Tube

In the unetched condition the alloy contained numerous non-metallic inclusions and some fine porosity. Etching showed that the tube was in the as-cast condition with dendrites growing in from the inner and outer surfaces. The XRF normalised data (Table A3.5.12) and the SEM data (Table A3.5.13), which are compared in Table A3.5.14, show that in contrast to the analyses of the previous artefacts, the XRF over-estimated the zinc content, and at *c* 32% Zn the microstructure should comprise alpha grains, with finely dispersed lead particles. The analysis of a dendrite and an interdendritic phase (Table A3.5.15) show that there is little difference in composition between the two phases. Again, the retention of a dendritic microstructure is due to the presence of the lead droplets, the sand particles and porosity.

SFN 1324 Strip 1

Two cross-sections and a longitudinal section were cut from the artefact; all three sections displayed the same microstructures. In the unetched condition the sections showed evidence of inter-dendritic corrosion; the most severe is black in colour, and the less severe corrosion is grey. When etched the sections displayed fine alpha dendrites with solid solution, plus lead droplets at the inter-dendritic boundaries. The XRF data (Table A3.5.16) and the SEM data (Table A3.5.17) are compared in Table A3.5.18, and demonstrate that the alloy is low in zinc with a low tin content. The values of zinc and lead obtained by the different techniques do differ, but this reflects corrosion mechanisms. Analysis of a dendrite and interdendritic phase (Table A3.5.19) show again that there is only a very slight difference between the two phases, which indicates that the residual dendritic microstructure is formed by the lead droplets (analysis shown in Table A3.5.19), the porosity and the sand inclusions. The strip is in the as-cast condition which is unexpected as it would have been cold-worked and annealed or hot-worked, which should have removed the as-cast microstructures and evened out any compositional differences though diffusion. To investigate the effect of corrosion phenomena on archaeological and historical brasses, an area of corrosion noted in the unetched micrograph was investigated by SEM. The area investigated was examined using BSE imaging, and starkly demonstrates the changing composition at the metal/corrosion interface. On the right of Plate 13 is the un-corroded metal displaying the remnant dendritic microstructure, the finely dispersed lead particles (white) and the sand grains and porosity (black). The area and phase analyses that are presented in Table A3.5.20. The un-corroded metal shows the low zinc content (9%). In contrast, the corroded metal is in oxide form, and the zinc has been lost. The area analysis of the corroded area is characterised by the high sulphur content and low zinc values. The sulphur is associated with the lead rich phase, where it is occurring as lead sulphide. It is also clear how the lead is retained in the corroded zone and forms a larger volume of the corrosion hence it is enhanced in the XRF analyses.

SFN 1406 Disc 9

The disc had been cleaned back during the XRF programme to compare the corroded surface with the bright metallic surface. Hence the flat surface was mounted and prepared for metallographic analysis as well as a cross-section.

In the unetched condition the cross-section only showed a few small non-metallic inclusions present in the sample. When etched (the microstructure showed equi-axed alpha grains with twins, and beta prime at the grain boundaries. Within the microstructure there were some lighter coloured phases, indicative of either light corrosion or non-alloyed copper. The XRF data (Table A3.5.21) and the SEM data (Table A3.5.22) are compared in Table A3.5.23, and show close correspondence, especially the SEM data and the XRF data of the cleaned surface. The alloy contains *c* 35% zinc, which should result in a final microstructure of alpha grains. However, the optical microscopy is confirmed by the SEM study, which shows that the grain boundary is beta prime (Table 3.5. A3.5.24). The lighter phase identified in the optical study is confirmed as copper (Table A3.5.24). The microstructure deviates from the phase diagram because under equilibrium conditions the phase diagram predicts that the microstructure should just comprise alpha grains. The manufacturing process induces the formation of the beta prime phase, perhaps by heating to above 250°C and quenching. The presence of twins is expected as they result from cold-working and annealing, or hot-working.

The flat section is has the same microstructure as the cross-section, the only slight difference is that the beta prime is segregated towards the centre of the section. There are no twins apparent.

SFN 1406 Sheet, offcut 3

Three sections, flat, longitudinal-section, and a cross-section were cut from the sample.

In the unetched condition the flat section appeared to contain either a large number of non-metallic inclusions or corrosion was present. Both the long-section and the cross-section did not display the non-metallic component and so it is believed that the flat section displayed corrosion. In the etched condition the longitudinal section displayed elongated alpha grains with finely dispersed lead particles. The XRF data (Table A3.5.25) and the SEM data (Table A3.5.26) area compared in Table A3.5.27, and show that the XRF analysis reflected the corrosion, with the lower zinc value and enhanced lead. The 'cleaned surface' still clearly retains some corrosion as indicated by the lead content. This is confirmed by the iron values obtained from the XRF analysis, which shows a high value present in the corroded surface, but still present on the 'cleaned surface'. Significantly, the SEM analysis did not detect silica, which has been present in the other specimens, especially those displaying the remnant dendritic microstructures. This sample was rich in zinc indicating that little had been lost during the working process, and displayed the expected microstructure of distorted alpha grains with finely dispersed lead droplets.

SFN 1406, Strip and wire

This artefact was a strip that had been drawn to wire, so that one end was classed as strip and the other as wire. The metallography should reflect these differences. The XRF analyses show marked differences between the two ends (Table A3.5.28), which is unlikely but may represent differences in corrosion.

In the unetched condition (the strip end was very clean with very few non-metallic inclusions present. When etched the microstructure comprises grains of alpha with grain boundary beta prime. There is no evidence for distortion or twins.

In the unetched condition the wire end was also very clean. When etched the microstructure comprises grains of alpha with grain boundary beta prime, with finely distributed lead particles. The grain boundary phase is concentrated in the centre of the wire, but is not distorted. The XRF data (Table A3.5.28) shows a difference between the two ends, with a very high zinc value for the strip end (58% Zn). The SEM data (Table A3.5.29) show consistent values between the two ends and comparison with the XRF data (Table A3.5.30) that the wire end XRF data was more representative. An alloy with a zinc content of *c* 37% should result in a microstructure of alpha grains; hence, the presence of grain boundary beta prime indicates deviation from the phase diagram, despite extensive working, especially of the wire end (A3.5.31).

SFN 1406 Wire 5

A longitudinal and cross-section were removed from the artefact. In the unetched condition the both sections were clean with a few non-metallic inclusions but with evidence of corrosion penetration. In the etched condition (the alpha grains are delineated by the inter-granular corrosion, which will have removed the lead droplets. The grains display heavily distorted features due to drawing process. The XRF data (Table A3.5.32) and the SEM data (Table A3.5.33) are compared in Table A3.5.34, and show that the XRF data radically under-estimated the zinc content. This is due to the thinness of the wire and the corrosion effect combing to produce this result.

SFN 58 Wire

A longitudinal and a cross-section sample were prepared. In the unetched condition both the cross-section and the longitudinal section show evidence of apparent segregation. When etched the microstructure showed alpha grains with some annealing twins were present. The XRF data (Table A3.5.35) and the SEM data (Table A3.5.36) are compared in Table A3.5.37 and show that the XRF over-estimated the zinc content, which is in contrast to SFN 1406 Wire, which under-estimated the zinc content. The wire was cast, annealed and hot-worked

TABLE A3.5.1 XRF ANALYSES OF SFN 58 BAR INGOT (WEIGHT %)

SFN 58 ingot	Cu	Zn	As	Sn	Sb	Pb
Corroded Surface	66.2	32.9	0	0	0	1.2
Cleaned Surface	69.5	29.9	0	0	0	0.6

TABLE A3.5.2 SEM ANALYSES OF BAR 58

Element	Area 1	Area 2	Area 3	Mean
Ni	0.0	0.2	0.0	0.1
Cu	66.1	66.6	66.5	66.4
Zn	33.5	33.3	32.9	33.3
As	0.3	0.3	0.5	0.3
Sn	0.1	0.0	0.0	0.0
Sb	0.1	0.0	0.0	0.0
Pb	0.0	0.0	0.1	0.1
Totals	100.0	100.3	100.09	100.1

TABLE A3.5.3 COMPARISON OF SFN58 INGOT AND SEM DATA

	XRF (corroded Surface)	XRF (Cleaned Surface)	SEM
Cu	66.2	69.5	66.4
Zn	32.9	29.9	33.3
Sn	0.0	0	0.0
Sb	0.0	0	0.0
Pb	1.2	0.6	0.1

TABLE A3.5.4 XRF ANALYSIS OF SFN 1245 BAR/INGOT 1 (WEIGHT %)

SFN 1245 Bar 1					
Cu	Zn	As	Sn	Sb	Pb
71.1	24.1	0	0.2	0	4.2

TABLE A3.5.5 SEM ANALYSES OF SFN 1245 BAR 1 (WEIGHT %)

	Area 1	Area 2	Area 3	Average
Ni	0.1	0.2	0.3	0.2
Cu	65.6	65.5	65.1	65.4
Zn	33.1	33.2	32.9	33.0
As	0.2	0.0	0.1	0.1
Sn	0.3	0.3	0.4	0.3
Sb	0.0	0.1	0.2	0.1
Pb	0.7	1.0	1.1	0.9

TABLE A3.5.6 COMPARISON OF SFN 1245 BAR 1 XRF AND SEM DATA

	XRF	SEM
Cu	71.1	65.4
Zn	24.1	33.0
Sn	0.2	0.3
Sb	0.0	0.1
Pb	4.2	0.9

TABLE A3.5.7 ANALYSIS OF A LEAD INCLUSION IN SFN 1245 BAR 1

	Pb incl
Ni	0.2
Cu	42.6
Zn	25.4
As	0.0
Sn	0.7
Sb	0.2
Pb	31.2

TABLE A3.5.8 XRF ANALYSIS OF SFN 1245 BAR/INGOT 2 (WEIGHT %)

SFN 1245 Bar 2					
Cu	Zn	As	Sn	Sb	Pb
89.8	8.1	0.2	0	0	1.8

TABLE A3.5.9 SEM ANALYSES OF SFN 1245 BAR 2 (WEIGHT %)

	Area 1	Area 2	Area 3	Average
Ni	0.1	0.1	0.1	0.1
Cu	72.7	72.2	71.7	72.2
Zn	26.4	26.8	26.7	26.7
As	0.2	0.0	0.2	0.1
Sn	0.0	0.2	0.3	0.2
Sb	0.0	0.1	0.1	0.0
Pb	0.9	0.8	0.9	0.8

TABLE A3.5.10 COMPARISON OF SFN 1245 BAR 2 XRF AND SEM DATA

	XRF	SEM
Cu	89.8	72.2
Zn	8.1	26.7
Sn	0.0	0.2
Sb	0.0	0.0
Pb	1.8	0.8

TABLE A3.5.11 SEM ANALYSES OF EXAMPLES OF THE DENDRITES IN SFN 1245 BAR 2 (WEIGHT %)

	Dendrite 1	Dendrite 2	Mean
Ni	0.1	0.2	0.2
Cu	71.2	73.6	72.4
Zn	27.8	26.0	26.9
As	0.2	0.0	0.1
Sn	0.4	0.0	0.2
Sb	0.2	0.0	0.1
Pb	0.3	0.5	0.4

Table A3.5.12 XRF Data SFN 1245 tube (Weight %, n.d- not detected)

Cu	Zn	As	Sn	Sb	Pb
58.9	37.7	n.d.	0.2	n.d.	3.4

Table A3.5.13 SEM analyses of SFN 1245 Bar 2 (Weight %)

	Area 1	Area 2	Area 3	Average
Ni	0.0	0.0	0.0	0.0
Cu	67.7	67.0	66.8	67.2
Zn	31.3	31.8	32.0	31.7
As	0.2	0.4	0.0	0.2
Sn	0.3	0.0	0.2	0.2
Sb	0.0	0.0	0.1	0.0
Pb	0.8	0.7	0.8	0.8

Table A3.5.14 Comparison of SFN 1245 tube XRF and SEM data

	XRF	SEM
Cu	58.9	67.2
Zn	37.7	31.7
Sn	0.2	0.2
Sb	n.d.	0.0
Pb	3.4	0.8

Table A3.5.15 SEM analyses of the dendritic and interdendritic phases in SFN 1245 Tube. (Weight %)

	Dend	Inter-dend
Ni	0.0	0.1
Cu	66.6	67.3
Zn	32.8	32.0
As	0.2	0.1
Sn	0.1	0.1
Sb	0.0	0.1
Pb	0.4	0.3

Table A3.5.16 XRF analysis of SFN 1324 Strip 1 (Weight %, Fe removed and normalised)

SFN 1324 Strip 1					
Cu	Zn	As	Sn	Sb	Pb
86.1	4.4	0.4	1.1	0	8

Table A3.5.17 SEM analyses of SFN 1324 Strip 1 (Weight %)

	Area 1	Area 2	Area 3	Average
Ni	0.1	0.0	0.1	0.1
Cu	88.0	87.8	87.7	87.8
Zn	9.7	9.9	9.7	9.7
As	0.0	0.0	0.1	0.1
Sn	1.5	1.3	1.5	1.4
Sb	0.0	0.0	0.0	0.0
Pb	0.8	1.1	1.0	1.0

Table A3.5.18 Comparison of SFN 1324 strip 1 XRF and SEM data

	XRF	SEM
Cu	86.1	87.8
Zn	4.4	9.7
Sn	1.1	1.4
Sb	0.0	0.0
Pb	8.0	1.0

Table A3.5.19 SFN 1324 Strip 1, dendrite and interdendritic phase. Lead inclusion (Weight %)

	Dendrite	Interdendritic	Lead Inclusion
Ni	0.2	0.0	0.1
Cu	84.3	87.2	34.3
Zn	8.6	9.9	3.8
As	0.1	0.2	0.0
Sn	6.4	2.3	1.1
Sb	0.4	0.1	0.2
Pb	0.1	0.4	60.5

Table A3.5.20 SFN 1324 Strip 1 analyses of corroded area (Weight %)

	Corroded Area	Corroded Alloy			Lead Phase	Uncorroded Metal
SO3	2.6	0.1		S	14.2	0.0
NiO	0.2	0.0		Ni	0.0	0.1
CuO	81.7	99.1		Cu	4.3	90.6
ZnO	2.7	0.2		Zn	0.2	9.0
As2O3	0.3	0.0		As	0.0	0.0
SnO2	4.9	0.1		Sn	0.9	0.4
Sb2O3	0.0	0.1		Sb	0.0	0.0
PbO	7.7	0.4		Pb	80.9	0.1

Table A3.5.21 XRF analysis of SFN 1406 Disc 9 (weight %, Fe removed and data normalised)

	Cu	Zn	As	Sn	Sb	Pb
Corroded surface	65.7	33.9	0	0	0	0.4
Cleaned surface	64	35.7	0	0	0	0.3

Table A3.5.22 SEM analyses of SFN 1406 Disc 9 (weight %)

	Area 1	Area 2	Area 3	Average
Ni	0.0	0.1	0.0	0.1
Cu	62.1	66.4	62.7	63.7
Zn	36.8	33.3	36.3	35.5
As	0.3	0.2	0.5	0.3
Sn	0.2	0.1	0.0	0.1
Sb	0.2	0.0	0.1	0.1
Pb	0.4	0.1	0.5	0.3

Table A3.5.23 Comparison of SFN 1406 Disc 9 XRF and SEM data

	XRF (corroded surface)	XRF (cleaned surface)	SEM
Cu	65.7	64.0	63.7
Zn	33.9	35.7	35.5
Sn	0.0	0.0	0.1
Sb	0.0	0.0	0.1
Pb	0.4	0.3	0.3

Table A3.5.24 Phase analyses of Sample 1406 Disc 9 (Weight %)

	Alpha Phase	Alpha Phase 2	Beta Prime	Cu Rich Phase	Cu Rich Phase 2	Cu Rich Phase 3
Ni	0.1	0.1	0.1	0.2	0.1	0.1
Cu	63.2	63.6	53.9	98.3	99.2	98.3
Zn	35.9	35.9	45.7	1.4	0.9	1.4
As	0.1	0.2	0.4	0.0	0.1	0.1
Sn	0.4	0.1	0.0	0.0	0.0	0.0
Sb	0.0	0.1	0.0	0.0	0.0	0.0
Pb	0.3	0.0	0.2	0.2	0.2	0.1

Table A3.5.25 XRF analysis of SFN 1406 Offcut 3 (weight %, Fe removed and normalised, Fe value given after total)

SFN 1406 Sheet, offcut 3	Cu	Zn	As	Sn	Sb	Pb	Total	Fe%
Offcut 3 side 1 corroded surface	68.4	30.0	0.0	0.0	0.0	1.6	100.0	3.8
Offcut 3 side 1 cleaned	64.5	34.3	0.0	0.0	0.0	1.1	100.0	0.1

Table 3.5.26 SEM analyses of SFN 1406 Offcut 3 (weight %)

	Area 1	Area 2	Area 3	Average
Ni	0.2	0.0	0.1	0.1
Cu	63.8	63.5	62.3	63.2
Zn	35.6	36.0	36.9	36.1
As	0.1	0.3	0.3	0.2
Sn	0.2	0.2	0.1	0.2
Sb	0.0	0.0	0.1	0.1
Pb	0.1	0.0	0.2	0.1

Table A3.5.27 Comparison of SFN 1406 Offcut 3 XRF data and SEM data

	XRF (corroded surface)	XRF (cleaned surface)	SEM
Cu	68.4	64.5	63.2
Zn	30.0	34.3	36.1
Sn	0.0	0.0	0.2
Sb	0.0	0.0	0.1
Pb	1.6	1.1	0.1

Table A3.5.28 XRF analyses of strip and wire ends of SFN 1406 (weight %, Fe removed and data normalised)

SFN 1406	Cu	Zn	As	Sn	Sb	Pb
Strip end	41.8	58.1	0	0	0	0.1
Wire end	64.2	34.6	0	0	0	1.1

Table A3.5.29 SEM analyses of SFN 1406 strip and wire (weight %)

	Strip end						Wire end			
	Area 1	Area 2	Area 3	Area 4	Average		Area 1	Area 2	Area 3	Average
Ni	0.1	0.2	0.0	0.1	0.1	Ni	0.1	-0.1	0.1	0.0
Cu	62.1	62.6	63.0	63.0	62.7	Cu	62.7	62.8	62.1	62.5
Zn	37.6	37.1	36.8	37.3	37.2	Zn	36.5	37.5	37.4	37.1
As	0.2	0.2	0.2	0.1	0.1	As	0.3	0.1	0.3	0.2
Sn	0.0	0.0	0.1	0.0	0.0	Sn	0.1	0.0	0.1	0.1
Sb	0.0	0.1	0.0	0.0	0.0	Sb	0.2	0.0	0.0	0.1
Pb	0.1	0.0	0.0	0.0	0.0	Pb	0.1	0.0	0.1	0.1

TABLE A3.5.30 COMPARISON OF SFN 1406 STRIP AND WIRE ENDS XRF AND SEM DATA

	XRF Strip End	SEM Strip End		XRF Wire End	SEM Wire End
Cu	41.8	62.7		64.2	62.5
Zn	58.1	37.2		34.6	37.1
Sn	0.0	0.0		0.0	0.1
Sb	0.0	0.0		0.0	0.1
Pb	0.1	0.0		1.1	0.1

TABLE A3.5.31 SEM PHASE ANALYSIS OF SFN 1406 STRIP AND WIRE (WEIGHT %)

	Strip End		Wire End				
	Alpha Phase	Beta Prime	Alpha Phase	Beta Prime	Beta Prime 2	Pb incl	Si incl
Ni	0.0	0.2	0.2	0.1	0.1	0.0	0.2
Cu	62.6	54.5	63.0	53.9	54.8	45.0	54.5
Zn	37.4	45.5	37.0	45.5	44.8	26.9	31.7
As	0.2	0.0	0.1	0.4	0.1	0.2	0.0
Sn	0.0	0.0	0.0	0.1	0.1	0.0	0.1
Sb	0.0	0.0	0.0	0.0	0.1	0.1	0.2
Pb	0.0	0.2	0.1	0.2	0.0	27.9	0.1
Si							13.2

TABLE A3.5.32 XRF ANALYSIS OF SFN 1406 WIRE 5 (WEIGHT %, FE REMOVED AND DATA NORMALISED)

SFN 1406 Wire 5					
Cu	Zn	As	Sn	Sb	Pb
85.9	12.1	0.2	0.2	0	1.7

TABLE A3.5.33 SEM ANALYSES OF SFN 1406 WIRE 5 (WEIGHT %)

	Area 1	Area 2	Area 3	Area 4	Average
Ni	0.1	0.1	0.1	0.1	0.1
Cu	65.2	65.3	65.3	65.3	65.3
Zn	34.5	34.2	34.4	33.6	34.2
As	0.0	0.2	0.1	0.4	0.2
Sn	0.1	0.0	0.0	0.3	0.1
Sb	0.1	0.1	0.1	0.0	0.1
Pb	0.0	0.2	0.2	0.2	0.1

TABLE A3.5.34 COMPARISON OF SFN 1406 WIRE 5 XRF AND SEM DATA

	XRF	SEM
Cu	85.9	65.3
Zn	12.1	34.2
Sn	0.2	0.1
Sb	0	0.1
Pb	1.7	0.1

TABLE A3.5.35 XRF ANALYSIS OF SFN 58 WIRE (WEIGHT %, FE REMOVED AND DATA NORMALISED)

SFN 58 Wire					
Cu	Zn	As	Sn	Sb	Pb
38.8	58.9	0	0	0	2.3

TABLE A3.5.36 SEM ANALYSES OF SFN 58 WIRE (WEIGHT %)

	Area 1	Area 2	Area 3	Average
Ni	0.1	0.1	0.0	0.0
Cu	64.1	64.4	64.2	64.2
Zn	35.1	35.6	35.2	35.3
As	0.5	0.1	0.2	0.3
Sn	0.1	0.1	0.2	0.1
Sb	0.0	0.0	0.1	0.0
Pb	0.2	0.0	0.1	0.1

TABLE A3.5.37 COMPARISON OF SFN 58 WIRE XRF AND SEM DATA

	XRF	SEM
Cu	38.8	64.2
Zn	58.9	35.3
Sn	0.0	0.1
Sb	0.0	0.0
Pb	2.3	0.1

APPENDIX 3.6. DETAILED RESULTS FROM METALLOGRAPHIC ANALYSIS OF THE SMITHING SLAG SAMPLES

SFN 1109 36a

Morphologically, the slag was atypical smithing slag lump, randomly shaped, black in colour and an agglomerated texture, ie not flowed. The reflected light microscope study showed that the microstructure comprises fine crystal laths and very fine dendritic phase in a glassy matrix. There are no copper alloy prills present. The SEM examination confirmed the optical study and the bulk analyses (Table A3.6.1) showed that the slag is rich in alumina, silica and lime and relatively low in iron oxide compared to archaeological smithing slags. The bulk analyses confirm that this slag does not contain elevated levels of copper or other non-ferrous metals and hence has not been used for working copper alloy. However the low iron content (*c* 5%) suggests that the slag was not formed during ironworking, as most slags contain excess iron oxide in the form of free iron oxide dendrites. The fine skeletal dendrites appeared to be iron oxide dendrites but the phase analyses (Table A3.6.2) clearly demonstrate that this is not the case, and is instead identified as hercynite and iron oxide, and alumina mineral. The grey lath crystals are aluminium lime silicates. The material attached to the slag is sand (SiO2). The slag is predominantly aluminium lime silicate with some crystals of alumina iron oxide dendrites.

SFN 1411 SSL-Cin

Morphologically similar to SFN 1109 36 a, but considered to have a higher silica content due to its lighter colour. The microstructure is dominated by a silicate or glassy matrix, with some possible silicate crystals, occasional crystals of iron oxide and a very few metal or metal oxide prills. The bulk analysis (Table A3.6.3) shows the slag is dominated by alumina and silica with varying quantities of lime and iron oxide. The analyses of the iron oxide crystals confirm them as iron oxides (Table A3.6.4). The possible silicate crystals were small variations of concentration of the bulk analysis. On one side of the section was an area of iron oxide adjacent to which was a silicate phase (Table A3.6.4). Overall the slag was an aluminium silicate of varying composition particularly in the levels of lime and iron oxide.

SFN 1255 Cu-SSL

This sample had a varied microstructure ranging from long laths in a silicate matrix (Plate 3) to areas with squat light/dark grey crystal in a glassy matrix containing fine dendrites. Copper alloy prills are common throughout the microstructure especially in the glassy/dendritic areas. Two large metallic inclusions were present. The bulk analysis (Table A3.6.5) shows a composition dominated by alumina and silica with lime, and higher levels of iron oxide than were present in the ironworking slags. Copper was present at significant levels with minor levels of tin, lead and zinc. The phase analyses (Table A3.6.6) show that one crystal phase is alumino-silicates solid solutions with various quantities of lime or iron oxide present; others are hercynites (principal oxides alumina and iron oxide). The analysis of one of the oxides metal prills and one of the large metallic inclusions show that they are copper, occurring as either oxides or chlorides.

SFN 1324 Cu-SSL

The microstructure comprises fine dendrites in a silicate matrix with copper alloy prills present throughout the section. The bulk analysis (Table A3.6.7) shows that the slag is dominated by alumina and silica with a significant iron oxide content. The lime content is low and although copper oxide and zinc oxide are present, they are at low levels but the area analyses avoided the large metallic prills. For example Area 4 (Plate 5) includes some prills, they contribute only a small level to the overall analysis. The phase analyses (Table A3.6.8) show that the dendrites are hercynite (2FeOAl2O3) in an alumina-silicate matrix which at high magnification may occur as squat lath crystals. The analyses of metallic prills (Table A3.6.9) show that they are copper.

SFN 1329 Cu-SSL

The microstructure comprised silicate crystals in a glassy or silicate matrix, with numerous copper alloy prills, dispersed throughout the section. The bulk analyses (Table A3.6.10) show that five of the areas (Areas 1–5) are broadly similar and show that the slag is an alumino-silicate slag, with a reasonable level of iron oxide. The analyses are distinguished from other slags by the high zinc oxide content. A sixth area was taken from part of the slag section that displayed a different microstructure and was dominated by low Z number phase and was silica rich, and the phase analyses (Table A3.6.11) confirm the particles as sand. The phase analyses confirm the presence of zinc oxide in all phases, the two crystalline phases (light and dark grey phases in Table A3.6.11), are zinc-rich, one being a silicate, the other an alumina-iron oxide - zinc oxide mineral silicates. The analyses of the prill and associated corrosion products (Table A3.6.12) show that both copper and lead prills are present. The copper prill does not contain zinc.

SFN 1369 Lining

Under reflected light microscopy there was little discrimination of phases. There were no copper alloy prills or evidence of copper corrosion present in the sample. In the SEM BSE imaging showed that the lining section showed a variable microstructure ranging from an amorphous uniform low Z number (grey) matrix with occasional high Z number (white) crystals to a multiple crystalline microstructure. The black phase in Plate 9 is either charcoal or more probably coal due to the presence of a sulphur peak in the spectrum obtained from a similar phase in another part of the slag. The bulk area analyses (Table A3.6.13) show that alumina and silica dominate the composition with a significant level of lime. The phase analyses (Table A3.6.14) show that the phases identified in the different areas analysed vary in composition, but all are rich in alumina.

TABLE A3.6.1 BULK ANALYSES OF SFN 1109 36 A (WEIGHT %)

	Area 1	Area 2	Area 3	Area 4	Area 5	Area 6	Average
Na2O	3.7	1.9	1.9	2.3	2.8	2.9	2.5
MgO	0.5	0.4	0.3	0.6	0.5	1.1	0.5
Al2O3	30.6	32.3	31.4	29.4	29.9	29.8	30.7
SiO2	42.7	46.0	45.8	42.8	43.6	40.7	44.2
P2O5	0.9	0.2	0.3	0.6	0.6	0.6	0.5
SO3	0.2	0.1	0.9	0.6	0.2	0.6	0.4
K2O	0.9	0.4	0.4	0.7	0.6	0.8	0.6
CaO	13.0	14.4	14.7	14.7	15.1	14.3	14.4
TiO2	1.4	0.6	0.6	0.9	0.8	1.2	0.8
V2O5	0.1	0.1	0.0	0.2	0.2	0.0	0.1
Cr2O3	0.0	0.0	0.0	0.0	0.2	0.0	0.0
MnO	0.0	0.2	0.0	0.0	0.1	0.2	0.0
FeO	5.5	3.4	3.5	5.8	5.3	7.8	4.7
CoO	0.0	0.1	0.2	0.1	0.0	0.1	0.0
NiO	0.2	0.0	0.0	0.2	0.1	0.0	0.1
CuO	0.1	0.2	0.1	0.3	0.1	0.0	0.1
ZnO	0.0	0.0	0.0	0.1	0.1	0.0	0.0
SnO2	0.4	0.3	0.1	0.7	0.0	0.1	0.3
PbO	0.0	0.1	0.0	0.1	0.0	0.0	0.0

TABLE A3.6.2 - SAMPLE SFN 1109 36 A PHASE ANALYSES

	White dendrite	Grey matrix around white dendrite	Grey lath crystal	Silica
Na2O	0.5	0.9	1.0	0.1
MgO	2.1	0.4	0.1	0.0
Al2O3	15.8	33.6	31.5	0.0
SiO2	8.3	49.5	49.7	100.3
P2O5	0.5	0.9	0.1	0.0
SO3	0.1	0.1	0.1	0.2
K2O	0.1	0.2	0.3	0.0
CaO	1.9	10.4	15.3	0.0
TiO2	4.9	0.5	0.1	0.0
V2O5	0.5	0.0	0.0	0.0
Cr2O3	0.0	0.0	0.0	0.0
MnO	0.2	0.1	0.1	0.1
FeO	64.2	3.3	1.6	0.1

CoO	0.3	0.0	0.0	0.1
NiO	0.2	0.0	0.0	0.0
CuO	0.0	0.1	0.1	0.1
ZnO	0.3	0.0	0.1	0.0
SnO2	0.0	0.1	0.5	0.0
PbO	0.1	0.0	0.0	0.0

TABLE A3.6.3 BULK ANALYSES OF SFN 1411 SSL-CIN

SFN 1411 SSL-Cin						
	Area 1	Area 2	Area 3	Area 4	Area 5	Mean
Na2O	1.2	0.2	0.8	0.0	0.6	0.6
MgO	7.8	0.7	6.9	6.5	10.3	6.4
Al2O3	23.8	35.5	26.6	20.3	27.0	26.6
SiO2	49.4	54.5	40.7	35.9	46.3	45.4
P2O5	0.1	0.0	0.1	0.2	0.1	0.1
SO3	0.3	0.4	0.4	1.8	0.5	0.7
K2O	1.2	2.3	0.9	0.3	1.1	1.2
CaO	5.8	0.4	4.4	1.1	5.7	3.5
TiO2	0.9	1.7	0.6	0.7	1.2	1.0
V2O5	0.0	0.0	0.1	0.0	0.0	0.0
Cr2O3	0.1	0.3	0.0	0.0	0.0	0.1
MnO	0.8	0.0	0.8	0.5	1.1	0.6
FeO	9.0	3.8	17.9	32.4	6.4	13.9
CoO	0.0	0.1	0.1	0.0	0.0	0.0
NiO	0.0	0.0	0.1	0.0	0.0	0.0
CuO	0.0	0.0	0.0	0.3	0.0	0.1
ZnO	0.0	0.0	0.0	0.0	0.0	0.0
SnO2	0.0	0.2	0.0	0.2	0.1	0.1
PbO	0.0	0.1	0.1	0.7	0.0	0.2

TABLE A3.6.4 PHASE ANALYSES OF SFN 1411 SSL-CIN

	FeOx Area 6	FeOx Area 2	Silicate phase in corrosion
Na2O	0.3	0.2	0.2
MgO	0.0	0.0	3.0
Al2O3	0.4	0.5	0.4
SiO2	2.8	0.4	28.1
P2O5	0.4	1.8	0.2
SO3	0.6	0.0	0.5
K2O	0.0	0.0	0.0
CaO	0.0	0.2	0.4
TiO2	0.2	0.0	0.1
V2O5	0.0	0.0	0.2

Cr2O3	0.0	0.0	0.1
MnO	0.0	0.1	0.4
FeO	95.5	96.4	66.1
CoO	0.5	0.4	0.0
NiO	0.0	0.0	0.2
CuO	0.0	0.2	0.2
ZnO	0.0	0.1	0.1
SnO2	0.2	0.0	0.2
PbO	0.0	0.1	0.0

TABLE A3.6.5 SEM BULK ANALYSES OF SAMPLE SFN1255 CU-SSL (WEIGHT %)

SFN 1255 Cu-SSL						
	Area 1	Area 2	Area 3	Area 4	Area 5	Mean
Na2O	1.7	1.5	1.3	1.6	0.8	1.4
MgO	1.6	2.1	1.8	1.4	1.4	1.7
Al2O3	21.2	19.9	19.5	19.9	22.1	20.5
SiO2	44.6	47.1	49.2	49.1	48.0	47.6
P2O5	0.5	0.4	0.5	0.6	0.3	0.4
SO3	0.3	0.1	0.1	0.2	0.1	0.2
K2O	1.9	1.7	1.6	2.0	1.9	1.8
CaO	7.6	6.7	4.9	7.2	4.5	6.2
TiO2	1.0	1.2	1.1	1.4	1.2	1.1
V2O5	0.0	0.2	0.2	0.0	0.1	0.1
Cr2O3	0.0	0.0	0.0	0.0	0.0	0.0
MnO	0.3	0.2	0.1	0.3	0.2	0.2
FeO	17.1	16.8	18.4	14.5	16.5	16.6
CoO	0.1	0.0	0.1	0.1	0.0	0.1
NiO	0.1	0.0	0.0	0.1	0.0	0.0
CuO	1.4	1.1	0.6	0.9	2.2	1.2
ZnO	0.6	0.5	0.4	0.8	0.7	0.6
SnO2	0.3	0.4	0.3	0.0	0.4	0.3
PbO	0.0	0.1	0.0	0.1	0.0	0.0

TABLE A3.6.6 SEM PHASE ANALYSES OF SAMPLE SFN 1255 CU-SSL (WEIGHT %)

	Metal prill (oxide)	Dark grey silicate	Light grey silicate	Matrix	Grey lath Area 4	Matrix Area 4	Hercynite Area 5	Hercynite dendrite	Area 5 matrix
Na2O	1.9	1.4	1.5	1.4	2.4	1.3	0.9	0.6	0.6
MgO	0.0	-0.1	2.2	2.2	0.0	2.4	5.5	4.5	1.4
Al2O3	0.2	31.2	19.1	17.5	26.9	19.0	52.6	43.4	22.7
SiO2	1.4	48.6	44.4	44.1	54.4	48.8	5.9	1.9	49.6
P2O5	0.1	0.2	0.4	0.6	0.1	0.6	0.0	0.0	0.3
SO3	0.1	0.0	0.1	0.2	0.3	0.1	0.1	0.0	0.1
K2O	0.0	0.4	2.9	3.0	1.2	1.7	0.0	0.0	2.0

CaO	0.0	16.2	6.1	5.2	11.9	7.3	0.1	0.2	4.9
TiO2	0.1	0.1	0.9	1.3	0.3	1.4	0.3	0.6	1.0
V2O5	0.1	0.0	0.1	0.0	0.0	0.0	0.3	0.3	0.0
Cr2O3	0.0	0.0	0.0	0.0	0.1	0.0	0.8	0.0	0.0
MnO	0.0	0.1	0.3	0.2	0.0	0.4	0.1	0.2	0.3
FeO	2.4	1.1	20.1	22.8	2.0	15.9	26.3	40.9	15.3
CoO	0.0	0.0	0.1	0.3	0.0	0.2	0.1	0.3	0.0
NiO	0.0	0.1	0.0	0.0	0.0	0.0	0.2	0.3	0.0
CuO	93.3	0.1	0.5	0.2	0.1	0.4	0.2	0.7	1.1
ZnO	0.3	0.2	0.7	1.0	0.0	0.5	7.1	6.7	0.5
SnO2	0.2	0.4	0.6	0.2	0.2	0.5	0.0	0.0	0.1
PbO	0.1	0.1	0.1	0.0	0.1	0.0	0.0	0.0	0.2

TABLE A3.6.6 CONT METALLIC AND CHLORIDE PHASE ANALYSIS AND OXIDISED METAL ANALYSIS (WEIGHT %)

	Prill	Chloride			Cu oxide	Cu oxide 2
Si	3.2	0.3		SiO2	0.0	0.2
S	0.1	0.0		SO3	0.1	0.0
l	0.0	27.7		FeO	0.6	0.0
Fe	2.2	0.0		NiO	0.0	0.1
Ni	0.0	0.1		CuO	99.7	99.7
Cu	94.4	65.3		ZnO	0.0	0.4
Zn	0.4	0.0		As2O3	0.0	0.1
As	0.2	0.0		SnO2	0.0	0.0
Sn	0.0	0.0		Sb2O3	0.0	0.0
Sb	0.0	0.0		PbO	0.0	0.2
Pb	0.0	0.5				

TABLE A3.6.7 SFN 1324 CU-SSL BULK AREA ANALYSES (WEIGHT %)

SFN 1324 Cu-SSL						
	Area 1	Area 2	Area 3	Area 4	Area 5	Average
Na2O	0.6	0.4	0.5	0.5	0.6	0.5
MgO	0.9	0.9	1.0	0.7	0.9	0.9
Al2O3	22.1	19.7	21.5	21.9	21.7	21.4
SiO2	48.7	50.8	48.9	49.5	49.4	49.5
P2O5	0.2	0.2	0.2	0.2	0.3	0.2
SO3	0.1	0.3	0.1	0.1	0.0	0.1
K2O	2.3	2.5	2.2	2.4	2.2	2.3
CaO	3.1	3.1	3.1	3.0	2.5	3.0
TiO2	1.0	1.0	0.9	1.0	0.9	1.0
V2O5	0.0	0.1	0.0	0.0	0.0	0.0
Cr2O3	0.1	0.0	0.1	0.0	0.0	0.0
MnO	0.0	0.2	0.0	0.1	0.0	0.1
FeO	19.5	18.4	20.2	18.8	19.5	19.3

CoO	0.1	0.1	0.2	0.0	0.2	0.1
NiO	0.0	0.0	0.0	0.1	0.2	0.1
CuO	0.1	1.4	0.2	0.6	0.4	0.5
ZnO	1.1	0.7	1.0	0.8	1.0	0.9
SnO2	0.3	0.2	0.0	0.4	0.0	0.1
PbO	0.0	0.1	0.3	0.1	0.0	0.1

Table A3.6.8 SFN 1324 Cu-SSL phase analyses (weight %)

	Xtalite Area 1	Xtalite Area 5	Xtalite Area 6	Dark phase Area 1	Matrix Area 1	Matrix Area 5	Matrix Area 6
Na2O	0.6	0.5	0.4	0.5	0.7	0.7	0.4
MgO	2.4	2.2	2.3	0.6	0.6	0.5	0.6
Al2O3	54.3	52.6	55.6	15.4	14.7	15.5	15.8
SiO2	2.3	0.7	0.6	57.1	61.0	59.3	56.6
P2O5	0.0	0.0	0.1	0.3	0.4	0.4	0.3
SO3	0.0	0.0	0.0	0.0	0.2	0.1	0.1
K2O	0.2	0.1	0.0	2.5	2.9	2.7	2.4
CaO	0.2	0.1	0.1	4.5	3.7	4.6	4.3
TiO2	0.5	0.8	0.5	1.2	1.0	1.1	1.2
V2O5	0.3	0.3	0.4	0.0	0.1	0.0	0.0
Cr2O3	0.0	0.0	0.0	0.0	0.0	0.1	0.0
MnO	0.0	0.2	0.0	0.1	0.1	0.0	0.2
FeO	34.5	38.7	36.2	17.0	13.7	14.0	17.4
CoO	0.0	0.2	0.2	0.1	0.0	0.3	0.1
NiO	0.0	0.0	0.0	0.0	0.0	0.1	0.0
CuO	0.1	0.1	0.0	0.4	0.2	0.3	0.1
ZnO	4.4	3.4	3.6	0.2	0.4	0.3	0.3
SnO2	0.3	0.3	0.0	0.2	0.3	0.0	0.3
PbO	0.0	0.1	0.3	0.2	0.1	0.2	0.0

Table A3.6.9 SFN 1324 Cu-SSL analyses of metallic prills (weight %)

	Prill	Chloride
Si	3.2	0.3
S	0.1	0.0
Cl	0.0	27.7
Fe	2.2	0.0
Ni	0.0	0.1
Cu	94.4	65.3
Zn	0.4	0.0
As	0.2	0.0
Sn	0.0	0.0
Sb	0.0	0.0
Pb	0.0	0.5

	Cu oxide	Cu oxide 2
SiO2	0.0	0.2
SO3	0.1	0.0
FeO	0.6	0.0
NiO	0.0	0.1
CuO	99.7	99.7
ZnO	0.0	0.4
As2O3	0.0	0.1
SnO2	0.0	0.0
Sb2O3	0.0	0.0
PbO	0.0	0.2

TABLE TABLE A3.6.10 - SFN 1329 CU-SSL BULK AREA ANALYSES (WEIGHT %)

	Area 1	Area 2	Area 3	Area 4	Area 5	Mean	Mean of Area 1-4
Na2O	2.2	3.3	3.0	2.5	0.8	2.3	2.7
MgO	1.6	0.8	0.9	1.7	0.3	1.0	1.2
Al2O3	18.0	14.4	11.6	17.0	7.1	13.6	15.2
SiO2	35.6	35.6	32.5	35.3	76.4	43.1	34.8
P2O5	0.3	0.1	0.2	0.3	0.3	0.2	0.2
SO3	0.1	0.0	0.3	0.3	0.7	0.3	0.2
K2O	1.3	0.8	0.9	1.2	3.3	1.5	1.0
CaO	4.6	2.8	4.9	4.2	0.2	3.3	4.1
TiO2	0.7	0.7	0.5	0.5	0.5	0.6	0.6
V2O5	0.0	0.1	0.1	0.0	0.0	0.0	0.0
Cr2O3	0.1	0.0	0.0	0.0	0.0	0.0	0.0
MnO	0.0	0.1	0.1	0.1	0.0	0.1	0.1
FeO	11.9	9.4	11.2	11.4	2.6	9.3	11.0
CoO	0.2	0.0	0.0	0.0	0.0	0.0	0.0
NiO	0.2	0.0	0.2	0.1	0.1	0.1	0.1
CuO	3.6	5.4	6.5	3.9	3.6	4.6	4.9
ZnO	18.0	24.2	24.3	19.7	4.2	18.1	21.6
SnO2	0.8	0.6	0.8	0.7	0.0	0.6	0.7
PbO	0.9	1.9	2.0	1.1	0.0	1.2	1.5

TABLE A3.6.11 - SFN 1329 CU-SSL PHASE ANALYSES (WEIGHT %)

	Light grey phase (Area 3)	Dark grey phase (Area 3)	Matrix Area 3	Sand inclusion
Na2O	4.9	2.7	1.8	0.1
MgO	0.9	2.0	0.7	0.0
Al2O3	1.3	43.7	8.4	0.0
SiO2	26.6	0.0	42.0	99.9
P2O5	0.1	0.2	0.2	0.0
SO3	0.1	0.0	0.0	0.2
K2O	0.1	0.0	1.3	0.0
CaO	0.4	0.0	8.5	0.0
TiO2	0.2	0.1	0.8	0.0
V2O5	0.0	0.0	0.0	0.0
Cr2O3	0.1	0.3	0.0	0.0
MnO	0.1	0.1	0.1	0.0
FeO	4.1	14.8	13.5	0.1
CoO	0.2	0.0	0.1	0.1
NiO	0.1	1.1	0.1	0.0
CuO	0.6	8.8	0.2	0.1
ZnO	59.9	26.3	18.9	0.2
SnO2	0.2	0.0	1.0	0.1
PbO	0.1	0.0	2.5	0.0

TABLE A3.6.12 - SFN 1329 CU-SSL METAL PRILL AND CORROSION PHASE ANALYSES (WEIGHT %)

	Metal Prill			Corrosion 1	Corrosion 2	Lead Prill
Si	3.0		SiO2	0.0	0.2	0.2
S	1.1		SO3	32.9	0.7	8.8
Cl	0.0			0.0	0.0	0.0
Fe	0.1		FeO	0.0	0.1	0.0
Ni	0.2		NiO	0.1	0.0	0.1
Cu	93.9		CuO	66.6	98.0	6.0
Zn	0.0		ZnO	0.4	0.3	0.0
As	0.1		As2O3	0.0	0.1	0.0
Sn	1.2		SnO2	0.0	0.3	0.0
Sb	0.3		Sb2O3	0.0	0.1	0.0
Pb	0.3		PbO	0.2	0.0	85.2

TABLE A3.6.13 - SEM BULK ANALYSES OF SAMPLE SFN 1369 SLAGGED LINING (WEIGHT %)

SFN 1369 Slagged lining								
	Area 1	Area 2	Area 3	Area 4	Area 5	Area 6	Average of first five	Overall average
Na2O	3.7	1.9	1.9	2.3	2.8	2.9	2.5	2.6
MgO	0.5	0.4	0.3	0.6	0.5	1.1	0.5	0.6
Al2O3	30.6	32.3	31.4	29.4	29.9	29.8	30.7	30.6
SiO2	42.7	46.0	45.8	42.8	43.6	40.7	44.2	43.6
P2O5	0.9	0.2	0.3	0.6	0.6	0.6	0.5	0.5
SO3	0.2	0.1	0.9	0.6	0.2	0.6	0.4	0.4
K2O	0.9	0.4	0.4	0.7	0.6	0.8	0.6	0.6
CaO	13.0	14.4	14.7	14.7	15.1	14.3	14.4	14.4
TiO2	1.4	0.6	0.6	0.9	0.8	1.2	0.8	0.9
V2O5	0.1	0.1	0.0	0.2	0.2	0.0	0.1	0.1
Cr2O3	0.0	0.0	0.0	0.0	0.2	0.0	0.0	0.0
MnO	0.0	0.2	0.0	0.0	0.1	0.2	0.0	0.1
FeO	5.5	3.4	3.5	5.8	5.3	7.8	4.7	5.2
CoO	0.0	0.1	0.2	0.1	0.0	0.1	0.1	0.1
NiO	0.2	0.0	0.0	0.2	0.1	0.0	0.1	0.1
CuO	0.1	0.2	0.1	0.3	0.1	0.0	0.1	0.1
ZnO	0.0	0.0	0.0	0.1	0.1	0.0	0.0	0.0
SnO2	0.4	0.3	0.1	0.7	0.0	0.1	0.3	0.3
PbO	0.0	0.1	0.0	0.1	0.0	0.0	0.0	0.0

TABLE A3.6.14 - SEM PHASE ANALYSES OF SAMPLE SFN 1369 SLAGGED LINING (WEIGHT %)

	Xtalite 1	Xtalite 2	Matrix 1	Matrix 2	Matrix dark	Matrix light	Dark phase in Area 4	Dark phase in Area 2
Na2O	0.2	0.0	9.9	0.7	0.8	7.8	11.0	1.2
MgO	11.2	12.1	0.8	0.0	0.0	1.5	1.4	0.0

Al2O3	40.6	45.1	25.6	33.8	34.4	23.1	22.6	30.6
SiO2	1.7	0.1	42.5	46.8	46.1	40.0	41.2	46.8
P2O5	0.0	0.0	1.8	0.0	0.0	1.3	2.2	0.1
SO3	0.0	0.2	0.1	0.0	0.3	0.2	0.3	2.7
K2O	0.0	0.0	2.5	0.2	0.1	2.2	2.4	0.3
CaO	0.3	0.1	7.9	16.9	17.2	14.2	7.1	16.0
TiO2	0.6	0.2	2.7	0.1	0.2	2.7	2.8	0.2
V2O5	0.1	0.0	0.3	0.0	0.0	0.2	0.2	0.2
Cr2O3	0.2	1.1	0.0	0.0	0.1	0.1	-0.1	0.0
MnO	0.6	0.5	0.2	0.0	0.0	0.3	0.3	0.0
FeO	42.3	36.9	5.3	0.7	1.1	6.2	8.5	1.4
CoO	0.5	0.3	0.0	0.0	0.0	0.0	0.1	0.1
NiO	0.9	3.0	0.0	0.1	0.0	0.0	0.0	0.0
CuO	0.4	0.0	0.0	0.0	0.2	0.2	0.2	0.1
ZnO	0.2	0.3	0.0	0.1	0.0	0.0	0.0	0.3
SnO2	0.2	0.1	0.4	0.7	0.2	0.3	0.4	0.2
PbO	0.0	0.1	0.1	0.2	0.0	0.0	0.0	0.0

TABLE A3.6.15 - MEAN VALUES OF BULK AREA ANALYSES OF ALL SAMPLES

	Iron working slags		Copper working slags			Slagged lining	
	SFN 1109 36a	SFN 1411	SFN 1255 Cu-SSL	SFN 1324	SFN 1329	SFN 1369	Mean of the slags
Na2O	2.5	0.6	1.4	0.5	2.7	2.6	1.5
MgO	0.5	6.4	1.7	0.9	1.2	0.6	2.1
Al2O3	30.7	26.6	20.5	21.4	15.2	30.6	22.9
SiO2	44.2	45.4	47.6	49.5	34.8	43.6	44.3
P2O5	0.5	0.1	0.4	0.2	0.2	0.5	0.3
SO3	0.4	0.7	0.2	0.1	0.2	0.4	0.3
K2O	0.6	1.2	1.8	2.3	1.0	0.6	1.4
CaO	14.4	3.5	6.2	3.0	4.1	14.4	6.2
TiO2	0.8	1.0	1.1	1.0	0.6	0.9	0.9
V2O5	0.1	0.0	0.1	0.0	0.0	0.1	0.1
Cr2O3	0.0	0.1	0.0	0.0	0.0	0.0	0.0
MnO	0.0	0.6	0.2	0.1	0.1	0.1	0.2
FeO	4.7	13.9	16.6	19.3	11.0	5.2	13.1
CoO	0.0	0.0	0.1	0.1	0.0	0.1	0.1
NiO	0.1	0.0	0.0	0.1	0.1	0.1	0.1
CuO	0.1	0.0	1.2	0.5	4.9	0.1	1.4
ZnO	0.0	0.0	0.6	0.9	21.6	0.0	4.6
SnO2	0.3	0.1	0.3	0.1	0.7	0.3	0.3
PbO	0.0	0.2	0.0	0.1	1.5	0.0	0.3

TABLE A3.6.16 - PHASE ANALYSES OF HIGH Z DENDRITES

	Iron working slags		Copper working slags			Slagged lining	
	SFN 1109 36a	SFN 1411	SFN 1255 Cu-SSL	SFN 1324	SFN 1329	SFN 1369	Classic hercynite
Na2O	0.5		0.6	0.4			
MgO	2.1		4.5	2.3			
Al2O3	15.8		43.4	55.6			59.0
SiO2	8.3		1.9	0.6			
P2O5	0.5		0.0	0.1			
SO3	0.1		0.0	0.0			
K2O	0.1		0.0	0.0			
CaO	1.9		0.2	0.1			
TiO2	4.9		0.6	0.5			
V2O5	0.5		0.3	0.4			
Cr2O3	0.0		0.0	0.0			
MnO	0.2		0.2	0.0			
FeO	64.2		40.9	36.2			41.0
CoO	0.3		0.3	0.2			
NiO	0.2		0.3	0.0			
CuO	0.0		0.7	0.0			
ZnO	0.3		6.7	3.6			
SnO2	0.0		0.0	0.0			
PbO	0.1		0.0	0.3			

TABLE A3.6.17 - PHASE ANALYSES OF LATH PHASES

	Iron working slags		Copper working slags			Slagged lining
	SFN 1109 36a	SFN 1411	SFN 1255 Cu-SSL	SFN 1324	SFN 1329	SFN 1369
Na2O	1.0		2.4	0.5	4.9	1.2
MgO	0.1		0.0	0.6	0.9	0.0
Al2O3	31.5		26.9	15.4	1.3	30.6
SiO2	49.7		54.4	57.1	26.6	46.8
P2O5	0.1		0.1	0.3	0.1	0.1
SO3	0.1		0.3	0.0	0.1	2.7
K2O	0.3		1.2	2.5	0.1	0.3
CaO	15.3		11.9	4.5	0.4	16.0
TiO2	0.1		0.3	1.2	0.2	0.2
V2O5	0.0		0.0	0.0	0.0	0.2
Cr2O3	0.0		0.1	0.0	0.1	0.0

MnO	0.1		0.0	0.1	0.1	0.0
FeO	1.6		2.0	17.0	4.1	1.4
CoO	0.0		0.0	0.1	0.2	0.1
NiO	0.0		0.0	0.0	0.1	0.0
CuO	0.1		0.1	0.4	0.6	0.1
ZnO	0.1		0.0	0.2	59.9	0.3
SnO2	0.5		0.2	0.2	0.2	0.2
PbO	0.0		0.1	0.2	0.1	0.0

TABLE A3.6.18 - MATRIX COMPOSITIONS OF THE SLAGS

	Iron working slags		Copper working slags			Slagged lining
	SFN 1109 36a	SFN 1411	SFN 1255 Cu-SSL	SFN 1324	SFN 1329	SFN 1369
Na2O	0.9		1.3	0.4	1.8	0.8
MgO	0.4		2.4	0.6	0.7	0.0
Al2O3	33.6		19.0	15.8	8.4	34.4
SiO2	49.5		48.8	56.6	42.0	46.1
P2O5	0.9		0.6	0.3	0.2	0.0
SO3	0.1		0.1	0.1	0.0	0.3
K2O	0.2		1.7	2.4	1.3	0.1
CaO	10.4		7.3	4.3	8.5	17.2
TiO2	0.5		1.4	1.2	0.8	0.2
V2O5	0.0		0.0	0.0	0.0	0.0
Cr2O3	0.0		0.0	0.0	0.0	0.1
MnO	0.1		0.4	0.2	0.1	0.0
FeO	3.3		15.9	17.4	13.5	1.1
CoO	0.0		0.2	0.1	0.1	0.0
NiO	0.0		0.0	0.0	0.1	0.0
CuO	0.1		0.4	0.1	0.2	0.2
ZnO	0.0		0.5	0.3	18.9	0.0
SnO2	0.1		0.5	0.3	1.0	0.2
PbO	0.0		0.0	0.0	2.5	0.0

ABBREVIATIONS

ABG	Aris's Birmingham Gazette
BA	Birmingham Archaeology
BBP	Birmingham Building Plan, Birmingham Archives and Heritage
BC	Birmingham Corporation
BCA	Birmingham City Archives, Birmingham Archives and Heritage
BDG	Birmingham Daily Gazette
BDP	Birmingham Daily Post
BG	Birmingham Gazette
BJ	Birmingham Journal
BUFAU	Birmingham University Field archaeology Unit
BWC	Boulton and Watt Collection, Birmingham Archives and Heritage
IMechE	The Institution of Mechanical Engineers
LG	London Gazette
MLC	Michael Lane Collection
NA	National Archives, Kew, London
WMAFS	Warwickshire Museum Archaeological Field Services
WRO	Warwickshire Records Office

BIBLIOGRAPHY

SECONDARY SOURCES

Aitken, W C, 1866 Brass and Brass Manufacturers, in Timmins 1866, 225–381

Allen, G C, 1927 *The Industrial Development of Birmingham and the Black Country 1860–1927*, London: G Allen and Unwin Ltd.

Amin, A, 2000 Industrial Districts, in E Sheppard and T J Barnes (eds) *A Companion To Economic Geography*, Oxford: Blackwell, 149–68

American Institute of Chemical Engineers, 1977 *Twenty-Five Years of Chemical Engineering Progress*, New Hampshire: Ayer Publishing

Anon, 1882 Instillation of the Rolling Mill by Boulton & Watt in 1852 at the Cambridge Street Works, *The Electrician* **1**, 9th September 1882

Bayley, J (ed) 2008 Science in Archaeology an Agenda for the Future, London: English Heritage

Becker, B H, 1833 *Brass Work at Birmingham*, London

Behagg, C, 1998 Mass Production without the Factory: Craft Producers, Guns and Small Firm Innovation, 1790–1815, *Business History* **40/3**, 1–15

Belford, P, Demidowicz, G, Hislop, M, 2003 *An Historic Environment Assessment (with Ecological Statement) of the Soho Foundry, Foundry Lane, Smethwick, West Midlands*, unpubl rep

Belford, P, 2011 'The archaeology of everything' - grappling with post-medieval, industrial and contemporary archaeology, in S Watt *The Archaeology of the West Midlands: a framework for research*, Oxford: Oxbow Books, 211–229

Benton, J H, 1914 *John Baskerville: type founder and printer, 1706–1775*, Birmingham: DB Updike

Berg, M, 1991 Commerce and Creativity in 18th century Birmingham, in M Berg, *Markets and Manufacture in Early Industrial Europe*, London: Routledge, 173–201

Berg, P, and Berg, T (trans) 2001 *R R Angerstein's Illustrated Travel Diary, 1753–1755: industry in England and Wales from a Swedish perspective*, London: British Museum

Bevan, L, Mould, Q and Rátkai, S, 2009 Worked bone, ivory and horn, in C Patrick and S Rátkai, *The Bull Ring Uncovered: excavations at Edgbaston Street, Moor Street, Park Street and The Row, Birmingham, 1997–2001*, Oxford: Oxbow Books, 179–80

Billing, M (ed), 1858 *The New Illustrated Directory Entitled Men and Things of Modern England*, Birmingham: M Billing

Birmingham Archaeology 2009 *Written Scheme of Investigation for archaeological excavations at the site of the Library of Birmingham, Cambridge Street, Birmingham*, unpub BA rep

Booth, D T N, 1978 *Warwickshire Watermills*, Smethwick: Midland Wind and Watermills Group

Buchanan, RA, and Watkins, GW, 1976 *The Industrial Archaeology of the Stationary Steam Engine*, London: Penguin

Cadbury, E M, Matheson, C, and Shann, G, 1907 *Women's Work and Wages: a phase of life in an industrial city*, Chicago: University Press

Campanella, L, Colacicchi Alessandri, O, Ferretti, M, and Plattner, S H, 2009 The effect of tin on dezincification of archaeological copper alloys, *Corrosion Science* **51**, 2183–2191

Cattell, J, Sheila, E, and Jones, B, 2002 *The Birmingham Jewellery Quarter*, Swindon: English Heritage

Chadwick, E, 1842 Report from the Poor Law Commissioners on an Inquiry into the Sanitary Conditions of the Labouring Population of Great Britain, London: Poor Law Commission

Chamberlain, J, 1866 Manufacture of Iron-wood screws, in Timmins 1866, 604–609

Cook, M 1936 The development of the non-ferrous metal industries in Birmingham, *Metal Industry* **18**, 485–490

Cook, M, 2008 *Watching brief at land at Blakesley Hall School, Yardley Green Road, Stetchford*, unpub rep

Cooksey, M, 2003 *Brickyards of the Black Country, Forgotten Industries, Refractories*, Cradley Heath: M Cooksey

Conway, M, and Litherland, S, 2003 *Survey of Nos. 28–37, Sheepcote Street, Birmingham*, BA rep 792

Court, W H B, 1938 *The Rise of the Midland Industries 1600–1838*, Oxford: University Press

Crowe, N, 1994 *English Heritage Book of Canals*, Northampton, Belmont Press

Davis, W J, 1892 *A Short History of the Brass Trade*, London

Day, J, 1995 Trade and Innovation in Non-Ferrous Metals at Bristol, in D R Hook and D R M Gaimster *Trade and Discovery: the scientific study of artefacts from post-medieval Europe and beyond*, British Museum Occasional Paper 109, London: British Museum Press, 205–220

Day, J, 1973 *Bristol Brass: a history of the industry*, Newton Abbott: David and Charles

Day, J, and Tylecote R F, (eds) 1991 *The Industrial Revolution in Metals*, London: Institute of Metals

Demidowicz, G, 1991 *The Watermills of Birmingham*, Draft Copy of unpub doc

Demidowicz, G, 2008 *Medieval Birmingham: the borough rentals of 1296 and 1344–5*, Dugdale Society Occasional Papers 48

Dent, R K, 1894 *The Making of Birmingham*, Birmingham: J L Allday

Dilworth, D, 1976 *The Tame Mills of Staffordshire*, London: Phillimore

DoE (Department of the Environment) 1990 *Policy and Planning Guidance Note (PPG) 16*, London

Dring, G, 1940 'Bakelite', *Royal Institute of Chemistry Jnl* **36**, 201–202

Driver, L, 2005 *Eastside, Birmingham: an archaeological desk-based assessment*, BA rep 1272

Driver, L, and Hislop, M, 2004 *Edgbaston Reservoir Edgbaston Reservoir and Icknield Port Loop, Birmingham: an archaeological desk-based assessment*, BA rep 1241

Driver, L, Hislop, M, and Litherland, S, 2004 The north service range, Aston Hall, Birmingham: excavation and recording, *Post-Medieval Archaeology* **42:1**, 104–129

Drozda, T, and Wick, C, (eds) 1983, *Tool and Manufacturing Engineers Handbook. Forming 2*, Michigan: Society of Manufacturing Engineers

Duncan, M, Hewitson, C, and Krawiec, K, forthcoming Excavations at Rea Street, Birmingham 2008, in C Hewitson and S Rátkai

Duncan, M, and Jones, A, 2010 *Excavations at Site 1, Couchman Road, Saltley, Birmingham 2010,* BA rep 2038

English Heritage 1991 *The Management of Archaeological Projects*, London: English Heritage

English Heritage 2006 *Management of Research Projects in the Historic Environment: the MoRPHE project manager's guidelines*, London: English Heritage

Evans, C and Rydén, G, 2007 *Baltic Iron and the Atlantic World in the Eighteenth Century*, Boston: Brill

Fleming, J, Honour, H, and Pevsner, N, 1998 *Dictionary of Architecture and Landscape Architecture*, 4th edn, London: Penguin

Floud, R, and McCloskey, D (eds), 1994a *The Economic History of Britain Since 1700, Vol 1: 1700–1860*, 2nd ed, Cambridge: Cambridge University Press

Floud, R, and McCloskey, D (eds), 1994b *The Economic History of Britain Since 1700, Vol 2: 1860–1940*, 2nd ed, Cambridge: Cambridge University Press

Foster, A, 2005 *Birmingham: Pevsner Architectural Guides*, London: Yale University Press.

Forster, A, and Rátkai, S, 2008 *'...that great dirty leviathan': archaeology and development in Birmingham City Centre, AD 1100–1900*, BA rep 1610

Gale, W K J V, 1946 *Soho Foundry*, Birmingham: W and T Avery Limited

Gale, W K J V, 1968 *Boulton and Watt and the Soho undertakings,* Birmingham: Museum of Science and Industry

Gale, W K J V, 1976 Ignorance of metallurgy led Bessemer to his process, in Mortimer 1976, 160–167

Grainger, R D, 1840 *Parliamentary Papers, Evidence for Employment of Children collected, 30th November 1840*

Greenslade, MW 1976 Smethwick: Economic history, *A History of the County of Stafford*, Vol 17: Offlow hundred (part) 107–118. URL: http://www.british-history.ac.uk/report.aspx?compid=36177 Date accessed: 29 October 2011.

Grenville, J 2004 The archaeology of the late and post-medieval workshop - a review and proposal for a research agenda, in P S Barnwell, M Palmer and M Airs (eds), *The Vernacular Workshop: from craft to industry, 1400–1900*, CBA Res Rep 140, York: Council for British Archaeology

Gwyn, D, and Palmer, M (eds), 2005 Understanding the workplace: a research framework for industrial archaeology in Britain, *Industrial Archaeology Review* **27**

Hadfield, C, 1985 *The Canal Age*, 3rd edn, New York: Praeger

Hall, P G, 1960 The location of the clothing trades in London, 1861–1951, *Transactions of the Institute of British Geographers* **28**, 155–178

Hamilton, H, 1967 *The English Brass and Copper Industries to 1800*, 2nd edn, London: Frank Cass and Co Ltd

Hannoford, H R, 1989, *Lifford Mill, Kings Norton: an archaeological evaluation*, BUFAU rep 73

Harley, K, 1994 Foreign trade: comparative advantage and performance, in R Floud and D McCloskey (eds) 1994a, 300–331

Harris, J R, 1964 *The Copper King: a biography of Thomas Williams of Llanidan*, Liverpool: Liverpool University Press

Hayes, G, 1983 *Stationary Steam Engines*, 2nd edn, Princes Risborough: Shire Publications

Herbert, L, 1849 *The Engineer's and Mechanic's Encyclopedia*, London: Kelly

Hewitson, C, 2002 *An Archaeological Watching Brief of Soft Stripping at 24 and 25 Sheepcote Street, Birmingham, 2000* BUFAU rep 615.02

Hewitson, C, 2005 *Investigation and Building Recording at the Former Garage Site, Wolverhampton Road, Walsall Waterfront*, BA rep 1268

Hewitson, C, 2011 *Before the Lunar Society: industrial development in early post-medieval Birmingham*, Proceedings of the Fe09 Conference 2009, BAR Brit Ser 523, Oxford: BAR Publishing.

Hewitson, C and Rátkai, S (eds), forthcoming Birmingham on the Edge of Industry: the archaeology of industrialisation in Digbeth, Deritend and Bordesley, BAR Brit Ser

Hiley, EN 1957 *Brass Saga: the story of the brassfoundry industry*, London: Earnest Benn

Hills, RL 1989 *Power from Steam: a history of the stationary steam engine*, Cambridge: Cambridge University Press

Hislop, M, Demidowicz, G, and Price, S, 2011 *'Northeton...a praty uplandyshe towne...' Building Recording, Excavation and Documentary Research in King's Norton, Birmingham, 2005–2007*, BAR Brit Ser 529, Oxford: BAR Publishing

Hodder, M, 2004 *Birmingham: the hidden history*, Stroud: Tempus

Hopkins, E, 2002 Industrial Change and Life at Work in Birmingham 1850–1914, *Midland History* **27**, 112–129

Hughes, S R, 2000 *Copperopolis: landscapes of the early industrial period in Swansea*, Cardiff: CADW

Hughes, S R, 2004 Social Archaeology: A possible methodology of the study of workers' settlements based on the 18th- and 19th-century copper industry in Swansea, in D Barker and D Cranstone (eds) *The Archaeology of Industrialization,* London: Maney

IMechE, 1897 Excursions and notices of works visited, Institution of Mechanical Engineers' summer meeting in Cardiff and Newport, Messrs Winfield's Cambridge Street works and rolling mill, *Proceedings of the Institute of Mechanical Engineers* **53**, 403–404

Jones, E, 1987 *A History of GKN: innovation and enterprise, 1759–1918*, Basingstoke: Macmillan

Jones, P M, 2008 *Industrial Enlightenment: science, technology, and culture in Birmingham and the West Midlands*, 1760–1820 Manchester: Manchester University Press

Joyce, P, 1980 *Work, Society and Politics: the culture of the factory in late Victorian England*, New Brunswick: Rutgers University Press.

Keate, R J, 1910 A description of the oldest English-speaking brass town, *Metal Industry* **8**

King, P W, 2003 *The Iron Trade in England and Wales 1500–1815: the charcoal iron industry and its transition to coke*, unpubl PhD thesis, Wolverhampton University

Krawiec, K, 2009 *Geotechnical test pits, Cambridge Street, Birmingham, archaeological watching brief, 2009*, BA rep 1905.

Lievers, M, 2010 Archaeological investigations at Digbeth Coach Station, Birmingham, *Transactions of the Birmingham and Warwickshire Archaeological Society* **114**, 111-114.

Linnane, S J, 1998 *Early Gasworks, Gas Street, Birmingham: architectural recording and analysis,* BUFAU rep 550

Litherland, S, 1995 *An Archaeological Assessment of the Digbeth Economic Regeneration Area and Cheapside Industrial Area, Birmingham*, BUFAU rep 337

Litherland, S, and Winter, C, 1999 *Archaeological Evaluation at 24/25 Sheepcote Street, Birmingham: building recording and desk-top assessment,* BUFAU rep 615.01

Lloyd, N, 1925 *A History of English Brickwork* (reprinted 1983), London: A. Montgomery

Lobb, M, 2006 *Proposed New Library, Cambridge Street, Birmingham: an archaeological desk-based assessment*, BA rep 1517

McMillan, A, 1917 An advantage for the gas-annealing furnace, *Industrial Engineering Chemistry* **9**, 984

Martineau, RF 1866 Cut Nails, in Timmins, 613–617

McCloskey, D, 1994 1780–1860: a survey, in R Floud and D McCloskey (eds) 1994a, 242–270

McDonnell, G, 2009 *Assessment Report of the Potential of Archaeo-metallurgical Residues from Cambridge Street, Birmingham*, unpubl rep

McKenna, J, 2005a *Birmingham: The Building of a City*, Stroud: Tempus

McKenna, J 2005b *Birmingham Breweries*, Studley: Brewin Books

McLain, L, 1976 Labouring man was an engine in that output depended on input, in Mortimer 1976, 222–227

Morgan, C E, 2001 *Women Workers and Gender Identities, 1835–1913: The Cotton and Metal Industries in England*, London: Routledge

Mortimer, J (ed), 1976 *The Engineer, Highlights of 120 years*, London: The GKN Group.

Morton, J, 1983 *Thomas Bolton and Sons Ltd, 1783–1983: the bi-centenary history of a major copper and brass manufacturer*, Ashbourne: Moorland Publishing

Mould, C, 1999 *An Archaeological Desk-Based Assessment of part of the Digbeth Millennium Quarter, Birmingham City Centre*, BUFAU rep 575

Nash, T R, 1799 *Collections for a History of Worcestershire, 1781–99*

Nicholas, P, 2009 Metal-working Debris from Park Street, in C Patrick and S Rátkai, 227–235

Page, M 1976 Aluminium ousting copper: a journalistic exaggeration, in Mortimer 1976, 170–175

Palmer, M, 2005 Understanding the Workplace: a research framework for industrial archaeology in Britain, *Industrial Archaeological Review* **27**, 9–17

Patrick, C, and Rátkai, S, 2009 *The Bull Ring Uncovered: excavations at Edgbaston Street, Moor Street, Park Street and The Row, Birmingham, 1997–2001*, Oxford: Oxbow Books

Pelham, R A, 1953 Establishment of the Willoughby Ironworks in North Warwickshire in the 16th century, *University of Birmingham Historical Journal* **4,** 18-29.

Peyton, E, 1866 Manufacture of iron and brass bedsteads, in Timmins 1866, 624–629

Phipson, T, 1866, The pin trade, in Timmins 1866, 601–603

Plot, R, 1686 *A Natural History of Staffordshire*, Oxford: Theater

Pollard, S, 1994 Entrepreneurship 1870–1914, in R Floud and D McCloskey (eds) 1994b, 62-89.

Ramsey, E, 2008 *Cambridge Street, Birmingham, Archaeological Evaluation, 2008*, BA rep 1788

Rimmer, W G, 1960 Castle Foregate Flax Mill, Shrewsbury, *Trans Shropshire Archaeol Soc* **56,** 49

Rose, R B, 1960 The Priestley Riots of 1791, *Past and Present* **18/1**, 68-88

Rowell, H W, 1939 Plastics as applied to engineering *Proceedings of the Birmingham Association of Mechanical Engineers* **35**

Rowlands, M B, 1975 *Masters and Men in the West Midland Metalware Trades before the IndustrialR revolution,* Manchester: Manchester University Press

Shill, R, 2002a *The Birmingham Canal Navigations: at the heart of the British canal system*, Stroud: Tempus

Shill, R, 2002b *Birmingham's Industrial Heritage 1900–2000*, Stroud: Sutton

Shill, R, 2006a BCN branches and byeways, *Boundary Post: The Birmingham Canal Network Society Journal*

Shill, R, 2006b *Workshop of the World: Birmingham's industrial legacy*, Stroud: Sutton

Skipp, V, 1970 *History of Yardley London*, Chichester: Phillimore

Skipp, V, 1980 *A History of Greater Birmingham down to 1830*, Birmingham: Skipp.

Smith, A, 1776 *An Inquiry into the Nature and Causes of the Wealth of Nations*, London: Methuen

Smith, D, 1982 *Conflict and Compromise: class formation in English society 1830–1914: a comparative study of Birmingham and Sheffield*, London: Routledge and Keegan Paul

Society for the Diffusion of Useful Knowledge 1833 *The Penny Cyclopedia*, Vol 4, London: Charles and Knight

Sommerville, L, forthcoming, The shell, in A Forster, C Hewitson and S Rátkai (eds) Birmingham on the Edge of Industry: the archaeology of industrialisation in Digbeth, Deritend and Bordesley, BAR Brit Ser

Staples, W G, and Staples, C L, 2001 *Power, Profits and Patriarchy: the social organisation of work at a British metal trades firm*, 1791–1922, Oxford: Rowman and Littlefields

Stephens, W B, 1964 Economic and social history: industry and trade, 1500–1880, and economic and social history: industry and trade, 1880–1960, in WB Stephens (ed) *The Victoria History of the County of Warwick. Vol 7 The City of Birmingham*, Oxford: Oxford University Press, 81–139, and 140–208

Strauss, G L M, Quin, C W, Brough, J C, Archer, T, Tegetmeier, W B, and Prowse, W J, 1864 *England's Workshops*, London: Groombridge and Sons

Stuart, I, 1995 The Hoffman Brick and Tile Company, Melbourne, Australia, *Industrial Archaeological Review* **17**, 129-144

Taylor, M J, and Wood, P A, 1973 Industrial linkage and local agglomerations in the West Midlands metal industries, *Transactions of the Institute of British Geographers* **59**, 127–154

Thring, O, 2010 Consider the brown source, *The Guardian*, 4th May 2010

Timmins, S, (ed) 1866 *Birmingham and Midland Hardware District*, London: Robert Hardwicke

Toomey, R R, 1985 *Vivian and Sons: a study of the firm in the copper and related industries*, London: Garland

Trinder, B, 1992 Ditherington flax mill, Shrewsbury: A Re-evaluation, *Textile History* **23–22**, 189–223

Turner, J P, 1866 The Birmingham Button Trade, in Timmins 1866, 432–451

Van Laun, J, and Day, J, 1995 *The Warmley Brass Works of William Champion*, Hereford: City of Hereford Archaeology Unit

Vance, J E, 1967 Housing the Worker: determinative and contingent ties in 19th century Birmingham, *Economic Geography* **43:2**, 95–127

Vince, C A, and Bunce, J T, 1902 *History of the Corporation of Birmingham*, Birmingham: Kessinger Publishing

WMAFS, 2002 *Archaeological evaluation at Lifford Hall*, WMAFS unpubl rep

Watts, H D, Wood, A M, and Wardle, P, 2003 Making friends or making things?': Interfirm transactions in the Sheffield metal-working cluster *Urban Studies* **40:3**, 615–630

White, D P, 1977 The Birmingham button industry, *Post-medieval Archaeology* **11**, 67–79

Wise, M J, 1949 On the evolution of the jewellery and gun quarters in Birmingham, *Transactions of the Institute of British Geographers*, **15**, 57–72

Wise, M J, and Johnson, B L C 1950, The changing regional pattern during the 18th century, in Kinvig *et al*, 161–186

Wise, M J, and Thorpe, P O, 1950 The growth of Birmingham 1800–1950, in R H Kinvig, J G Smith and M J Wise, *Birmingham and its Regional Setting*, Birmingham: British Association for the Advancement of Science, 161–228

PRIMARY SOURCE DOCUMENTS

BBP 21502, 30th May 1910 Single-storey warehouses were built on the former Baskerville House Mill site

BBP 22091, Application dated 6th April 1911

BBP 22163, Application dated 10th May 1911

BBP 23159, Application dated 30th May 1912

BBP 26436, Application for works and offices for Player and Mitchell on the corner of Doris Road and Garrison Lane, dated February 23rd 1915

BBP 26514A, Application for the Richardson and Boiler House premises formed part of the new extension for the No 3 die casting shop, dated 23rd March 1915

BBP 28058 The old showrooms fronting Cambridge Street became a 'Dispatch and Finished' shop, dated 4th December 1916

BBP 28315 Application for the Richardson and boiler house premises formed part of the new extension for the No 3 die casting shop, dated 12th March 1917

BBP 29213, Mathews and Timings premises, Tube Works, application dated 9th September 1918

BBP 6354, Architect John P Osborne Winfield's time the buildings in this area rose to four storeys and one had the upper storey rebuilt after the fire of 1888, dated 30th October 1888

BC, 1920 Birmingham Corporation General Purposes Committee Minutes 10690, 16th February 1920

BC, 1922a Birmingham Corporation Staff and Accommodation Sub Committee, 1922

BC, 1922b Birmingham Corporation General Purposes Committee, 10th April 1922

BC, 1936 Birmingham Corporation General Purposes Committee Minutes 15156, minutes referred to in civic centre sub-committee, January 1936

BCA MS 86, Leonard Leigh user of Gibson's Lock brought loads of coal up to the mills and took away rubbish to a tip. Traffic forwarded by canal, dated September 1934

BCA MS 119 Inventory of Peyton and Peyton's Factory, dated 1884

BCA MS 1342/376 R W Winfield, General and ornamental brassfounders; Cambridge Street works, Birmingham, engraved letterhead: coat of arms, medals, dated 1874

BCA MS 322/1 Lease of a piece of land in Baskerville Place. Thomas Gibson to Daniel Ledsam, Joseph Ledsam, William Potts, Matthew Dixon and R W Winfield, dated 1st October 1824

BCA MS 322/8 Assignment of leasehold mill in Baskerville Place Daniel Ledsam and others to Robert Walter Winfield, 27th September 1853

BCA MS 322/9 Mortgage of property in Cambridge Street. Robert Winfield to Thornley, Rotton and Bartleet, 28, September 1853, Transfer of mortgage, Henry Thornley to Thornley, Bartleet and Rotton, 17th January 1861. Reassignment endorsed, 21 March 1868

BCA MS 322/10 Premises in Cambridge Street, George Crowther to Robert Walter Winfield, dated 31st October 1857

BCA MS 322/12 Copy recital in abstracted deed 10th March 1879 re land in Cambridge St, n.d.

BCA MS 322/15 to MS322/23 Various regarding formation of Winfield's Rolling Ltd. First chairman was Francis Mitchell (Steel Pen Maker), assisted by directors George and William Dugard (partners in the firm of Dugard Brothers). Includes conditional agreement for sale of Lot 1 (Rolling, Wire and Tube Mills) exclusive of lacquer and varnish trade, dated 15 January 1898, also draft underlease of rolling wire and tube mills in Cambridge Street, and sale particulars of the mills, for sale by auction, dated 27th January 1898

BCA MS 322/29 Plan of Plot 1, Rolling Tube and Wire Mills, Cambridge Street, dated November 1897

BCA MS 322/30 Plan and schedule of rolling, wire and tube mills and lacquer and staining varnish trade. Winfield's Ltd, dated 27th January 1898

BCA MS 322/31 Plan of Cambridge Street Works, undated

BCA MS 322/32 Report on rolling mill, Winfield and Co, by James Watt and Co, dated February 1897

BCA MS 322/34 Specification of electrical equipment for new rolling mills in course of erection for Winfields Rolling Mills Ltd. Cambridge Street, undated

BCA MS 322/35, 36 Prospectus of Winfield Rolling Mills Ltd. Issue of shares, dated February 1898

BCA MS 322/ 45B Winfield's Rolling Ltd Cost analysis book, dated 1899–1915

BCA MS 322/ 46-155 Miscellaneous file of letters concerning Winfields Ltd, Dugard Bros. Ltd, and others, dated 1898–1950

BCA MS 322/159-166 Notes concerning Robert Walter Winfield, n.d.

BCA MS 322/173 Paper read before Birmingham Philosophical Society, entitled 'The cause of imperfections in the colour of sheet brass', by Professor Thomas Turner 1891

BCA MS 322/175 *The Efficiency Magazine*, Special Avery edition. Includes article concerning Watt beam engine in use at Winfield's Mills, dated October edition 1932

BCA MS 322/176-180 News cuttings concerning the demolition of Winfield's mills and the scrapping of the James Watt beam engine, dated –1936

BCA MS 322/181-185 Engraved illustrations and photographs of Winfield's Rolling Mills, dated 1856–*c*1934

BCA MS 322/186 Extract from *Illustrated London News* for June 28th, 1851, concerning the Royal Commissioners of the Great Exhibition's visit to Mr Winfield's factory in Birmingham, dated 28th June 1851

BCA MS 322/ 192 Pamphlet issued after the visit of the British Association to Winfield's mills, dated 1865

BCA MS 322/197 Typescript extract from *Official Illustrated Guide to the Great Western Railway*, June 1861, concerning Winfield's mills, dated June 1861.

BCA MS 322/199 Plan of lot 9, the Metallic Bedspread Dept. of the Cambridge St. works of Winfield's Ltd, undated.

BCA MS1342/376 RW Winfield, general and ornamental brassfounders: Cambridge Street works, Birmingham, dated 1874

BCA MS1422 60/1/7/1 Repair of lock gates on Gibson's Lock. Birmingham Canal Navigation Company commissioned to replace the lock gates and fit new paddle gear, dated May 1907

BCA MS1422 60/1/7/1 Winfield's Rolling Ltd directors' minutes, various entries, dated 16th May 1911 through to 21st May 1912

BCA MS1422 60/1/7/2 17th April 1917 building plans show they took over the former Sperryn, Player and Mitchell and Richardson premises in 1915 and in 1917 to have the whole former bedstead works. They also took over the former Winfield's boiler house next to Sperryns in 1917

BCA MS 3375/411595 Plan of premises in Essington Street, Sheepcote Street and Grosvenor Street West *c* 1890

BWC, Letter Book No.63, Letter dated 21st February 1848, (Boulton and Watt Corporation)

BWC, Letter Book No.123, Letter Book BW3147/3/134, (Boulton and Watt Corporation)

NA BT 31/32228/145098 Peyton's factory at Bordesley was equipped to supply both the domestic and foreign markets.

NA BT31/3815/23936 Manufactured products included chandeliers, gas fittings, bedstead tubes, art metal, stained glass and electric light fittings

NA BT31/7060/49740 Offices transferred to London and by 1901 the title was changed to the Birmingham Aluminium Castings Ltd

NA BT31/7986/57388 new company was formed in May 1898 that had the lengthy title J and A Tonks (Late Winfield's Brass Foundry) Ltd

NA Rail 810 The link with the Newhall Branch was made during 1812, BCN proprietors minute- March 6th and July 24th 1812

WRO QS/11/c/61 Survey of the Smiths' Hearths in Edgbaston, Erdington and Little Bromwich December 1683

GAZETTES AND NEWSPAPERS

ABG, 1780 A serious address to Birmingham merchants and manufacturers of hardware, 9th October 1780

ABG, 1788 Sale of Baskerville House, dated 18th April 1788

ABG, 1810 Advert, property to let on the Baskerville estate, dated 22nd August 1810

ABG, 1811 Advert, Spoil and earth from cutting of Gibson's private branch, dated 24th June 1811

ABG, 1814 Notice of plans for the Crescent, dated 4th July and 25th June 1814

ABG, 1823 Notice, Thomas Gibson leaves merchant, rolling and slitting business to his sons, dated 6th October 1823

ABG, 1835 Sale Notice John Stephens, steam engine maker, dated 7th February 1835

ABG, 1845 Sale notice, Baskerville Mill, dated 12th May 1845

ABG, 1846 Union Rolling Mills advertised room space suitable for wire drawing, dated 21st May 1846

BDG, 1888, Serious fire broke out in the bedstead works, dated 6th September 1888

BDP 1877, Notice Marrian Works, Millward Street, Small Heath bedstead, fireguard and fancy ironwork makers filed for bankruptcy, dated October 23rd 1877

BDP, 1861 Stoppage announced of Mr Thomas Dowling, timber merchant of Gresham House, Old Broad Street, Rotherhithe and Birmingham, dated 1st May 1861

BDP, 1862 Sale of lease, Morewood and Co, dated 6th April 1862

BDP, 1864 Charles Weston and James Atkins join partnership, dated 18th August 1864

BDP, 1870a Notice Railway Ironworks commence work, dated 22nd January 1870

BDP, 1870b Railway Ironworks previous warehouse in Edgbaston Street, 21st May 1870

BDP, 1882 Winfield's supply electric power and lighting to the Birmingham Town Hall for the music festival, dated 28th August 1882

BDP, 1883, Advertisement Benjamin Cook, Green Street Factory, dated January 6th 1883

BDP, 1884 Notice with regards Edward Peyton settlement with creditors, dated June 14th 1884

BDP, 1887 Winfield's supplied a window to the Priory Church, Malvern, dated 16th December 1887

BDP, 1888a Serious fire broke out in the bedstead works, dated 7th September 1888

BDP, 1888b Baskerville House Mill demolition of the house was carried out in September 1888 and the materials were sold off in lots, dated 14th September 1888

BDP, 1890 Action was brought against Julia Snepp, Charles Torr, Winfield's Ltd and their solicitors Milward and Co, January 1890

BDP, 1891 Winfield's supplied stained glass windows to the refurbished Grand Hotel in Colmore Row, dated 15th June 1891

BDP, 1892 Winfield's premises in Baskerville Place leased out to the Birmingham Municipal Technical School, dated 28th January 1892 and 19th August 1892

BDP, 1895 Article in serial by W J Davis, 'A summary of the brass trades' dated 6th and 14th December 1895

BDP, 1898 Advertisement Cox and Luckman, Stanhope Street, dated July 16th 1898

BDP, 1901 Sales Notice of plant from J and A Tonks, Tube Works, dated 4th May 1901

BDP, 1901 J and A Tonks, Tube Works went into liquidation in 1901, dated 31st August 1901

BG, 1843 Lease by Thomas Crowther, timber merchant, box and cask maker, dated 7th August 1843

BG, 1847 Sale of first steam engine described as 'Three boilers and a 95 horse power steam engine', dated 25th January to 8th March 1847

Birmingham Mail, 1936, Description of the beam engine at Winfield's Rolling Mill, 5th September 1936

BJ, 1840 Thomas Gibson obituary, dated 4th July 1840

BJ, 1841 Advertisement for Baskerville House Mill 29th May 1841

BJ, 1852 Advertisement Thomas Bolton and Co, 18th December 1852

BJ, 1854 When Prosser died, the tube mill plant was put up for sale, dated 2nd September 1854

Freemans Journal, 1884 Sale notice Peyton and Peyton, Dublin premises, dated April 2nd 1884

Furniture Gazette, 1875 An account of the process of bedstead manufacture at S B Whitfield's factory, which specialised in making iron beds, dated April 3rd 1875

LG, 1845, dated 7th March 1845

LG, 1850 Birmingham Patent Welded Iron Tube Company. Bower left the partnership and for a time (until 1850) William Hodges of Great Bookham was a partner, dated 29th March 1850.

LG 1879, Notice, dated 14th February 1879

LG, 1883 Advertisement, Benjamin Cook and Son at 261–262 Bradford Street, dated May 22nd 1883

LG, 1884 Advertisement Edward Peyton and Henry Eagles, Birmingham, Liverpool, Manchester, London, Glasgow, Dublin, Paris and New York, dated March 11th 1884

LG, 1891 Sperryn and Co was established first at Hospital Street and comprised a partnership of E C Kemp, Clifford Kemp, H Herbert Wright and George Sperryn. The Kemps left the partnership in December 1890, dated June 5th 1891

LG, 1907 Player and Mitchell, John Player left this partnership in 1907, but the firm continued to be known as Player and Mitchell, dated May 17th 1907

Martineau and Smith's Hardware Trade Journal 1887 The works of Messrs R W Winfield and Co. in 1887 from 'The homes of our metal manufactures. Messrs R W Winfield and Co's Cambridge Street works and rolling mills, Birmingham', dated January 31st, 1887

Metal Industry 1911, High pressure gas fired melting furnace first of its type to be employed in Britain, dated May 1911

Metal Industry 1922, Extruded Brass Rod, dated May 12th 1922

Metal Industry 1923, Birmingham Aluminium Castings supplied many thousands of cylinders and pistons for the aircraft, dated 1st June 1923

Metal Industry 1936, Article by M Cook 'The development of the non-ferrous metal industries in Birmingham', dated April 1936

Morning Post 1810, Advertisement, Pryer, Steains and Mackenzie who sold the patent Brass-Screw Bedsteads, dated January 18th 1810

Morning Chronicle 1820, Advertisement, John Thomas Thompson, travelling equipage maker, dated May 20th 1820

Morning Post 1821, John Thomas Thompson, travelling equipage maker sale notice to take place November 14th, dated November 5th 1821

The Times 1803, Advert, T Butler, Catherine Street, London, patent bedstead factory, dated October 3rd 1803

The Times 1910, Birmingham Aluminium Casting Company town gas from the Birmingham Gas Department was used to supply a high-pressure gas fired melting furnace in 1910, dated 26th October 1910

Times, 1919 Harper Bean and Co, car makers of Dudley & Tipton, had purchased £100,000 of shares in the Birmingham Aluminium Casting Company and contracted with them to supply aluminium parts for 2000 cars per week, dated 19th December 1919

Trade Directories

Chapman, 1803, 1808 *Chapman's Birmingham directory; or, Alphabetical List of the Merchants, Tradesmen and Principal Inhabitants of the Town of Birmingham, and its Vicinity*, Birmingham: T Chapman

Kelly 1883, 1892 *Kelly's Directory of Birmingham*, London: Kelly's Directory

Pigot 1829, 1842 *Pigot and Cos Commercial Directory of Birmingham and its Surrounds*, Birmingham: Pigot and Co.

PO 1845, 1854, 1864, 1871, 1875, 1881 *Post Office Directory of Birmingham, Warwickshire, and part of Staffordshire*, London: Kelly's Directory

Slater 1852 *Slater's Directory of Birmingham and its Vicinity*, Slater's Directory Limited

White 1855 *History, Gazetteer and Directory of Birmingham*, Sheffield: Francis White and Company

Wrightson, 1821, 1824, 1833, 1835, 1839 *Birmingham Directory*, Birmingham: Wrightson

Wrightson and Webb, 1843 *Birmingham Directory*, Birmingham: Wrightson and Webb

PATENTS

Patent 3460, Benjamin Cook and Thomas Attwood, Combining and connecting different metals, or metals and wood, so as to make the combination have the same appearance, dated 27th June 1811

Patent 3560, John Thomas Thompson, travelling equipage maker, produced and sold metallic bedsteads manufactured, dated 30th April 1812

Patent 3609, Benjamin Cook, The cased rod in the manufacture of various items of furniture including bedsteads, dated 31st October 1812

Patent 3677, John Bennett, The dove-tail joint, 1813

Patent 4798, Thomas Attwood, Making cylinders for printing cottons, calicoes and other articles, dated 3rd June 1823

Patent 5109, Certain improvements in manufacturing tubes for gas and other purposes. February 26th, 1825

Patent 5573, Robert Walter Winfield, Tubes or rods produced by a new method of manufacture, and manufacturing the same into part of Bedstead or other articles dated 4th December 1827

Patent 6206, Robert Walter Winfield, Construction of Bedstead or other articles, dated 22nd December 1831

Patent 9187, William Church and Jonathan Harlow, Manufacturing metal tubes and mode of joining the or other tubes or pieces, dated 16th December 1841

Patent 11705 Jonathan Harlow, Bedstead patents, dated 18th May 1847.

Patent 12838 Jonathon Harlow with Richard Peyton, Manufacturing bedsteads, dated 10th November 1849

Patent No 12268 Manufacture of metallic tubes by Robert Walter Winfield, 1848

Patent No 12302 Construction and manufacture of metallic bedsteads by Robert Walter Winfield, 1848

Patent No 13576 Bedsteads by Robert Walter Winfield with T Woods, Portsea, 1848.

Patent No 1452 Furnaces for boilers furnaces produced by Charles Torr and J Johnstone, managers of Robert Walter Winfield and Co dated 1873

Patent No 1942 Tube design produced by Charles Torr and J Atkins, managers of Robert Walter Winfield and Co dated 1873

Patent No 2624 New furnaces produced by Charles Torr and J Johnstone, managers of Robert Walter Winfield and Co dated 1873

Patent No 2724 Ornamenting metallic bedsteads by Robert Walter Winfield, 1861Patent 3677, John Bennett, metal dovetail joint, for furniture and any kind of framework, dated 7th April 1813

Patent No 4489 Bedstead design by Charles Torr, 1876

Patent No 5686 Centre Light by Winfields Ltd, Cambridge Street works, dated 1895

Patent No 8891 Metallic Bedsteads by Robert Walter Winfield, 1841

Patent No 9888 Bedstead design by Charles Torr, 1886

Rate Books

Rate Books of Birmingham, Ladywood ward, 1834/ 19*, 1840/ 36-37*, 1845/70*, 1846/ 75*, 1860/ 180-181*, 1871/ 246*, 1876/ 257*, 1881/ 268-269*, Birmingham Archives and Heritage

*Microfiche Number

Cartographic

Bickley and Hill 1891 *Conjectural map of Birmingham, based on the written records of the 1553 Borough Survey*. *It relies on later historical maps, predominantly Westley's 1731/2 map and prospect for its topographic basis of the town.

Bradford, Samuel 1750 *Plan of Birmingham surveyed in MDCCL*, Scale 100ft to 1 inch, Birmingham Archives and Heritage

Earl of Dartmouth 1824-1825 *Map of Birmingham, Engraved from a minute trigonometrical survey made in the years 1824 and 1825, Dedicated to The Right Honourable William Earl of Dartmouth, Viscount Lewisham.* *produced by Pigott-Smith, Birmingham Archives and Heritage

Hanson, Thomas 1778 *Birmingham survey'd by Thomas Hanson*, Birmingham Archives and Heritage

Kempson, John 1810 *Map of Birmingham,* Birmingham Archives and Heritage

Pigott-Smith 1855–1857 *Map of Birmingham*, Birmingham Archives and Heritage

Snape 1796 *Snape's Survey of the Gooch Estate*, BCA 669754, Birmingham Archives and Heritage

Westley, William 1731/2 *The Plan of Birmingham, Survey'd in the Year 1731*, Birmingham Archives and Heritage

Ordnance Survey 1887–1889 Birmingham: Warwickshire, Sheet 14, 1st edition

Ordnance Survey 1901–1903 Birmingham: Warwickshire, Sheet 14, 1st revision

Ordnance Survey 1917–1918 Birmingham: Warwickshire, Sheet 14, 2nd revision

Ordnance Survey 1936 Birmingham: Warwickshire, Sheet 14, 3rd revision

ACKNOWLEDGEMENTS

The project was commissioned by Carillion Building on behalf of Birmingham City Council and without their sponsorship this project would not have been possible. Particular thanks are due to Ed Gardner, groundworks site manager and the staff of Carillion Building and Armoury Demolition for their co-operation and assistance during the fieldwork stage of the project. Thanks also go to Mike Hodder, Planning Archaeologist, who monitored the project on behalf of Birmingham City Council. The archaeological work on site was undertaken by Elisabeth Bishop, Mark Charles, Mary Duncan, Emma Collins, Samantha Hepburn, Kristina Krawiec, Erica Macey-Bracken, Phil Mann and William Mitchell who supervised the work and Chris Hewitson who managed the work for Birmingham Archaeology. The specialists who contributed to the project include Rod MacKenzie, Gerry McDonnell (Archaeo-metallurgy), David Dungworth (Glass Residues), Emma Collins, Samantha Hepburn (Finds) and Ray Shill (Historic Research). Chris Hodrien and Jim Williams provided advice on site with regard to Stationary Steam Engines. Chris Hewitson and William Mitchell produced the written report with major contributions by Ray Shill and Gerry McDonnell. Special thanks must go to Malcolm Dick of the University of Birmingham who kindly reviewed the publication in his own time. The publication was illustrated by Nigel Dodds and Chris Hewitson. Special thanks must be given to the staff of Birmingham Archives and Heritage who assisted in the original research that highlighted the significance of the site and have kindly allowed permission to reproduce many of the illustrations presented within the report (individual credits given with the illustrations).